Large Grid-Connected Wind Turbines

Large Grid-Connected Wind Turbines

Special Issue Editors

S M Muyeen
Frede Blaabjerg

MDPI • Basel • Beijing • Wuhan • Barcelona • Belgrade

MDPI

Special Issue Editors
S M Muyeen
Curtin University
Australia

Frede Blaabjerg
Aalborg University
Denmark

Editorial Office
MDPI
St. Alban-Anlage 66
4052 Basel, Switzerland

This is a reprint of articles from the Special Issue published online in the open access journal *Applied Sciences* (ISSN 2076-3417) from 2018 to 2019 (available at: https://www.mdpi.com/journal/applsci/special_issues/wind_turbines)

For citation purposes, cite each article independently as indicated on the article page online and as indicated below:

LastName, A.A.; LastName, B.B.; LastName, C.C. Article Title. *Journal Name* **Year**, *Article Number*, Page Range.

ISBN 978-3-03897-756-8 (Pbk)
ISBN 978-3-03897-757-5 (PDF)

Cover image courtesy of pexel.com user pixabay.

Contents

About the Special Issue Editors

S M Muyeen received his B.Sc. Eng. Degree from Rajshahi University of Engineering and Technology (RUET), Bangladesh, formerly known as Rajshahi Institute of Technology, in 2000 and his M. Eng. and Ph.D. Degrees from Kitami Institute of Technology, Japan, in 2005 and 2008, respectively, all in Electrical and Electronic Engineering. At present, he is working as an Associate Professor in the Electrical and Computer Engineering Department at Curtin University, Perth, Australia. He is serving as Editor/Associate Editor for many prestigious Journals from IEEE, IET, and other publishers, e.g., IEEE Transactions of Sustainable Energy, IEEE Power Engineering Letters, IET Renewable Power Generation, and IET Generation, Transmission and Distribution. He is the Editor-in-chief for the Smart Grid Section of Frontier in Energy Research. He has served as Guest Editor-in-chief/Leading Editor for many special issues. He has been the recipient of many awards, including the Petroleum Institute Research/Scholarship Award 2012, which was the only research award for the entire university until 2013. He is the author/co-author of about 200 scientific articles, including 80+ journals and 6 Books as an Author/Editor. In his short career, he has secured many prestigious research grants at national and international levels. Dr. Muyeen has been given many Keynote and Invited speeches to International Conferences. His research interests are renewable energy, smart grid, and power system stability. Dr. Muyeen is a Senior Member of IEEE and Fellow of Engineers Australia (FIEAust).

Frede Blaabjerg was with ABB-Scandia, Randers, Denmark, from 1987 to 1988. From 1988 to 1992, he studied for his PhD degree in Electrical Engineering at Aalborg University, which he received in 1995. He became an Assistant Professor in 1992, an Associate Professor in 1996, and a Full Professor of power electronics and drives in 1998. In 2017, he became a Villum Investigator. He is honoris causa at University Politehnica Timisoara (UPT), Romania and Tallinn Technical University (TTU) in Estonia. His current research interests include power electronics and its applications, such as in wind turbines, PV systems, reliability, harmonics, and adjustable speed drives. He has published more than 600 journal papers in the field of power electronics and its applications. He is the co-author of four monographs and editor of ten books in power electronics and its applications. He has received 29 IEEE Prize Paper Awards, the IEEE PELS Distinguished Service Award in 2009, the EPE-PEMC Council Award in 2010, the IEEE William E. Newell Power Electronics Award 2014, and the Villum Kann Rasmussen Research Award 2014. He was the Editor-in-Chief of the IEEE TRANSACTIONS ON POWER ELECTRONICS from 2006 to 2012. He was a Distinguished Lecturer for the IEEE Power Electronics Society from 2005 to 2007 and for the IEEE Industry Applications Society from 2010 to 2011, as well as 2017 to 2018. He is currently serving s as President of IEEE Power Electronics Society for the 2019–2020 tenure. He also serves as Vice-President of the Danish Academy of Technical Sciences. He was nominated in 2014, 2015, 2016 and 2017 by Thomson as among the 250 most cited researchers in Engineering in the world.

applied
sciences

MDPI

Editorial

Special Issue on 'Large Grid-Connected Wind Turbines'

S. M. Muyeen [1],* and Frede Blaabjerg [2],*

[1] School of Electrical Engineering, Computing, and Mathematical Sciences, Curtin University,
    Perth 6845, Australia
[2] Department of Energy Technology, Aalborg University, Pontoppidanstraede 111, DK9220 Aalborg, Denmark
* Correspondence: sm.muyeen@curtin.edu.au (S.M.M.); fbl@et.aau.dk (F.B.)

Received: 25 February 2019; Accepted: 26 February 2019; Published: 6 March 2019

1. Introduction

The renewable energy penetration rate to the power grid is rapidly increasing these days. As per the statistics available in 2018 [1], the total renewable power capacity is 1,081 GW, excluding hydropower. Wind, solar, biomass, tidal and geothermal are considered as the prime renewable sources in generating electric power, while wind energy is the most attractive one dominating the energy market. In 2017, 52 GW of wind power was added globally, making a total of 539 GW, which is more than 50% of the total power generated from various renewable resources. China, the United States, Germany, India and Spain are the top five countries investing most in the wind energy sector. Among many wind turbine manufacturers, Vestas, Siemens-Gamesa and GE have captured the majority of the wind market in 2017, having a market share of 16.7%, 16.6%, and 7.6%, respectively [1].

Because of the growing nature of wind energy penetration, large scale wind turbines are getting popular, especially in offshore wind farms. In June 2017, MHI Vestas, a joint venture between Vestas and Mitsubishi Heavy Industries, launched a 9.5 MW offshore turbine, which is currently the world's largest wind turbine as of today, and it has been scheduled for 10 MW. Standing 187 meters tall and with 80-meter blades, this wind turbine is an upgraded version of MHI Vestas' 8 MW V164 turbine, which is already in operation at the Burbo Bank Extension and Blyth offshore wind farms in Britain. Seeing the bright future of the large scale wind turbine, GE has recently introduced the Haliade-X 12 MW, which will be the most powerful offshore wind turbine in the world to date, featuring a 12 MW capacity, 220-meter rotor, and a 107-meter blade. The Haliade-X will also be the most efficient of wind turbines in the ocean having a capacity factor of 63%—which of course will be dependent on wind conditions. This turbine is expected to be commissioned for trial by the middle of 2019 [2].

2. Challenges and Opportunities of Large Wind Turbines

Considering the growing nature of large wind turbines, many technological challenges are to be resolved. Wind power intermittency is a well-known problem. To address this issue, it is important to exercise well how to predict the wind speed, and only then can this intermittency problem be handled precisely. The smoothing of the variable wind power is the next step in which the energy storage system and other advanced control approaches may play vital roles [3,4]. The intermittency of wind power may also lead to voltage and frequency instabilities. Many FACTS devices and energy storage systems can help in augmenting these stabilities. A modern approach to address the frequency dip problem is to use the synthetic inertial [5]. Grid Interfacing, fault ride through and power electronic converters' reliability are other important issues to be addressed [6] in order to get a reliable grid operation.

The wind energy sector is blessed by the contribution from many engineering and science disciplines in the last few decades, mainly from mechanical, electrical, electronic, computer and aerospace. The individual and joint efforts from scientists within different disciplines are prime drivers for the wind industry to reach maturity. In this special issue, the present and future development schemes of wind turbine generator systems are depicted based on the contribution from many renowned scientists and engineers in different disciplines. To make this special issue collection useful for students and researchers, a wide variety of research outcomes are merged together, putting focus on the variability, stability and scalability issues of wind energy conversion systems. To address the challenges mentioned above, 10 different articles are included in this special issue collection, selected from many submissions, and a brief description for each article is given below.

The first paper [7], authored by Arman Oshnoei, Rahmat Khezri, SM Muyeen and Frede Blaabjerg, presents a pure technical review on Wind Farm Automatic Generation Control (AGC), focusing on frequency regulation. In this article, the contribution of wind farms in the supplementary/load frequency control of AGC is overviewed first. Then the authors proposed a fractional order proportional-integral-differential (FOPID) controller to regulate the speed of the turbine rotor in the participating AGC. The second paper [8] provides another review on Fault Current Limiting Devices to augment the fault ride through the capability of large wind turbines using doubly fed induction generators, authored by Seyed Behzad Naderi, Pooya Davari, Dao Zhou, Michael Negnevitsky and Frede Blaabjerg, giving a focus mainly on fault current limiters and series dynamic braking resistors. This is indeed a timely reporting.

There are three papers addressing wind power variability and intermittency issues. In the first paper [9], authored by Yanxia Shen, Xu Wang and Jie Chen computed the potential uncertainties of wind power in constructing prediction intervals (PIs) and prediction models using a wavelet neural network (WNN). In order to optimize the forecasting model, the authors have proposed a multi-objective artificial bee colony (MOABC) algorithm combining multi-objective evolutionary knowledge, called EKMOABC. It appears that in this way, a better short-time wind power forecasting is very much possible. In the second paper [10], the novel output power smoothing control strategy for a wind farm based on the allocation of wind turbines is proposed by the Authors Ying Zhu, Haixiang Zang, Lexiang Cheng and Shengyu Gao. The wind turbines in the wind farm are divided into control wind turbines (CWT) and power wind turbines (PWT), separately. The rotor inertia based power smoothing method is adopted in that study for the sake of better performance. This is another timely reporting in addressing wind power variability problem. In the third paper [11], authored by Andrzej Tomczewski and Leszek Kasprzyk, the wind power intermittency problem is addressed by using Flywheel Energy Storage System (FESS).

In paper [12], authored by Abdullah Bubshait and Marcelo G. Simões, is presented a new control approach for wind turbines in regulating system frequency. The prime focus was to design a control method to maintain the reserved power of the wind turbine, by simultaneously controlling the blade pitch angle and rotor speed. The transient stability augmentation method of a wind farm populated with both fixed and variable speed wind generators is presented in [13], by Md. Rifat Hazari, Mohammad Abdul Mannan, S. M. Muyeen, Atsushi Umemura, Rion Takahashi and Junji Tamura. In that work, it has been shown that a fuzzy based control approach works well to stabilize the squirrel cage induction generators of a wind farm through the use of doubly fed induction generators.

Condition monitoring of equipment and accessories including sensors are very important tasks for large wind turbines to reduce any downtime. This is critical because the shutdown of a megawatt class wind turbine causes considerable revenue loss. The current sensor fault diagnosis and isolation problem for a permanent magnet synchronous generator (PMSG) based wind system is presented in [14], by Zhimin Yang, Yi Chai, Hongpeng Yin and Songbing Tao. In [15], Bin Li, Junyu Liu, Xin Wang and Lili Zhao discussed the protection issues of wind turbines using the doubly fed induction generator. Feasibility and economic studies of wind farm using FACTS devices in a distribution network have been conducted in [16] by Lina Wang, Kamel Djamel Eddine Kerrouche, Abdelkader Mezouar, Alex Van Den Bossche, Azzedine Draou and Larbi Boumediene.

3. Concluding Remarks

It is obvious that although wind energy technology has reached a stage of maturity, more in-depth research should be carried out in this area to welcome 10+ MW class wind turbines onto the market. The main problems identified so far are the intermittency, stability and reliability issues of large scale wind turbines. This requires further improvements in existing control and protection mechanisms and hence, combined efforts from scientists and researchers from multiple disciplines are very much essential and most appreciated, in order to drive down the cost of energy.

References

1. Renewables 2018—Global Status Report. Available online: http://www.ren21.net/ (accessed on 12 February 2019).
2. Haliade-X Offshore Wind Turbine Platform. Available online: https://www.ge.com/renewableenergy/wind-energy/turbines/haliade-x-offshore-turbine (accessed on 20 February 2019).
3. Muyeen, S.M.; Tamura, J.; Murata, T. *Stability Augmentation of a Grid-Connected Wind Farm*; Springer: London, UK, 2008; ISBN 978-1-84800-315-6.
4. Blaabjerg, F.; Ma, K. Wind Energy Systems. *Proc. IEEE* **2017**, *105*, 2116–2131. [CrossRef]
5. Hazari, M.R.; Mannan, M.A.; Muyeen, S.M.; Umemura, A.; Takahashi, R.; Tamura, J. Transient Stability Augmentation of Hybrid Power System Based on Synthetic Inertia Control of DFIG. In Proceedings of the Australasian Universities Power Engineering Conference 2017 (AUPEC2017), Melbourne, Australia, 19–22 November 2017.
6. Blaabjerg, F.; Ionel, D.M. *Renewable Energy Devices and Systems with Simulations in MATLAB®and ANSYS®*; CRC Press: Boca Raton, FL, USA, 2017.
7. Oshnoei, A.; Khezri, R.; Muyeen, S.M.; Blaabjerg, F. On the Contribution of Wind Farms in Automatic Generation Control: Review and New Control Approach. *Appl. Sci.* **2018**, *8*, 1848. [CrossRef]
8. Naderi, S.; Davari, P.; Zhou, D.; Negnevitsky, M.; Blaabjerg, F. A Review on Fault Current Limiting Devices to Enhance the Fault Ride-Through Capability of the Doubly-Fed Induction Generator Based Wind Turbine. *Appl. Sci.* **2018**, *8*, 2059. [CrossRef]
9. Shen, Y.; Wang, X.; Chen, J. Wind Power Forecasting Using Multi-Objective Evolutionary Algorithms for Wavelet Neural Network-Optimized Prediction Intervals. *Appl. Sci.* **2018**, *8*, 185. [CrossRef]
10. Zhu, Y.; Zang, H.; Cheng, L.; Gao, S. Output Power Smoothing Control for a Wind Farm Based on the Allocation of Wind Turbines. *Appl. Sci.* **2018**, *8*, 980. [CrossRef]
11. Tomczewski, A.; Kasprzyk, L. Optimisation of the Structure of a Wind Farm—Kinetic Energy Storage for Improving the Reliability of Electricity Supplies. *Appl. Sci.* **2018**, *8*, 1439. [CrossRef]
12. Bubshait, A.; GSimões, M. Optimal Power Reserve of a Wind Turbine System Participating in Primary Frequency Control. *Appl. Sci.* **2018**, *8*, 2022. [CrossRef]
13. Hazari, M.; Mannan, M.; Muyeen, S.; Umemura, A.; Takahashi, R.; Tamura, J. Stability Augmentation of a Grid-Connected Wind Farm by Fuzzy-Logic-Controlled DFIG-Based Wind Turbines. *Appl. Sci.* **2018**, *8*, 20. [CrossRef]
14. Yang, Z.; Chai, Y.; Yin, H.; Tao, S. LPV Model Based Sensor Fault Diagnosis and Isolation for Permanent Magnet Synchronous Generator in Wind Energy Conversion Systems. *Appl. Sci.* **2018**, *8*, 1816. [CrossRef]
15. Li, B.; Liu, J.; Wang, X.; Zhao, L. Fault Studies and Distance Protection of Transmission Lines Connected to DFIG-Based Wind Farms. *Appl. Sci.* **2018**, *8*, 562. [CrossRef]
16. Wang, L.; Kerrouche, K.; Mezouar, A.; Van Den Bossche, A.; Draou, A.; Boumediene, L. Feasibility Study of Wind Farm Grid-Connected Project in Algeria under Grid Fault Conditions Using D-Facts Devices. *Appl. Sci.* **2018**, *8*, 2250. [CrossRef]

applied
sciences

MDPI

Review

On the Contribution of Wind Farms in Automatic Generation Control: Review and New Control Approach

Arman Oshnoei [1], Rahmat Khezri [2,*], SM Muyeen [3] and Frede Blaabjerg [4]

[1] Faculty of Electrical and Computer Engineering, Shahid Beheshti University, Tehran 1983969411, Iran; a_shnoei@sbu.ac.ir
[2] College of Science and Engineering, Flinders University, Adelaide, SA 5042, Australia
[3] Department of Electrical and Computer Engineering, Curtin University, Perth, WA 6845, Australia; sm.muyeen@curtin.edu.au
[4] Department of Energy Technology, Aalborg University, 9220 Aalborg, Denmark; fbl@et.aau.dk
* Correspondence: rahmat.khezri@flinders.edu.au; Tel.: +61-451-645-568

Received: 18 September 2018; Accepted: 3 October 2018; Published: 9 October 2018

Abstract: Wind farms can contribute to ancillary services to the power system, by advancing and adopting new control techniques in existing, and also in new, wind turbine generator systems. One of the most important aspects of ancillary service related to wind farms is frequency regulation, which is partitioned into inertial response, primary control, and supplementary control or automatic generation control (AGC). The contribution of wind farms for the first two is well addressed in literature; however, the AGC and its associated controls require more attention. In this paper, in the first step, the contribution of wind farms in supplementary/load frequency control of AGC is overviewed. As second step, a fractional order proportional-integral-differential (FOPID) controller is proposed to control the governor speed of wind turbine to contribute to the AGC. The performance of FOPID controller is compared with classic proportional-integral-differential (PID) controller, to demonstrate the efficacy of the proposed control method in the frequency regulation of a two-area power system. Furthermore, the effect of penetration level of wind farms on the load frequency control is analyzed.

Keywords: large-scale wind farm; automatic generation control; load frequency control; fractional order proportional-integral-differential controller

---

1. Introduction

In the 21st century, electrical energy is needed more than ever, and the harmful effect of using fossil fuels to generate electrical energy, such as carbon dioxide emission, has become more serious. Accordingly, the demand on renewable energy sources to produce electricity from clean energies, such as wind, solar, hydro, biomass and geothermal, have globally increased. Renewable energies are salient choice to solve the air pollution problem, however the intermittent output power can create new challenges in the operation of power systems. Impacts of renewable energies on power systems operation cannot be ignored, and should be analyzed, along with developing effective mitigation strategies and technologies—especially for higher level of renewable penetrations.

In the territory of renewable energy sources, development of wind turbine as a source to produce electrical energy from wind is going on swiftly around the world. In 2017, the installed wind energy worldwide was more than 539 GW [1]. Such a high generation needs more and more attention, in order to address the intermittency issues produced by wind turbine itself. Wind turbines are divided into two groups: Fixed speed and variable speed. The first group, fixed speed, generally use an induction generator (IG) that is connected directly to the grid, and are known as fixed speed wind turbine (FSWT). The second group, variable speed, typically use permanent magnet synchronous generator (PMSG), or doubly-fed induction generator (DFIG) in their structure, and are known as variable speed wind turbines (VSWTs). Taking advantage of the power electronic converters, the PMSG is fully decoupled from the grid. It means that, the stator of PMSG is connected to the back-to-back fully rated power electronic converters in order to inject the power into the grid. In the other case, DFIGs have both direct connection and converter-based connection to the grid. The stator of DFIG is connected directly to the grid while the rotor is connected through partially rated back-to-back converters. The power electronic converter used in a variable speed wind turbine enables the wind turbine to regulate the output power over a wide range of wind speeds [2,3]. The variable type is the dominant and promising type of wind turbines for application in large-scale wind farms. In the domain of wind turbines, VSWT technology has attracted a lot of attention for integration in power networks, because of its salient features. The primary advantage of VSWT driven wind generators is that, they allow the amplitude and frequency of their output voltages to be maintained at a constant value, no matter the speed of the wind blowing on the wind turbine rotor. Therefore, it can be inferred that they can be directly connected to the ac power network and remain synchronized at all times with the ac power network. Other advantages include the ability to control the voltage at point of common coupling and power factor control (e.g., to maintain the power factor at unity). Furthermore, the VSWT-based wind generators can produce the maximum power at variable speeds of wind [4].

Penetration level of large scale wind farm is increased in power systems [5]. Since that the large-scale wind farms have high capacity, they should be investigated like conventional power plants as they are connected to the electrical power network. In Reference [6], static planning of wind farms in power networks is reported by optimal load flow. Contribution of wind farms in ancillary services of power system are investigated in Reference [7]. In this regard, wind farms can contribute to voltage and reactive power control [8,9], frequency stability and control [10–12], power system stability enhancement [13,14], harmonic mitigation [15], oscillation damping [16] etc. Since frequency control is directly related to active power control, and active output power of wind farms has intermittent characteristic, due to wind uncertainty; this aspect of grid ancillary capability (frequency control) for wind farms has more significance from the viewpoint of the power system. Frequency control in wind farms is analyzed in three levels: Inertial control, primary control, and secondary control or automatic generation control. The inertial and primary control systems are well addressed in the literature [17,18].

Imbalance between load and generated power by generators in the system are exhibited through frequency deviations. An indelible off-normal frequency deviation directly affects power system operation, security, reliability, and efficiency by damaging equipment, degrading load performance, overloading of the transmission lines, and triggering the protection devices. Frequency performance of the power system is regulated in three levels; the first two levels by generation units, and in the third level, by loads through load shedding in severe situations. These three levels are known as primary control or inertial response [19], secondary or supplementary control [20] and tertiary or emergency control or load shedding [21].

Automatic generation control (AGC) is the manner of action of participating generation units in frequency control. Supplementary frequency control, which is known as load-frequency control (LFC), is a major function of AGC systems as they operate online to control system frequency and power generation [22]. Frequency control can be directly dependent on the speed control of the turbines in the generation units, due to the fact that the frequency generated in the electric network is proportional to the rotational speed and mechanical power of the generator. In conventional generators (non-wind turbine), the issue is initially sensed by the governor of the generators, which can adjust the valve position to change the fuel amount and subsequently change the mechanical power for the electrical part to track the load change and to restore the frequency to be near the nominal value.

AGC responsibilities can be classified as a significant control process that operates constantly to balance the generation and load in the power system at a minimum cost, adjust the generation to minimize frequency deviation and regulate tie-line power flows. Briefly, the AGC system is responsible for frequency control and power interchange, as well as optimal economic dispatch. The AGC system realizes generation changes by sending signals to the under-control generation units. Due to newly advanced technologies in wind turbines and their controllability, the wind farms can be mentioned as an acceptable choice to contribute to the AGC. The AGC performance is highly dependent on how those generating units respond to the commands. The generation unit response characteristics is mainly affected by the control strategy that the unit utilizes, like robust control methods [23,24], intelligent algorithms [25,26] and optimization approaches [27,28]. This part can be solved by integrating appropriate and efficient controller in the wind turbine structure, which is demonstrated in this study. The control strategy that is utilized for wind farm contribution in the AGC is the main topic for this paper.

This paper firstly reviews the contribution of wind farms in load frequency control as a major function of AGC. Different aspects of such contribution, like different control strategies (optimization, model predictive control and intelligent methods) and coordination with other devices, such as conventional generation units, flexible AC (alternative current) transmission systems (FACTS) and energy storage systems are reviewed. As the second stage, a new control methodology by implementing fractional order proportional-integral-differential (FOPID) controller for supplementary loop of a variable speed wind generator-based wind farms is proposed. The parameters of the proposed controller are optimized by sine-cosine algorithm (SCA) to attain efficient control performance in multi-area power systems. The main contributions of this paper are:

(1)   Wind farms contribution to frequency regulation of power systems is studied.
(2)   Review on application of wind farms in AGC is presented.
(3)   The FOPID controller is applied to variable speed wind turbine.
(4)   Performance of FOPID controller for variable speed wind turbine-based wind farm is compared to classical controller.
(5)   Frequency variation effect of different penetration levels of wind farms in power systems is investigated.

2. Wind Farm Contribution in Frequency Regulation

Granted by new technologies and control methods of wind turbines, wind farms are worthy choice to contribute to the frequency regulation of the power system. In this regard, operation under maximum power point tracking, connecting to power system with AC/DC (direct current)/AC converters and coordination with other conventional power plants are the main issues that to be investigated for wind farms in order to able to contribute to frequency regulation. Figure 1 shows the frequency regulation process and related strategies for wind farms to contribute in AGC. This section generally explains the wind farms contribution in primary frequency control, and their inertial response to support frequency regulation. Wind farms contribution in AGC and its details are provided in the next section.

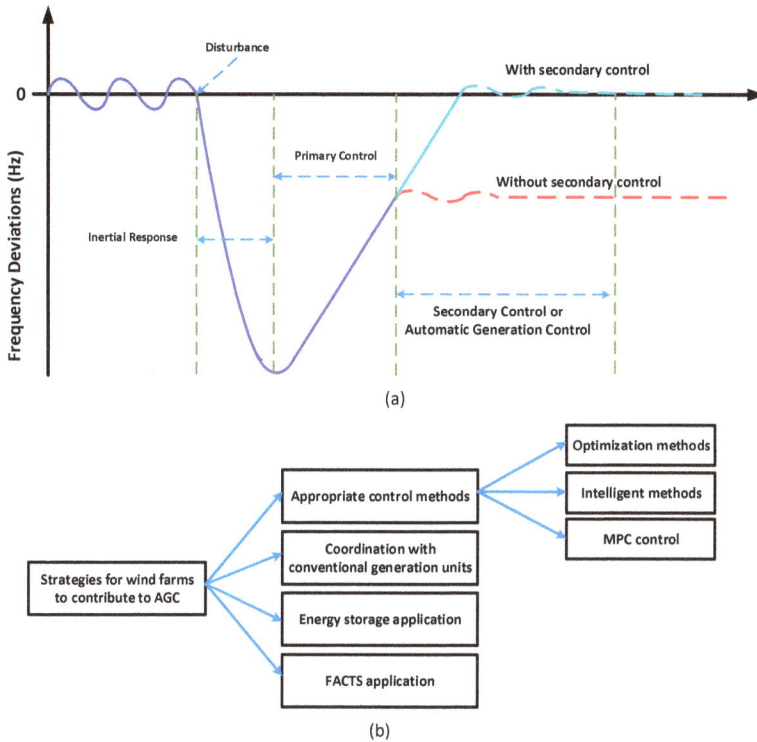

Figure 1. (**a**) Frequency regulation process; (**b**) control strategies for wind farms to contribute to the AGC. AGC, automatic generation control; FACTS, flexible AC transmission systems; MPC, model predictive control.

Normally, wind turbines operate in maximum power point of tracking (MPPT) mode to extract maximum possible power from the wind and convert it into electrical power. It is obvious that such models with MPPT operation mode is not appropriate in AGC studies. In Reference [29], the performance of wind farms by VSWTs were investigated when they are modeled under the set-point tracking control strategies for intermittent wind. The superiority of set-point tracking mode over MPPT mode was proven by the contribution of the wind farms in AGC. To reach high quality frequency performance, a new method is introduced in Reference [30] to preserve a certain amount of wind power reserve to contribute in frequency regulation of power system.

Inertial response of wind turbines is the other important issue for frequency regulation. The FSWTs can provide an inertial response to the frequency deviation, because of its direct coupling; however, this inertia is small compared to the synchronous generator. On the other hand, the VSWTs are connected to the grid by back-to-back voltage or current source converters, which decouples the complete wind turbine in the case of PMSG and the DFIG partially from the grid to support frequency regulation. In this situation, the wind turbine cannot contribute to the frequency regulation through inertial response with its original controller, since it is not connected directly to the grid. Therefore, it is needed to add supplementary controllers to converter part of VSWTs to prepare them for frequency change tracking and then do efficient control. In this regard, the VSWTs are equipped with efficient supplementary frequency control loops to support inertial response, primary frequency control and secondary frequency control [31–33]. A combination of inertial response and a droop active power support in VSWTs was shown in Reference [34], in which the wind turbine reached efficient frequency control for different levels of penetration. A new control strategy to coordinate the inertial control, rotor speed and pitch angle control in a DFIG are presented in Reference [35]. In Reference [36], a deloading strategy was proposed alongside with inertial control loop for the PMSGs to contribute to the frequency control, by increasing their active power production. Efficient damping of frequency oscillations and improving the frequency regulation capability were the main advantages for the strategy. In sharp contrast to MPPT, deloaded strategy is a method to limit the generation of wind turbine through power curve of turbine to make it feasible to contribute to frequency control [37]. In Reference [38] a novel strategy was proposed to strengthen the primary control capability of VSWTs for frequency control with fast response, which can improve the VSWTs contribution in the AGC.

Coordination between wind farms and conventional power plants in terms of frequency control is also investigated in order to reach more efficient frequency response through wind farms. In References [39,40], it is shown that the provided inertial response from DFIG wind farms is only efficient for low level of penetration. Authors have proposed an inertial coordination strategy between the wind farms and conventional power plants in a high level of penetration, which has shown improvement in frequency performance compared to uncoordinated strategy. In Reference [41], it is shown that the inertial response coordination between conventional power plants and wind farms can help the system to reach efficient frequency response in terms of smoothing tie-line fluctuations, reduced peak frequency excursion and settling time.

The literature shows also that the wind farms can contribute in the frequency regulation of the power systems by different strategies and at different stages of the frequency regulation. Most of the abovementioned studies covered by the inertial response and primary control of DFIGs to contribute in primary frequency control of power system. However, the secondary frequency control or LFC as a part of AGC is not well addressed for the wind farms.

3. Wind Farms Contribution in AGC

The AGC is responsible for minimization of power system cost, due to operation of different power plants, removing the steady-state and transient frequency deviations and smoothing the tie-line power between areas. As stated earlier, this study investigates the last items—which are known as secondary frequency control or LFC.

There is a vast body of literature on control strategies for wind farms to contribute in the LFC. Because of simplicity in structure and industry application of proportional integral (PI) controller, most of the studies have used the PI controller in VSWT structure to contribute to the secondary frequency control. However, a range of other control approaches can be examined for such contribution. Generally, over and undershoots, settling time, steady-state error and some other control indices are investigated to evaluate the performance of controllers in the LFC. A simple PI controller tuned by try and error for wind turbine was proposed for secondary frequency control in Reference [42], which was able to restore the frequency to its nominal value.

Applied control strategies for wind farms can be classified into optimization methods, intelligent methods, and model predictive control methods for PI controllers in wind turbine structure. Furthermore, contribution to the AGC issues through the wind farm by the integration of energy storage systems and FACTS devices are investigated.

3.1. Applied Control Methods

Optimization of the parameters in a PI controller structure is one of the basic methods to reach efficient control performance. Simple parameter optimization based-on integral of square error index was applied to the PI controller in the wind turbine structure to coordinate the wind turbine with AGC system of an isolated power system, as reported in Reference [43]. The efficiency of DFIG's contribution in AGC of isolated power system was demonstrated by a high level of wind farm penetration. Similar methods were used in Reference [44], to prove the DFIGs contribution to the AGC of multi-area power systems. Dynamic participation of DFIG-based wind farms in an optimal AGC is demonstrated in Reference [45], which is effective to reach higher a stability margin and smooth the frequency response with a suitable damping. These studies showed the contribution of the DFIGs in the AGC of power systems; however, the simple method for the optimization process is the deficiency.

A range of metaheuristic optimization methods are utilized in wind turbines to contribute to the AGC [46–52]. Ant colony optimization (ACO) approach [46], craziness-based particle swarm optimization (CRPSO) [47], improved particle swarm optimization (IPSO) [48], opposition learning based gravitational search algorithm (OGSA) [49], genetic algorithm (GA) [50], non-dominated sorting genetic algorithm-II (NSGAII) [51], and non-dominated Cuckoo search algorithm (NSCS) [52] are successfully applied to classic PI controller of VSWT wind turbines to contribute in the AGC. In Reference [53], ant lion optimization (ALO) is proposed to optimize the parameters of a new controller known as trajectory following controller in the structure of wind turbines. Efficiency of the proposed controller was shown over optimized PI and PID controllers in minimizing the settling time and peak overshoot of frequency performance.

Intelligent methods like fuzzy logic and artificial neural network (ANN) are applied to automatic generation control of wind integrated power systems [54–57]. Simplicity in the design process for fuzzy logic approach has made it impressive to apply in VSWT wind turbines. In Reference [54], a fuzzy-based PI controller is designed for DFIG-based wind farms to contribute in AGC of a two-area power system. It is shown that utilizing fuzzy approach can be more effective compared to a simple PI controller in the frequency regulation under different load changes and wind penetration levels. However, the fuzzy logic method is a suitable choice for frequency control of wind integrated power systems, the expert knowledge base design process of the fuzzy logic may deteriorate its performance. For this reason, in Reference [55], a combined Jaya algorithm (JA)-fuzzy logic based proportional integral differential (PID) controller is proposed for the LFC of three area power system integrated with wind farms. Trained for a wide range of operation conditions and load changes, a non-linear recurrent ANN is demonstrated in Reference [56] using off-line data for the wind farm contribution in the AGC of multi-area power system. Better frequency performance in terms of lesser undershoot and settling time, and faster oscillation damping compared to the conventional PI controller are the superiority of the proposed controller. As another class of supervised learning-based controller, a least square support vector machine method is proposed in Reference [57] for the AGC in a wind integrated multi-area power system. Compared to multi-layer perceptron neural network based AGC, the proposed controller is efficient for the frequency control purpose.

The model predictive control (MPC) approach has been utilized for an AGC system, incorporated with wind farms successfully [58–61]. The operation process through MPC is an optimization process that can be defined at each time instants. The important point of the optimization process of the MPC is to compute a new control input vector to be fed to the system and taking into account the system constraints at the same time. Considering the governor and turbine parameters variation, as well as load changes, an MPC approach is developed for the AGC in a single-area power system [58], and multi-area power system [59], in the presence of a DFIG-based wind farm. Robust performance of the proposed MPC, due to parameter variation is the main advantage compared to the classic controllers. To make the MPC more efficient for AGC studies in the presence of wind farms, a distributed MPC known as DMPC is employed by [60,61]. The main advantage of the DMPC application is dividing the whole system into some subsystems and controlling of each subsystem by a local MPC controller. The DMPC approach shows more robust and efficient performance in the AGC compared to the central or simple MPC.

3.2. BESS and FACTS Integration with Wind Farm

The intermittent and unpredictable output power of wind farms may cause the use of supplementary source of energies in the power system to reach a more efficient frequency control and AGC contribution in the power system. Energy storage system is one of the important sources of energy that can be installed in wind integrated power systems to absorb and release the energy.

Integration of a real 34 MW battery storage system in a 51 MW wind farm in Japan is shown in Reference [62]. Frequency control is one of the important aims in utilizing this battery storage.

As there are many types of energy storage technologies, the redox flow batteries (RFB), flywheels, capacitive energy storages and superconducting magnetic energy storages (SMES) have the most contribution for wind farms in the AGC based on the literature. Furthermore, there is high potential to improve the frequency control of multi-area power systems with FACTS devices, since they can control the tie-lines power. Regarding the mentioned features, energy storage systems and FACTS have the most contribution for wind farms to contribute in AGC of power systems.

Integration of flywheel storage systems with wind farms was investigated in Reference [63]; in which the wind-flywheel system reduced the settling time and smoothed the frequency deviations more effectively. In Reference [64], it was shown that the application of SMES system in addition to dynamic active power support from wind farm is an impressive solution to improve the transient performance of the frequency after some disturbances. Coordinated design of AGC and redox flow batteries to minimize the frequency deviation in the wind integrated multi-area power system was demonstrated in Reference [65].

The aforementioned studies have investigated the sole integration of energy storage systems for wind farms in the AGC issue. However, application of energy storages in one area may not affect the frequency performance of other areas; as well application of energy storage for all areas is not an economical solution. Therefore, it is rational to use FACTS devices alongside wind-storage system to better control of tie-line power. The efficiency of such system was proven in Reference [66] by coordinated design of thyristor-controlled phase shifter (TCPS) and SMES system (in one area) incorporated with dynamic participation of wind farms in a deregulated two-area power system. Similar system was investigated in Reference [67] by applying the SMES system to each areas of the two-area power system. Such applications increase the cost of system, since the electromagnetic energy storage systems are still expensive. Furthermore, coordinate application of RFB with static synchronous series compensators (SSSC), and capacitive energy storage (CES) with TCPS, in wind integrated power systems for better contribution in the AGC were addressed by the studies [68,69].

Furthermore, a detailed review of existing papers in Sections 3.1 and 3.2 is summarized in Table 1. Type of generation units contributed in the power systems, system configuration (energy storage and FACTS), control approach, and optimization techniques are addressed in Table 1. As shown, most of the studies are done in multi-area power system that demonstrates the capability of wind power to contribute to frequency control of large-scale power system.

Table 1. Review of papers in wind farm contribution in AGC.

Ref.	Generation Unit				System Configuration	Control Approach (AGC)	Optimization Techniques
	Wind	Thermal	Gas	Hydro			
[43]	√	√			Single area power system	Integral controller	–
[44]	√	√	√	√	Three area power system	Integral controller	–
[45]	√	√			Two area power system	Integral controller	–
[46]	√	√			Two area power system	PI controller	ACO
[47]	√	√		√	Two area power system	Integral controller	CRPSO
[48]	√	√			Three area power system	PI controller	IPSO
[49]	√	√	√	√	Four area power system	PID controller	OGSA
[50]	√	√		√	Deregulated two area power system	PI controller	GA
[51]	√	√			Two area power system	Integral controller	NSGAII
[52]	√	√			Two area power system	Integral controller	NSCS
[53]	√	√			Two area power system	Integral controller	ALO
[54]	√	√			Two area power system	Fuzzy logic-PI controller	–
[55]	√	√			Three area power system	Fuzzy logic-PID controller	JA
[56]	√	√			Two area power system	Non-linear recurrent ANN	–
[57]	√	√			Two area power system	Least squares vector machines	–

Table 1. *Cont.*

Ref.	Generation Unit				System Configuration	Control Approach (AGC)	Optimization Techniques
	Wind	Thermal	Gas	Hydro			
[58]	√	√			Single area power system	MPC	–
[59]	√	√			Two area power system	MPC	–
[60]	√	√			Three area power system	DMPC	–
[61]	√	√			Four area power system	DMPC	–
[62]	√	√			Two area power system with flywheel	PID controller	–
[63]	√	√		√	Deregulated two area power system with SMES	Integral controller	CRPSO
[64]	√	√		√	Two area power system with RFB	PID controller	Grey Wolf Optimizer (GWO)
[65]	√	√		√	Deregulated two area power system with SMES and TCPS	Integral controller	CRPSO
[66]	√	√		√	Two area power system with SMES and TCPS	Fuzzy logic-PID with derivative filter controller	Multi-Verse Optimizer (MVO)
[67]	√	√	√	√	Deregulated two area power system with CES and TCPS	Integral controller	–
[68]	√	√	√	√	Two area power system with RFB and SSSC	Integral controller	GA

4. Proposed Wind Farm Control Technique in AGC

4.1. Wind Turbine Modeling in AGC

DFIG is the most commercially used variable speed wind turbine and therefore it is used in this research to demonstrate the effect of wind turbine in AGC. The DFIG wind turbine participates in frequency control through releasing the stored kinetic energy in the turbine blades under sudden load changes. Extracting the stored kinetic energy and converting it into electric energy depends on the turbine inertia, and its control system. Figure 2 shows the block diagram of DFIG-based wind turbine used in this paper for the AGC issue.

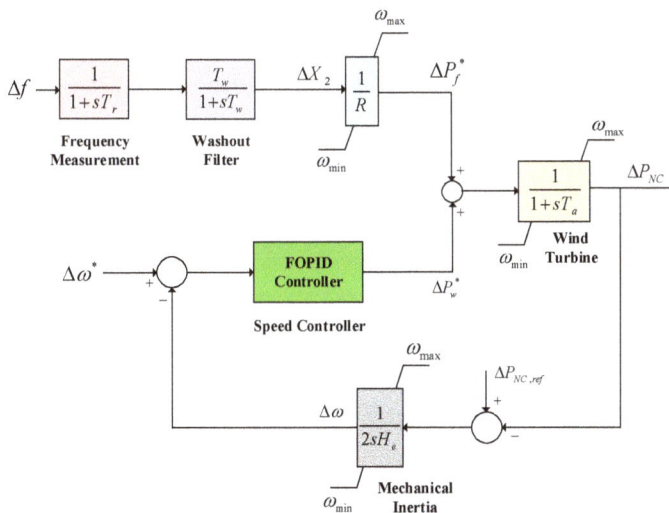

Figure 2. Block diagram of doubly-fed induction generator (DFIG) based wind turbine model.

In this model, the extra signal ΔP_f^* tries to adapt the set point power as a function of frequency deviation rate Δf. Another signal, ΔP_ω^*, attempts to maintain the speed of wind turbine at a desired value for producing the maximum output. In the considered model, e is error signal in speed of the wind turbine, $\Delta \omega^*$ and $\Delta \omega$ are the reference and actual deviations of wind turbine speed, respectively. In order to obtain $\Delta \omega$, a mechanical equation can be stated as follows [37]:

$$\frac{d\Delta\omega}{dt} = \frac{1}{2H}(T_m - T_e), \tag{1}$$

where T_m is the mechanical torque and T_e is the electrical torque. Equation (1) represents the swing equation, which shows that the change in generator speed result from a difference in the electrical torque and the mechanical torque. The equation can be rewritten based on the active power as follows:

$$2H \times \frac{d\Delta\omega}{dt} = \Delta P_{NC,ref} - \Delta P_{NC}, \tag{2}$$

where H and $\Delta P_{NC,ref}$ are the inertia constant of the generator and the desired wind source output obtained using power versus wind speed characteristics with the wind speed as its input. A description of the related calculations of $\Delta\omega$ is indicated in Figure 1. The wind turbine is specified by a time constant (T_a) as follows:

$$\frac{1}{1+sT_a}, \tag{3}$$

The effect of the conventional generators frequency changes on DFIG is determined by a filter with time constant T_r. A governor droop (R) is added, which is the rate of change of frequency with respect to generator power change. It shows a load sharing pattern of a particular generator. The activation of inertial and droop control loop determines the support to frequency regulation problem. The frequency change is an input to droop control loop and it provides additional active power support to the system. A washout filter with time constant T_w is added in the model to provide non-zero output during the transient periods only and is able to reject the steady-state frequency deviations. The power change of generator ΔP_f^* can be shown as follows:

$$\Delta P_f^* = \frac{1}{R}\Delta x_2, \tag{4}$$

where, R and Δx_2 are the parameter of speed regulation and sensed frequency changes, respectively. Equation (4) shows the droop control of the governor of the wind turbine as the primary frequency control [37]. Finally, the total injected power ΔP_{NC} into the power system can be written as:

$$\Delta P_{NC} = \Delta P_f^* + \Delta P_\omega^*. \tag{5}$$

It should be noted that an FOPID controller for supplementary loop of DFIGs is applied, which is known as speed controller. To tune the parameters of FOPID controller, the SCA method is employed in this study. Figure 3 illustrates the considered power system integrated with DFIG-based wind farms in each area. Detailed data about the system parameters and their description are accessible in Reference [26].

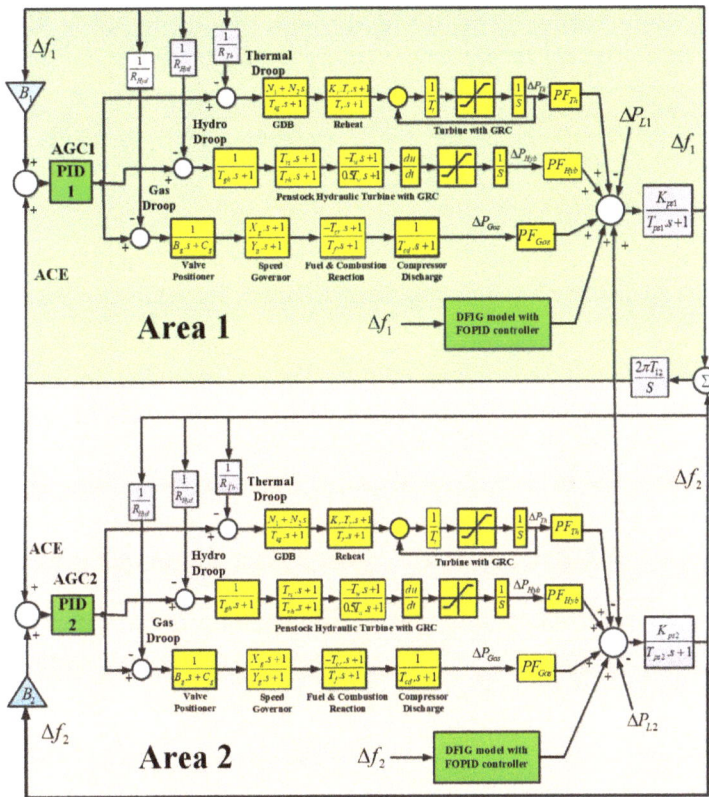

Figure 3. Two-area power system in the presence of wind farms.

4.2. Design Procedure of SCA-Based FOPID Controller

4.2.1. Fractional Order PID Controller

Benefit from a high degree of freedom (by selecting suitable values for λ and μ), fractional order controllers are regularly more adequate than usually used integer order models like PI and PID controllers. The general form of FOPID controller is depicted in Figure 3 and it is mathematically represented as:

$$PI^{\lambda}D^{\mu} = K_P + \frac{K_I}{S^{\lambda}} + K_D S^{\mu}. \tag{6}$$

Herein, λ and μ represent the fractional order operators often adjustable in the range of (0, 1) and K_P, K_I and K_D are the proportional, integral and differential gains of FOPID controller respectively. It can be seen from Figure 4 that, the FOPID controller can operate like simple classic controllers (P, I, PI, PID) by selecting 0 and 1 for λ and μ [70].

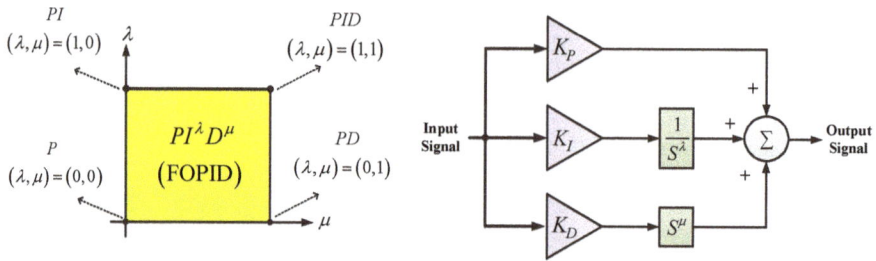

Figure 4. The general structure of the proposed fractional order proportional-integral-differential (FOPID) controller.

There are 5 parameters for the FOPID controller, which should be optimized by optimization algorithms. In this study, the SCA approach is utilized for optimization of FOPID controller. For this purpose and at the first step, the objective function and adjustable parameters of the controllers should be determined.

4.2.2. Objective Function Formulation and Employed Optimization Algorithm

In order to damp the frequency deviations and tie-line power oscillations effectively, considering a suitable objective function is essential. The considered objective function should be defined such that the output properties in the time domain, such as peak overshoot, peak time and settling time of the considered variables are minimized. In this paper, the integral of time multiplied squared error (ITSE) performance index is considered as the objective function.

$$ITSE = \int_{0}^{T_{sim}} t\,[\Delta f_1^2 + \Delta f_2^2 + \Delta P_{12}^2]\,dt, \tag{7}$$

where T_{sim} denotes the simulation time. Δf_1 and Δf_2 are the frequency deviations of area 1 and area 2, respectively. ΔP_{12} is the tie-line power deviation. The ITSE index uses advantages of both ISE and ITAE indices. The ITSE utilizes squared error and time multiplication to diminish large oscillations and decrease long settling time. The SCA optimization algorithm is employed here to minimize the ITSE index and optimize all the adjustable parameters subject to constraints. The constraints for a FOPID controller are as follows:

$$\begin{aligned} K_P^{\min} \le K_P \le K_P^{\max}, K_I^{\min} \le K_I \le K_I^{\max} \\ K_D^{\min} \le K_D \le K_D^{\max}, 0 \le \lambda \le 1, 0 \le \mu \le 1 \end{aligned} \tag{8}$$

4.2.3. Sine-Cosine Algorithm

Sine-cosine Algorithm (SCA) is a new population-based optimization approach for solving optimization problems [71]. The optimization process in SCA is based-on a set of random solutions that applies a sine and cosine functions based-on a mathematical model to fluctuate outwards or towards the best solution. The SCA has two phases known as exploration and exploitation in the optimization process. To establish exploration and exploitation of the search space to rapid achieve the optimal solution, this algorithm uses many random and adaptive variables. The position updating equation related to any search agent X_i can be written as follows:

$$X_i^{t+1} = \begin{cases} X_i^t + r_1 \times \sin(r_2) \times |r_3 P_i^t - X_i^t| & r_4 < 0.5 \\ X_i^t + r_1 \times \cos(r_2) \times |r_3 P_i^t - X_i^t| & r_4 \ge 0.5 \end{cases} \quad r_1 = c - t\frac{c}{T}, \tag{9}$$

$$r_1 = c - t\frac{c}{T}, \tag{10}$$

where X_i^t and P_i^t are the position of the current solution and the position of the best solution in j-th dimension at the t-th iteration, respectively. Furthermore, T is the maximum number of iterations, and c is a constant value. r_1, r_2, r_3 and r_4 are random numbers. r_1 is a control parameter that states the next position, which could be either in the space between the solution and destination or outside; r_2 expresses how far the movement should be towards or outwards the destination. r_3 is a random weighting parameter for emphasizing ($r_3 > 1$) or deemphasizing ($r_3 < 1$) the effect of destination to define the distance; and, r_4 is a switching parameter, which switches between the sine and cosine components, equally.

5. Simulation Results

To validate the performance of the DFIGs equipped with the SCA optimized FOPID controllers, fourth scenarios are considered and evaluated in the considered two-area power system. Simulations are accomplished in the MATLAB/Simulink environment. The scenarios show the performance of proposed controller for a step load change, sinusoidal load change, effects of different level of wind penetration on AGC, and sensitivity analysis for 25% deviation in the loading condition and the synchronizing coefficient (T_{12}). The optimal gains of the FOPID and PID controllers, which are optimized by the SCA method are given in Table 2.

Table 2. Optimal parameters of controllers obtained by the sine-cosine algorithm (SCA).

Controller Type	Areas	K_P	K_I	K_D	λ	μ
FOPID-DFIG	Area 1	0.2104	−0.2002	0.7112	0.4704	0.6387
PID-DFIG	Area 1	0.1509	−0.1807	0.7409	-	-
PID-AGC	Area 1	0.8544	0.2979	0.5840	-	-
FOPID-DFIG	Area 2	0.2194	−0.2233	0.2901	0.4418	0.5794
PID-DFIG	Area 2	0.1644	−0.1907	0.6987	-	-
PID-AGC	Area 2	0.4811	0.2977	0.5133	-	-

5.1. First Scenario: Step Load Change

As the first scenario, performance of the DFIGs equipped with the SCA optimized FOPID controllers is investigated under a 0.01 p.u. step load change in area 1. The area frequency and tie-line power deviations are illustrated in Figure 5. As shown in this figure, the performance of the SCA-based FOPID controller with 10% wind penetration is compared with the SCA-based PID controller, with 10% wind penetration and the condition that there is no DFIGs in any areas. Area frequency and tie-line power oscillations are remarkably damped by the proposed SCA-based FOPID controller compared with two other controllers. The ITSE performance index, peak overshoot, peak time and settling time are shown in Table 3. The SCA-based FOPID controller has the lowest value of the ITSE index compared to the other controllers for the first scenario. Therefore, it can be concluded that the proposed controller improves the dynamic responses for the studied power system more efficiently. Furthermore, the settling time and peak overshoot are reduced by the proposed controller. Note that the peak overshoot unit for the frequency deviation is Hz; and for the tie-line power deviations is pu. In this study, settling time is the time required for an output to reach and remain within a ±5% error band following some input stimulus. Furthermore, minimum damping ratio provides a mathematical means of expressing the level of damping in a system relative to critical damping.

15

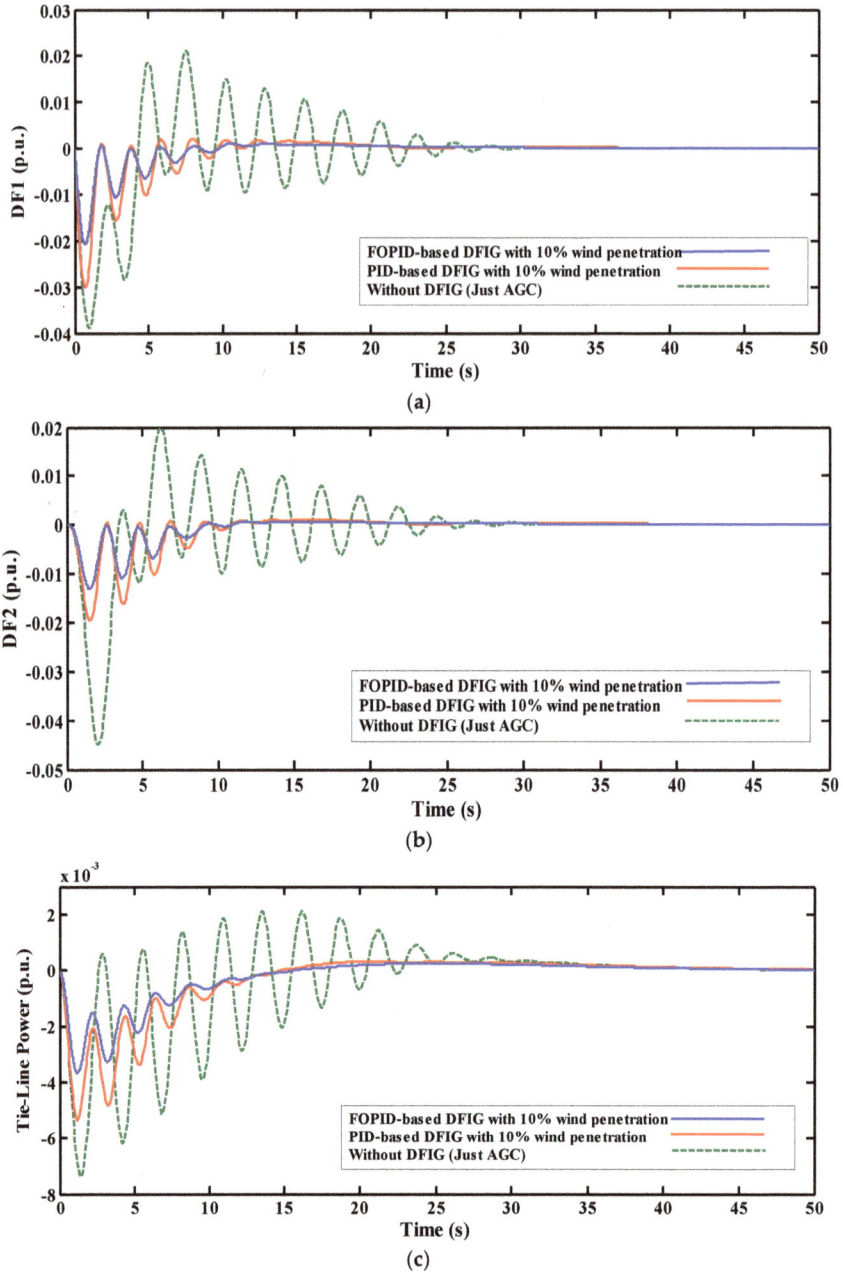

Figure 5. Frequency deviation and tie-line active power change in the first scenario, (**a**) frequency deviation in area 1; (**b**) frequency deviation in area 2; and (**c**) tie-line power deviation.

Table 3. Frequency deviation and tie-line power characteristics of FOPID, PID and just AGC.

Controller Type	Signal	MDR	PO	PT	ST	ITSE
FOPID-based DFIG & AGC	Δf_1		0.0208	0.6997	14.1210	
	Δf_2	0.1910	0.0131	1.5054	14.8400	0.0026
	ΔP_{12}		0.0036	1.1957	21.0161	
PID-based DFIG & AGC	Δf_1		0.0301	0.7311	17.0812	
	Δf_2	0.0658	0.0194	1.5536	18.3103	0.0041
	ΔP_{12}		0.0053	1.2089	23.4412	
Without DFIG (Just AGC)	Δf_1		0.0389	0.9309	28.1927	
	Δf_2	0.0498	0.0449	2.0776	27.1289	0.0333
	ΔP_{12}		0.0073	1.3934	32.5925	

MDR, minimum damping ratio; PO, peak overshoot; PT, peak time; ST, settling time.

5.2. Second Scenario: Comparison of Results in Different Level of Wind Penetration

In this scenario, the performance of the DFIGs equipped with the SCA optimized FOPID controller is investigated under different levels of wind power penetration. Herein, 10%, 15% and 20% penetration levels are considered for the wind power. The penetration level of wind power is increased through reducing the existing generator units by x%, i.e., an x% reduction in system inertia constant. In the other words, an x% increase in wind power is fulfilled through decreasing the inertia by x% [72]. In Figure 6, the system dynamic responses of 10%, 15% and 20% penetration levels of wind power are demonstrated. As can be seen, by increasing the wind power penetration level, although the inertia of wind system decreases, the proposed FOPID controller provides better frequency performance in high penetration level.

5.3. Third Scenario: Sinusoidal Load Change

The third scenario verifies the performance of the DFIGs equipped with the SCA optimized FOPID controllers in a sinusoidal load perturbation, which is applied in area 1. This sinusoidal load perturbation is formulated as follows:

$$\Delta P_d = -0.002 \sin(4t) + 0.002 \sin(4.7t) + 0.003 \sin(4.7t). \tag{11}$$

Under the supposed sinusoidal load perturbation, the area frequency and tie-line power oscillations are exhibited in Figure 7. The results illustrate that the oscillations are effectively damped by employing the FOPID controllers.

5.4. Fourth Scenario: Sensitivity Analysis

In the Fourth scenario, a sensitivity analysis is done to examine the robustness of the proposed controller under a wide variation in governor time constant of steam turbine (T_{sg}) and the synchronizing coefficient (T_{12}) separately. To do so, the T_{12} (as an indicator of active tie-line power between the areas) and T_{sg} are changed by ±25% of nominal value regarding the first scenario condition. The dynamic responses for the case of DFIGs equipped with the SCA optimized FOPID controllers after applying uncertainties in T_{12} and T_{sg} are depicted in Figures 8 and 9, respectively. Furthermore, the system damping characteristics for both changes are listed in Table 3.

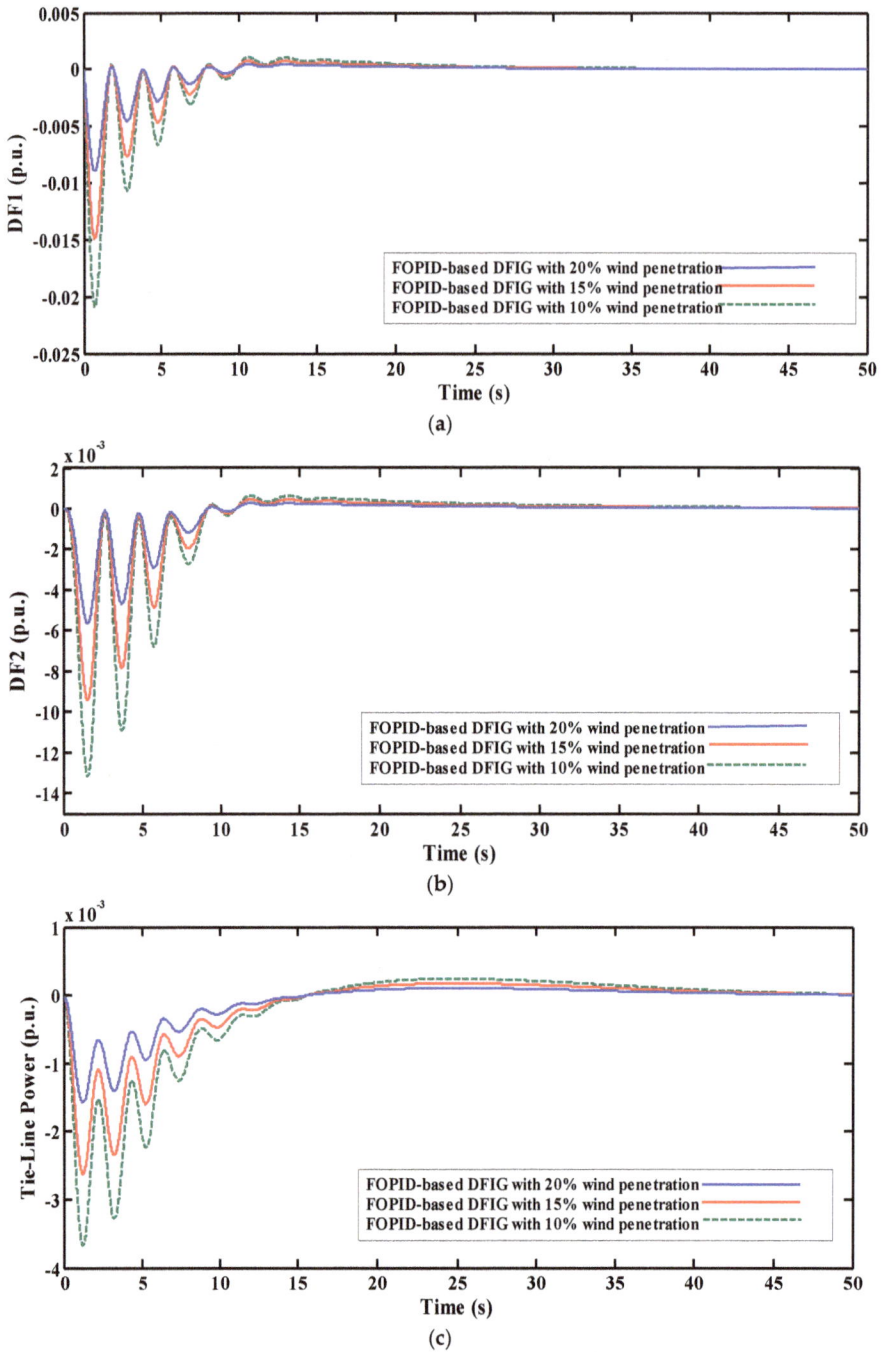

Figure 6. Frequency deviation and tie-line active power change in second scenario, (**a**) frequency deviation in area 1; (**b**) frequency deviation in area 2; and (**c**) tie-line power deviation.

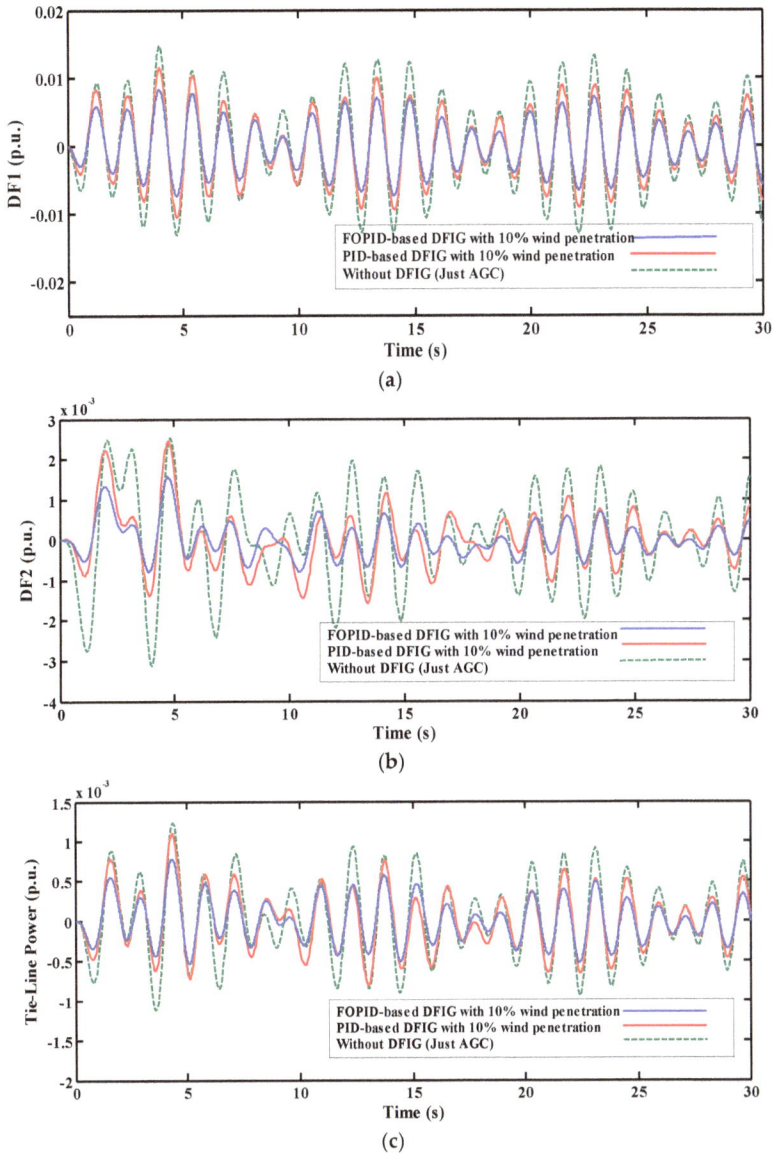

Figure 7. Frequency deviation and tie-line active power change in second scenario, (**a**) frequency deviation in area 1; (**b**) frequency deviation in area 2; and (**c**) tie-line power deviation.

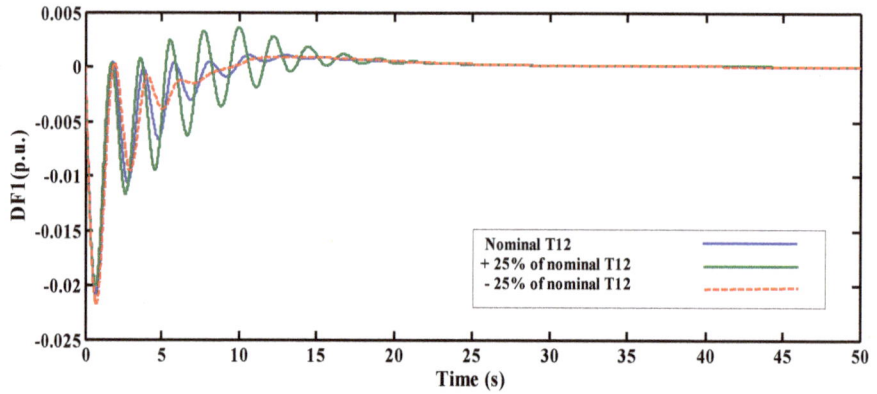

(a)

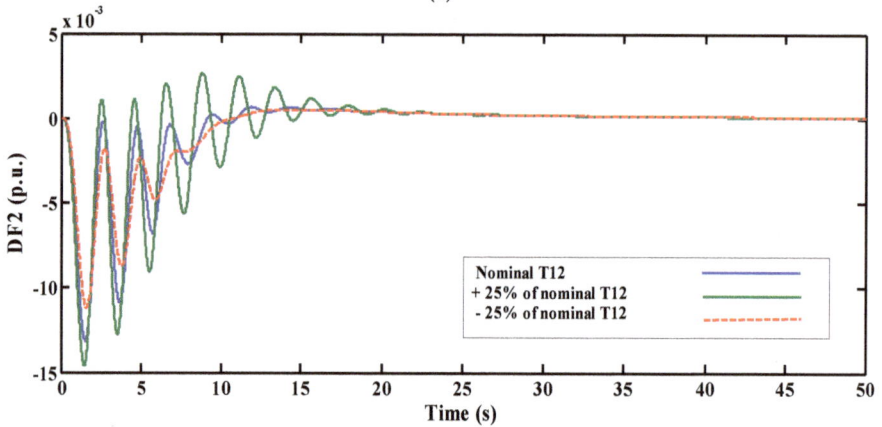

(b)

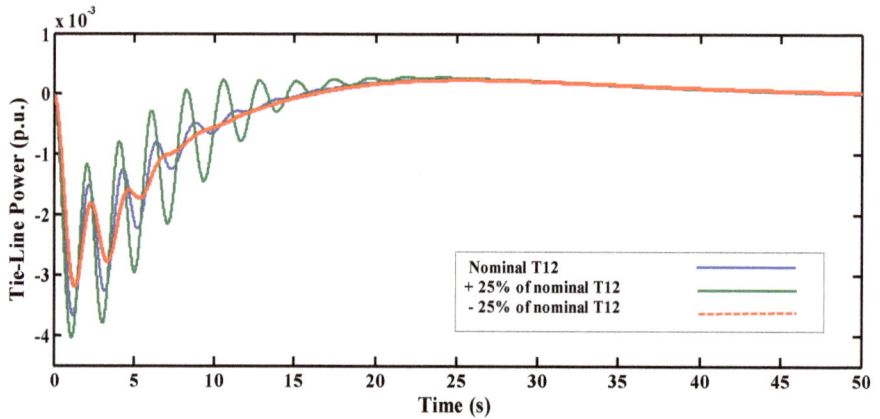

(c)

Figure 8. Simulation results for ±25% of synchronizing coefficient, (**a**) frequency deviation in area 1; (**b**) frequency deviation in area 2; and (**c**) tie-line power deviation.

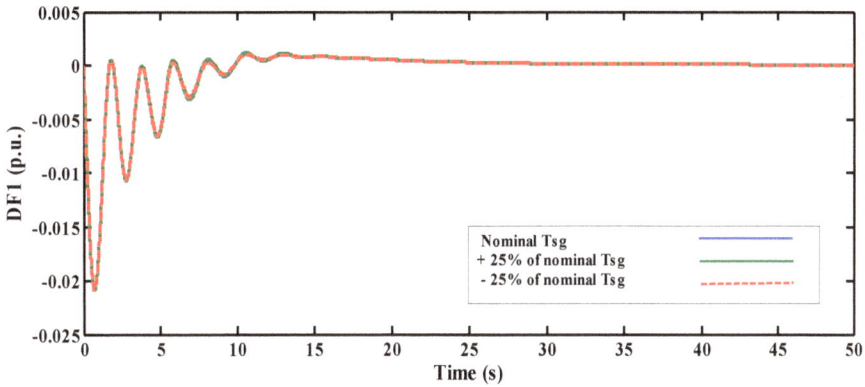

(a)

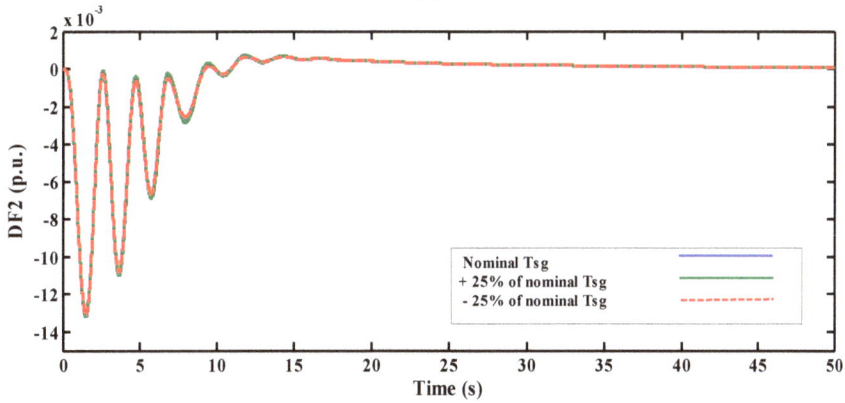

(b)

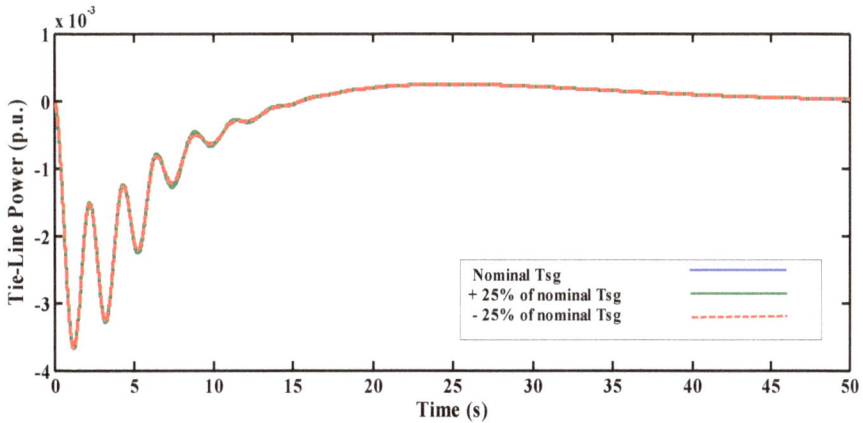

(c)

Figure 9. Simulation results for ±25% of time constant of steam turbine, (**a**) frequency deviation in area 1; (**b**) frequency deviation in area 2; and (**c**) tie-line power deviation.

It can be seen from the Figures 8 and 9, and in Table 4 that in comparison with the nominal responses demonstrated in Table 2 and Figure 5, the considered severe variations have negligible impacts on the system dynamic performance, since the damping ratios, performance indices and damping control measures deviate slightly from the nominal values so that the power system is still stable as before. Therefore, the adjustable parameters of the controllers are tuned for the nominal condition and it is not necessary to tune the parameters for the $\pm 25\%$ change in the T_{12} and T_{sg} again.

Table 4. Frequency deviation and tie-line power characteristics of FOPID controllers for $\pm 25\%$ variations in the time constant of steam turbine and synchronizing coefficient.

FOPID-based DFIG & AGC	Signal	MDR	PO	PT (s)	ST (s)	ITSE
$T_{12} + 25\%$	Δf_1		0.0201	0.6765	19.2995	
	Δf_2	0.0537	0.0146	1.4438	19.0713	0.0053
	ΔP_{12}		0.0040	1.1106	21.2130	
-25%	Δf_1		0.0217	0.7285	14.9600	
	Δf_2	0.1096	0.0112	1.5874	13.8402	0.0023
	ΔP_{12}		0.0031	1.3164	20.9504	
$T_{sg} + 25\%$	Δf_1		0.0209	0.6997	15.0300	
	Δf_2	0.0745	0.0132	1.5054	14.9200	0.0027
	ΔP_{12}		0.0036	1.1957	21.0041	
-25%	Δf_1		0.0208	0.6997	14.5210	
	Δf_2	0.0815	0.0131	1.5054	14.4467	0.0026
	ΔP_{12}		0.0036	1.1957	20.2161	

MDR, minim damping ratio; PO, peak overshoot; PT peak time; ST, settling time.

6. Conclusions and Future Directions

This paper presented a review on wind farms contribution to the automatic generation control of power systems. The applied control strategies and other alternatives for wind farms were investigated. Fractional order PID controller was deployed for DFIG wind turbines for more efficient contribution in the load frequency control of multi-area power systems. Four scenarios, including step and sinusoidal load changes, sensitivity analysis, and the effect of different levels of DFIG penetration, demonstrated that the proposed controller is efficient for the wind farms in the load frequency control. Minimizing the overshoot and settling time, and better oscillation damping were highlighted as the salient features for the SCA-based FOPID controller in the structure of DFIGs.

The role of wind farms to contribute in automatic generation control is still an active field of study for future research. The literature of control strategies in wind farm is limited to model predictive control, optimization and intelligent methods. On the other hand, application of lithium-ion battery energy storages, which nowadays are utilized in wind farms and other types of energy storage systems can be investigated in this area of research. Coordination between wind farms and photovoltaic farms as the frontier renewable energies to contribute in the automatic generation control can also be extended in the future.

Author Contributions: A.O. developed the main idea for this research study. R.K. collected the literature review. A.O. simulated the case studies. R.K., A.O., S.M. and F.B. analyzed the data and simulation results. R.K. and A.O. wrote the manuscript. S.M. and F.B. checked the manuscript and provided their comments on the paper.

Funding: This research received no external funding.

Conflicts of Interest: The authors declare no conflict of interest.

References

1. GWEC Report 2017. Available online: http://www.gwec.net/publications/global-wind-report-2/global-wind-report-2017-annual-market-update/ (accessed on 25 April 2017).

2.   Blaabjerg, F.; Ma, K. Wind energy systems. *Proc. IEEE* **2017**, *105*, 2116–2131.

3.   Li, H.; Chen, Zh. Overview of different wind generator systems and their comparisons. *IET Renew. Power Gener.* **2009**, *2*, 123–138. [CrossRef]

4.   Torkaman, H.; Keyhani, A. A review of design consideration for Doubly Fed Induction Generator based wind energy system. *Electr. Power Syst. Res.* **2018**, *160*, 128–141. [CrossRef]

5.   Khezri, R.; Bevrani, H. Voltage performance enhancement of DFIG-based wind farms integrated in large-scale power systems: Coordinated AVR and PSS. *Int. J. Electr. Power Energy Syst.* **2015**, *73*, 400–410. [CrossRef]

6.   Oshnoei, A.; Khezri, R.; Hagh, M.T.; Techato, K.; Muyeen, S.M.; Sadeghian, O. Direct probabilistic load flow in radial distribution systems including wind farms: An approach based on data clustering. *Energies* **2018**, *11*, 310. [CrossRef]

7.   Bousseau, P.; Belhomme, R.; Monnot, E.; Laverdure, N.; Boeda, D.; Roye, D.; Bacha, S. Contribution of wind farms to ancillary services. *CIGRE* **2006**, *21*, 1–11.

8.   Kayikci, M.; Milanovic, J.V. Reactive power control strategies for DFIG-based plants. *IEEE Trans. Energy Convers.* **2007**, *22*, 389–396.

9.   Gabash, A.; Li, P. On variable reverse power flow-part I: Active-reactive optimal power flow with reactive power of wind stations. *Energies* **2016**, *9*, 121. [CrossRef]

10.   Muyeen, S.M.; Tamura, J.; Murata, T. *Stability Augmentation of a Grid-Connected wind Farm*; Springer: Berlin, Germany, 2008.

11.   Golkhandan, K.R.; Aghaebrahimi, M.R.; Farshad, M. Control strategies for enhancing frequency stability by DFIGs in a power system with high percentage of wind power penetration. *Appl. Sci.* **2017**, *7*, 1140. [CrossRef]

12.   Jalali, M. DFIG Based Wind Turbine Contribution to System Frequency Control. Master's Thesis, University of Waterloo, Waterloo, ON, Canada, 2010; pp. 1–92.

13.   Tang, Y.; He, H.; Wen, J.; Liu, J. Power system stability control for a wind farm based on adaptive dynamic programming. *IEEE Trans. Smart Grid* **2015**, *6*, 166–177. [CrossRef]

14.   Hazari, M.R.; Mannan, M.A.; Muyeen, S.M.; Umemura, A.; Takahashi, R.; Tamura, J. Stability augmentation of a grid-connected wind farm by fuzzy-logic-controlled DFIG-based wind turbines. *Appl. Sci.* **2018**, *8*, 20. [CrossRef]

15.   Mishra, A.; Tripathi, P.M.; Chatterjee, K. A review of harmonic elimination techniques in grid connected doubly fed induction generator based wind energy system. *Renew. Sustain. Energy Rev.* **2018**, *89*, 1–15. [CrossRef]

16.   Liu, Y.; Garcia, J.R.; King, T.J.; Liu, Y. Frequency regulation and oscillation damping contributions of variable-speed wind generators in the U.S. eastern interconnection (EI). *IEEE Trans. Sustain. Energy* **2015**, *6*, 951–958. [CrossRef]

17.   Ma, H.; Chowdhury, B. Working towards frequency regulation with wind plants: Combined control approaches. *IET Renew. Power Gener.* **2010**, *4*, 308–316. [CrossRef]

18.   Yingcheng, X.; Nengling, T. Review of contribution to frequency control through variable speed wind turbine. *Renew. Energy* **2011**, *36*, 1671–1677. [CrossRef]

19.   Kim, M.-K. Optimal Control and Operation Strategy for Wind Turbines Contributing to Grid Primary Frequency Regulation. *Appl. Sci.* **2017**, *7*, 927. [CrossRef]

20.   Sun, Y.; Zhang, Z.; Li, G.; Lin, J. Review on frequency control of power systems with wind power penetration. In Proceedings of the 2010 International Conference on Power System Technology (POWERCON), Hangzhou, China, 24–28 October 2010; pp. 1–8.

21.   Khezri, R.; Golshannavaz, S.; Vakili, R.; Memar-Esfahani, B. Multi-layer under frequency load shedding in back-pressure smart industrial microgrids. *Energy* **2017**, *132*, 96–105. [CrossRef]

22.   Bevrani, H. *Robust Power System Frequency Control*, 2nd ed.; Springer: Berlin, Germany, 2014.

23.   Golshannavaz, S.; Khezri, R.; Esmaeeli, M.; Siano, P.L. A two-stage robust-intelligent controller design for efficient LFC based on Kharitonov theorem and fuzzy logic. *J. Ambient Intell. Hum. Comput.* **2017**, *9*, 1445–1454. [CrossRef]

24.   Mohamed, TH.; Bevrani, H.; Hassan, A.; Hiyama, T. Decentralized model predictive based load frequency control in an interconnected power system. *Energy Convers. Manag.* **2011**, *16*, 1208–1214. [CrossRef]

25. Khezri, R.; Golshannavaz, S.; Shokoohi, S.; Bevrani, H. Fuzzy Logic Based Fine-tuning Approach for Robust Load Frequency Control in a Multi-area Power System. *Electr. Power Compon. Syst.* **2016**, *44*, 2073–2083. [CrossRef]

26. Oshnoei, A.; Hagh, M.T.; Khezri, R.; Mohammadi-Ivatloo, B. Application of IPSO and fuzzy logic methods in electrical vehicles for efficient frequency control of multi-area power systems. In Proceedings of the 2017 Iranian Conference on Electrical Engineering (ICEE), Tehran, Iran, 2–4 May 2017; pp. 1349–1354.

27. Cam, E.; Gorel, G.; Mamur, H. Use of the Genetic Algorithm-Based Fuzzy Logic Controller for Load-Frequency Control in a Two Area Interconnected Power System. *Appl. Sci.* **2017**, *7*, 308. [CrossRef]

28. Khezri, R.; Oshnoei, A.; Tarafdar Hagh, M.; Muyeen, S.M. Coordination of Heat Pumps, Electric Vehicles and AGC for Efficient LFC in a Smart Hybrid Power System via SCA-Based Optimized FOPID Controllers. *Energies* **2018**, *11*, 420. [CrossRef]

29. Chang-Chien, L.; Chih-Che, S.; Yu-Ju, Y. Modeling of wind farm participation in AGC. *IEEE Trans. Power Syst.* **2014**, *29*, 1204–1211. [CrossRef]

30. Chang-Chien, L.R.; Lin, W.T.; Yin, Y.C. Enhancing frequency response control by DFIGs in the high wind penetrated power systems. *IEEE Trans. Power Syst.* **2011**, *26*, 710–718. [CrossRef]

31. Mauricio, J.M.; Marano, A.; Gómez-Expósito, A.; Ramos, J.L.M. Frequency regulation through variable-speed wind energy conversion systems. *IEEE Trans. Power Syst.* **2009**, *24*, 173–180. [CrossRef]

32. Hazari, R.M.; Mannan, M.A.; Muyeen, S.M.; Umemura, A.; Takahashi, R.; Tamura, J. Transient Stability Augmentation of Hybrid Power System Based on Synthetic Inertia Control of DFIG. In Proceedings of the Australasian Universities Power Engineering Conference 2017 (AUPEC2017), Melbourne, Australia, 19–22 November 2017; pp. 1–6.

33. Wang, Y.; Meng, J.; Zhang, X.; Xu, L. Control of PMSG-Based wind turbines for system inertial response and power oscillation damping. *IEEE Trans. Sustain. Energy* **2015**, *6*, 565–574. [CrossRef]

34. Ataee, S.; Khezri, R.; Feizi, M.R.; Bevrani, H. Investigating the impacts of wind power contribution on the short-term frequency performance. In Proceedings of the Smart Grid Conference (SGC), Tehran, Iran, 9–10 December 2014; pp. 1–6.

35. Zhang, Z.S.; Sun, Y.Z.; Lin, J.; Li, G.J. Coordinated frequency regulation by doubly fed induction generator based wind power plants. *IET Renew. Power Gen.* **2010**, *6*, 38–47. [CrossRef]

36. Kleftakis, V.; Rigas, A.; Papadimitriou, C.; Katsoulakos, N.; Moutis, P.; Hatziargyriou, N. Contribution to frequency control by a PMSG wind turbine in a Diesel-Wind Turbine microgrid for rural electrification. In Proceedings of the 9th Mediterranean Conference on Power Generation, Transmission Distribution and Energy Conversion (MedPower 2014), Athens, Greece, 2–5 November 2014; pp. 1–5.

37. Vidyanandan, K.V.; Senroy, N. Primary frequency regulation by deloaded wind turbines using variable droop. *IEEE Trans. Power Syst.* **2013**, *28*, 837–846. [CrossRef]

38. Yingcheng, X.; Nengling, T. System frequency regulation in doubly fed induction generators. *Int. J. Electr. Power Energy Syst.* **2012**, *43*, 977–983. [CrossRef]

39. Ataee, S.; Khezri, R.; Feizi, M.R.; Bevrani, H. Impacts of wind and conventional power coordination on the short-term frequency performance. In Proceedings of the 23rd Iranian Conference on Electrical Engineering (ICEE), Tehran, Iran, 10–14 May 2015; pp. 1–6.

40. Ataee, S.; Bevrani, H. Improvement of primary frequency control by inertial response: Coordination between wind and conventional power plants. *Int. Trans. Electr. Energy Syst.* **2017**, *27*, e2340. [CrossRef]

41. Verma, Y.P.; Kumar, A. Dynamic contribution of variable-speed wind energy conversion system in system frequency regulation. *Front. Energy* **2012**, *3*, 184–192. [CrossRef]

42. Badmasti, B.; Bevrani, H. On contribution of DFIG wind turbines in the secondary frequency control. In Proceedings of the 1st Conference on New Research Achievements in Electrical and Computer Engineering, Tehran, Iran, 12 May 2016; pp. 1–7.

43. Jalali, M.; Bhattacharya, K. Frequency regulation and AGC in isolated systems with DFIG-based wind turbines. In Proceedings of the IEEE Power & Energy Society General Meeting, Vancouver, BC, Canada, 21–25 July 2013; pp. 1–5.

44. Aziz, A.; Shafiullah, G.M.; Stojcevski, A.; Amanullah, M.T.O. Participation of DFIG based wind energy system in load frequency control of interconnected multi-generation power system. In Proceedings of the Australasian Universities Power Engineering Conference (AUPEC), Perth, Australia, 28 September–1 October 2014; pp. 1–6.

45. Ibraheem; Niazi, K.R.; Sharma, G. Study on dynamic participation of wind turbines in automatic generation control of power systems. *Electr. Power Compon. Syst.* **2015**, *43*, 44–55.

46. Abo-Elyousr, F.K. Load frequency controller design for two-area interconnected power system with DFIG-based wind turbine via ant colony algorithm. In Proceedings of the Eighteenth International Middle East Power Systems Conference (MEPCON), Cairo, Egypt, 27–29 December 2016; pp. 1–8.

47. Bhatt, P.; Roy, R.; Ghoshal, S.P. Dynamic participation of doubly fed induction generator in automatic generation control. *Renew. Energy* **2011**, *36*, 1203–1213. [CrossRef]

48. Oshnoei, A.; Khezri, R.; Ghaderzadeh, M.; Parang, H.; Oshnoei, S.; Kheradmandi, M. Application of IPSO algorithm in DFIG-based wind turbines for efficient frequency control of multi-area power systems. In Proceedings of the Smart Grids Conference (SGC), Tehran, Iran, 20–21 December 2017; pp. 1–6.

49. Preeti; Sharma, V.; Naresh, R.; Pulluri, H. Automatic generation control of multi-source interconnected power system including DFIG wind turbine. In Proceedings of the 1st IEEE International Conference on Power Electronics, Intelligent Control and Energy Systems (ICPEICES), Delhi, India, 4–6 July 2016; pp. 1–6.

50. Pappachen, A.; Fathima, A.P. Genetic algorithm based PID controller for a two-area deregulated power system along with DFIG units. In Proceedings of the IEEE Sponsored 2nd International Conference on Innovations in Information, Embedded and Communication systems (ICIIECS), Coimbatore, India, 19–20 March 2015; pp. 1–6.

51. Chaine, S.; Tripathy, M.; Satpathy, S. NSGA-II based optimal control scheme of wind thermal power system for improvement of frequency regulation characteristics. *Ain Shams Eng. J.* **2015**, *6*, 851–863. [CrossRef]

52. Chaine, S.; Tripathy, M.; Jain, D. Non dominated Cuckoo search algorithm optimized controllers to improve the frequency regulation characteristics of wind thermal power system. *Eng. Sci. Technol. Int. J.* **2017**, *20*, 1092–1105. [CrossRef]

53. Sahu, P.C.; Prusty, R.C.; Panda, S. ALO optimized NCTF controller in multi-area AGC system integrated with WECS based DFIG system. In Proceedings of the International Conference on Circuits Power and Computing Technologies (ICCPCT), Paris, France, 21–22 September 2017; pp. 1–6.

54. Kumar, A.; Sathans. Impact study of DFIG based wind power penetration on LFC of a multi-area power system. In Proceedings of the Annual IEEE India Conference (INDICON), New Delhi, India, 17–20 December 2015; pp. 1–6.

55. Azizipanah-Abarghooee, R.; Malekpour, M.; Zare, M.; Terzija, V. A new inertia emulator and fuzzy-based LFC to support inertial and governor responses using Jaya algorithm. In Proceedings of the IEEE Power & Energy Society General Meeting (PESGM), Boston, MA, USA, 17–21 July 2016; pp. 1–5.

56. Sharma, G.; Niazi, K.R.; Ibraheem. Recurrent ANN based AGC of a two-area power system with DFIG based wind turbines considering asynchronous tie-lines. In Proceedings of the IEEE International Conference on Advances in Engineering & Technology Research (ICAETR), Unnao, India, 1–2 August 2014; pp. 1–5.

57. Sharma, G.; Niazi, K.R.; Ibraheem; Bansal, R.C. LS-SVM based AGC of power system with dynamic participation from DFIG based wind turbines. In Proceedings of the 3rd Renewable Power Generation Conference (RPG), Naples, Italy, 24–25 September 2014; pp. 1–6.

58. Qudaih, Y.S.; Bernard, M.; Mitani, Y.; Mohamed, T.H. Model predictive based load frequency control design in the presence of DFIG wind turbine. In Proceedings of the 2nd International Conference on Electric Power and Energy Conversion Systems (EPECS), Sharjah, The United Arab Emirates, 15–17 November 2012; pp. 1–6.

59. Mohamed, T.H.; Morel, J.; Bevrani, H.; Hiyama, T. Model predictive based load frequency control design concerning wind turbines. *Int. J. Electr. Power Energy Syst.* **2012**, *43*, 859–867. [CrossRef]

60. Ma, M.; Liu, X.; Zhang, C. LFC for multi-area interconnected power system concerning wind turbines based on DMPC. *IET Gen. Transm. Distrib.* **2017**, *11*, 2689–2696. [CrossRef]

61. Zhang, Y.; Liu, X.; Qu, B. Distributed model predictive load frequency control of multi-area power system with DFIGs. *IEEE/CAA J. Autom. Sin.* **2017**, *4*, 125–135. [CrossRef]

62. Kawakami, N.; Iijima, Y.; Fukuhara, M.; Bando, M.; Sakanaka, Y.; Ogawa, K.; Matsuda, T. Development and field experiences of stabilization system using 34MW NAS batteries for a 51 MW wind farm. In Proceedings of the IEEE International Symposium on Industrial Electronics, Istanbul, Turkey, 1–4 June 2014; pp. 2371–2376.

63. Hui, W.; Wen, T. Load frequency control of power systems with wind turbines through flywheels. In Proceedings of the 27th Chinese Control and Decision Conference (CCDC), Qingdao, China, 23–25 May 2015; pp. 1–5.

64. Bhatt, P.; Roy, R.; Ghoshal, S.P. Dynamic contribution of DFIG along with SMES for load frequency control of interconnected restructured power systems. In Proceedings of the 10th International Conference on Environment and Electrical Engineering, Rome, Italy, 8–11 May 2011; pp. 1–6.
65. Kouba, N.Y.; Menaa, M.; Hasni, M.; Boudour, M. LFC enhancement concerning large wind power integration using new optimised PID controller and RFBs. *IET Gen. Transm. Distrib.* **2016**, *10*, 4065–4077. [CrossRef]
66. Bhatt, P.; Ghoshal, S.P.; Roy, R. Coordinated control of TCPS and SMES for frequency regulation of interconnected restructured power systems with dynamic participation from DFIG based wind farm. *Renew. Energy* **2012**, *40*, 40–50. [CrossRef]
67. Kumar, A.; Suhag, S. Effect of TCPS, SMES, and DFIG on load frequency control of a multi-area multi-source power system using multi-verse optimized fuzzy-PID controller with derivative filter. *J. Vib. Control* **2017**. [CrossRef]
68. Dhundhara, S.; Verma, Y.P. Evaluation of CES and DFIG unit in AGC of realistic multisource deregulated power system. *Int. Trans. Electr. Energy Syst.* **2017**, *27*, e2304. [CrossRef]
69. Shankar, R.; Pradhan, R.; Sahoo, S.B.; Chatterjee, K. GA based improved frequency regulation characteristics for thermal-hydro-gas & DFIG model in coordination with FACTS and energy storage system. In Proceedings of the 3rd Int'l Conf. on Recent Advances in Information Technology, Dhanbad, India, 3–5 March 2016; pp. 1–6.
70. Pan, I.; Das, S. Fractional order AGC for distributed energy resources using robust optimization. *IEEE Trans. Smart Grid* **2015**, *7*, 2175–2186. [CrossRef]
71. Mirjalili, S. SCA: A sine cosine algorithm for solving optimization problems. *Knowl.-Based Syst.* **2016**, *96*, 120–133. [CrossRef]
72. Dahiya, P.; Sharma, V.; Sharma, R.N. Optimal generation control of interconnected power system including DFIG-based wind turbine. *IETE J. Res.* **2015**, *61*, 285–300. [CrossRef]

applied
sciences

MDPI

Review

A Review on Fault Current Limiting Devices to Enhance the Fault Ride-Through Capability of the Doubly-Fed Induction Generator Based Wind Turbine

Seyed Behzad Naderi [1], Pooya Davari [2], Dao Zhou [2], Michael Negnevitsky [1,*] and Frede Blaabjerg [2]

[1]   School of Engineering, University of Tasmania, Hobart, TAS 7005, Australia;
      Seyedbehzad.Naderi@utas.edu.au
[2]   The Department of Energy Technology, Aalborg University, 9220 Aalborg, Denmark; pda@et.aau.dk (P.D.);
      zda@et.aau.dk (D.Z.); fbl@et.aau.dk (F.B.)
*   Correspondence: michael.negnevitsky@utas.edu.au; Tel.: +61-362-267-613

Received: 27 September 2018; Accepted: 23 October 2018; Published: 25 October 2018

Abstract: The doubly-fed induction generator has significant features compared to the fixed speed wind turbine, which has popularised its application in power systems. Due to partial rated back-to-back converters in the doubly-fed induction generator, fault ride-through capability improvement is one of the important subjects in relation to new grid code requirements. To enhance the fault ride-through capability of the doubly-fed induction generator, many studies have been carried out. Fault current limiting devices are one of the techniques utilised to limit the current level and protect the switches, of the back-to-back converter, from over-current damage. In this paper, a review is carried out based on the fault current limiting characteristic of fault current limiting devices, utilised in the doubly-fed induction generator. Accordingly, fault current limiters and series dynamic braking resistors are mainly considered. Operation of all configurations, including their advantages and disadvantages, is explained. Impedance type and the location of the fault current limiting devices are two important factors, which significantly affect the behaviour of the doubly-fed induction generator in the fault condition. These two factors are studied by way of simulation, basically, and their effects on the key parameters of the doubly-fed induction generator are investigated. Finally, future works, in respect to the application of the fault current limiter for the improvement of the fault ride-through of the doubly-fed induction generator, have also been discussed in the conclusion section.

Keywords: fault current limiters; doubly-fed induction generator; fault ride-through; superconductor; series dynamic braking resistor

1. Introduction

With the high penetration level of wind energy conversion systems in the grid, power system operators encounter new challenging issues, which could affect the stability of the power system. Therefore, keeping the wind turbines connected to the grid during faults with high wind power penetration is important, which is known as fault ride-through (FRT) capability [1].

Due to many significant characteristics of doubly-fed induction generator (DFIG) based wind turbines, they are widely employed in the power system, especially for the multi-megawatt applications [2]. In the configuration of the DFIG, the stator of the DFIG is directly connected to the grid, and the rotor circuit is linked to the network by partial-scale back-to-back voltage source converters. During the fault condition, a transient over-current goes through the rotor circuit towards the rotor side converter (RSC). The rotor over-current can either trip out the DFIG or damage the power electronic devices [3]. Therefore, keeping the DFIG based wind turbine connected to the utility and preventing the equipment from damage are important during the fault. For secure power system operation, the wind turbines should meet the grid requirements. The FRT requirement varies from country to country, following characteristics of the power system as shown in Figure 1 [4]. Among the different grid codes, which are regulated by the various operators, the E.ON grid code has the most severe FRT requirements [1,5]. With regard to E.ON grid code, even if the point of the common coupling (PCC) voltage drops to zero for 150 ms after the fault occurrence, the wind turbine must not be disconnected from the grid.

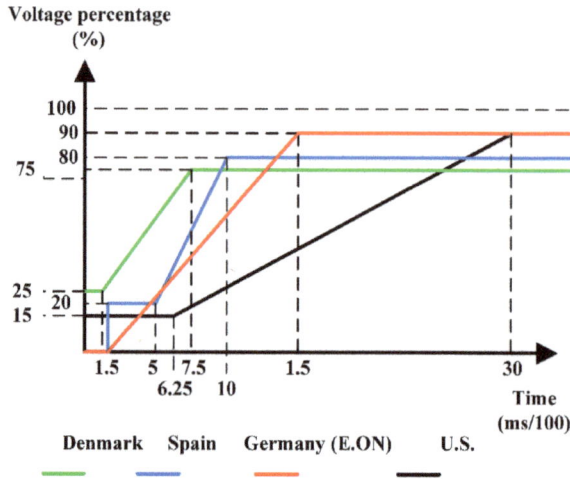

Figure 1. The grid code requirement for the fault ride-through of the wind turbine in the different countries [4]. Adapted with permission from [4], Copyright Publisher, 2015.

In the literature, different methods have been studied to enhance the FRT capability of the DFIG [6]. The previous introduced approaches can be classified as crowbar (active and passive) [7,8], DC braking chopper [9], new configurations for the DFIG [10,11], the advanced control techniques in back-to-back voltage source converters [12], and adding new hardware into the DFIG.

When adding new hardware into the DFIG, one of the common techniques is to employ fault current limiting devices to improve the FRT capability of the DFIG. Fault current limiters (FCLs) and series dynamic braking resistors (SDBRs) are placed in series connection, in different locations, in the DFIG (the stator side, the terminal, the DC link between the RSC and the DC-link capacitor, and the rotor side). In this paper, a review of most of the existing literature, which investigates the fault current limitation in the DFIG, is carried out. To improve the FRT capability of the DFIG, different configurations of the FCLs have been employed. Figure 2 shows the diagram of all employed FCLs to enhance the FRT capability of the DFIG. The operational behaviour of each configuration is briefly discussed in normal and fault conditions. The FCL structures can be categorized with respect to their impedance type during the fault condition. In the simulation section, the impedance type and the FCL's location will be discussed, which have various impacts on the key parameters of the DFIG. In this review, the detailed operation of the configurations is not considered, only the functionality is considered. Moreover, the steady state of the fault condition is discussed, which essentially depends on the type and size of the impedance in the fault current limiting devices.

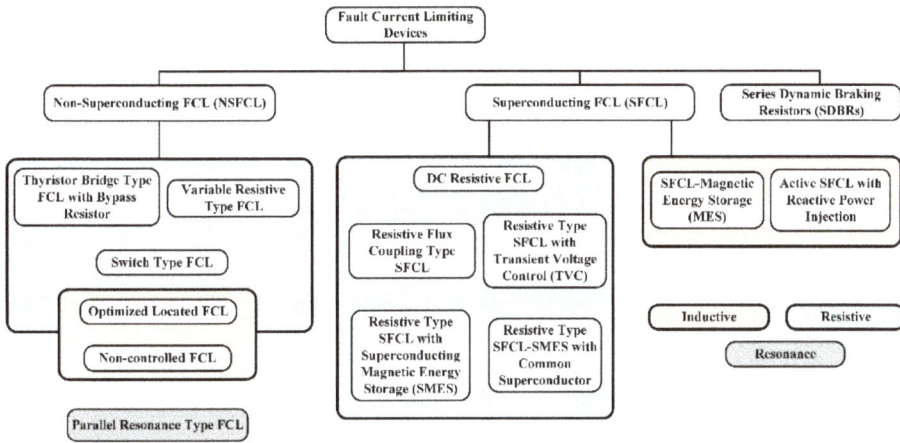

Figure 2. Diagram of the fault ride-through (FRT) capability enhancement by fault current limiting devices in the doubly-fed induction generator. FCL: Fault current limiter.

In the following, the operation of the DFIG is briefly expressed with the RSC's and grid side converter (GSC)'s control circuits. In Section 3, other approaches are mentioned with their pros and cons. In Section 4, an overview of almost all fault current limiting devices is studied, with their configuration and behaviour in normal condition and in the fault condition. Section 5 shows how the impedance and the location of the fault current limiting devices affect the operation of the DFIG in the steady state of the fault condition. Furthermore, a discussion is raised on the simulation results. Finally, a conclusion is given.

2. Doubly-Fed Induction Generator

The configuration of the DFIG is shown in Figure 3A. During the fault condition, a very high electromotive force (EMF) is induced in the rotor due to the electromagnetic interaction with the stator and the rotor of the DFIG, which leads to over-voltage and over-current in the rotor side. The over-voltage and the over-current could damage the RSC components if the semiconductor switches are continuously trigged in the fault condition. In this situation, the rotor high-transient over-current passes through the switches.

Meanwhile, the excess active power, which is not able to be delivered to the power system in the fault condition, charges the DC-link capacitor and rapidly increases the DC-link voltage. So, with regard to Figure 3A, the DC braking chopper typically operates to dissipate the excess energy and keep the DC-link voltage (V_{DC}) in a constant value. It should be noted that, because of the operation of the DC braking chopper and the time-varying characteristics of the electromagnetic interaction with the stator and the rotor, the response of the DFIG is very non-linear. However, during the fault, the DC-link voltage is almost kept constant when the DC braking chopper operates. In a static stator-oriented reference frame, the rotor and the stator fluxes, $\vec{\psi}_r$, $\vec{\psi}_s$, and the rotor and the stator voltages, $\vec{v}_r$, $\vec{v}_s$, are expressed as follows with respect to the Park model of the DFIG [13]:

$$\vec{v}_s = R_s \vec{i}_s + \frac{d\vec{\psi}_s}{dt} \tag{1}$$

$$\vec{v}_r = R_r \vec{i}_r + \frac{d\vec{\psi}_r}{dt} - j\omega_r \vec{\psi}_r \tag{2}$$

$$\vec{\psi}_s = L_s \vec{i}_s + L_m \vec{i}_r \tag{3}$$

$$\vec{\psi}_r = L_m \vec{i}_s + L_r \vec{i}_r \tag{4}$$

whereby $\vec{i}$, ω, R, and L represent current, angular frequency, resistance, and inductance, respectively. Additionally, stator, mutual, and rotor parameters are mentioned by subscripts of s, m, and r, respectively. By means of (3) and (4), $\vec{\psi}_r$ is computed with respect to $\vec{i}_r$ and $\vec{\psi}_s$. (5) expresses $\vec{\psi}_r$ as follows:

$$\begin{cases} \vec{\psi}_r = \frac{L_m \vec{\psi}_s}{L_s} + \sigma L_r \vec{i}_r \\ \sigma = 1 - \frac{L_m^2}{L_s L_r} \end{cases} \tag{5}$$

whereby the leakage coefficient is represented by σ. To calculate the rotor voltage with respect to (2) and (5), the following expression is deduced:

$$\vec{v}_r = \overbrace{\frac{L_m}{L_s}\left(\frac{d}{dt} - j\omega_r\right)\vec{\psi}_s}^{\vec{v}_{ro}} + \left(R_r + \sigma L_r(\frac{d}{dt} - j\omega_r)\right)\vec{i}_r \tag{6}$$

whereby $\vec{v}_{ro}$ denotes the rotor voltage when the rotor is in an open circuit condition and the rotor current is zero. If the reference frame is changed from a static stator-oriented reference frame to a rotor-oriented reference frame, the rotor voltage is expressed in (7):

$$\vec{v}_r^r = \vec{v}_{ro}^r + \left(R_r + \sigma L_r\frac{d}{dt}\right)\vec{i}_r^r \tag{7}$$

whereby $\vec{v}_r^r$, $\vec{v}_{ro}^r$, and $\vec{i}_r^r$ are the rotor voltage, the open circuit rotor voltage, and the rotor current in the rotor-oriented reference frame, respectively. With regard to (7), the rotor side electrical circuit is modelled by a three-phase voltage source of $\vec{v}_{ro}^r$, the rotor resistance (R_r), and the transient inductance (σL_r) during the fault condition. The level of rotor transient over-current will be changed with regard to the fault instant, the type of fault, and the voltage sag depth [14]. For instance, during a symmetrical grid fault, the open circuit rotor voltage ($\vec{v}_{ro}^r$) includes two expressions as follows:

$$\vec{v}_{ro}^r = (1-p)V_s\frac{L_m}{L_s}se^{js\omega_s t} - \frac{L_m}{L_s}\left(\frac{1}{\tau_s} + j\omega_r\right)\frac{pV_s}{j\omega_s}e^{-j\omega_r t}e^{\frac{-t}{\tau_s}} \tag{8}$$

whereby the pre-fault stator voltage magnitude, the voltage sag depth, and the slip are represented by V_s, p, and s during the symmetrical grid fault, respectively. Also, τ_s decaying time constant is equal to L_s/R_s.

The schematic of control circuits for the rotor side converter and the grid side converter are presented in Figure 3B [15]. In both control systems, proportional–integral (PI) controllers are employed for regulation. In the rotor side converter, the reference active power (P_{ref}) is computed in respect to maximum power point tracking and then the extracted active power ($P_{extract}$), which depends on the wind speed, is compared to P_{ref}. Consequently, the d-axis reference current of the rotor is achieved. Meanwhile, the reference value for the reactive power of the stator (Q_{s-ref}) is considered zero. Therefore, during the normal operation, the absorbed reactive power (Q_s) from the stator side of the DFIG will be equal to the reference value, and the required reactive power for the DFIG will be covered by the back-to-back converters. To maintain the DC-link voltage in constant value, the grid side converter provides the active power to the rotor side. Therefore, the d-axis reference current is obtained by comparing the DC-link voltage with the reference value (V_{DC-ref}). Meanwhile, the reactive power (Q) in the GSC and the RSC is adjusted by the reference value of Q_{ref}.

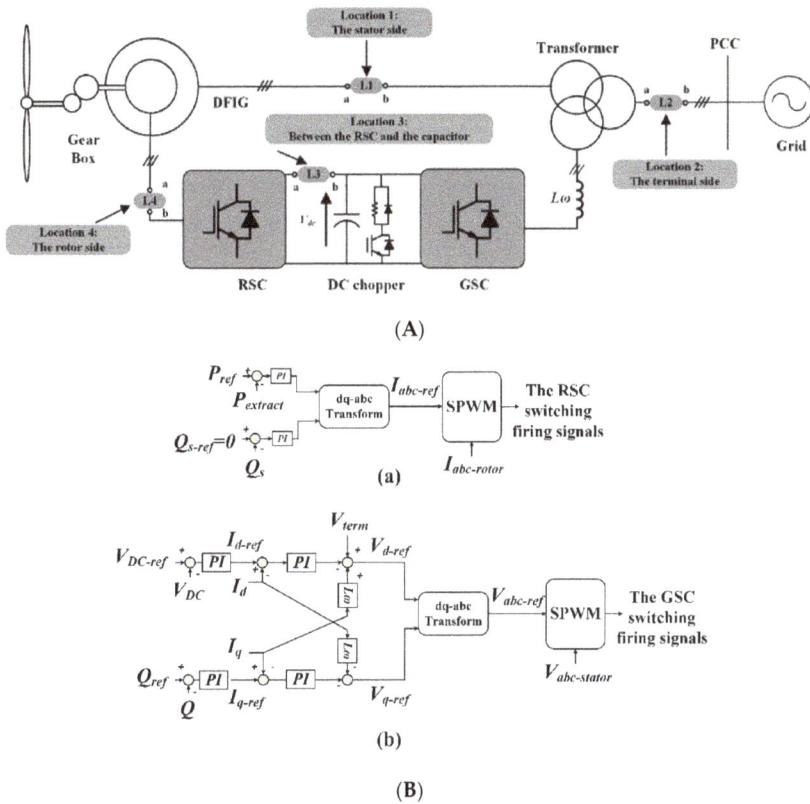

Figure 3. (**A**) The doubly-fed induction generator (DFIG) configuration, with all possible highlighted locations for the fault current limiters and series dynamic braking resistors, in the fault ride-through capability improvement. RSC: Rotor side converter; GSC: Grid side converter; PCC: Point of the common coupling. (**B**) The schematic of control circuits for: (**a**) The rotor side converter; and (**b**) the grid side converter [15]. Adapted with permission from [15], Copyright Publisher, 2017.

3. Fault Ride-Through

As mentioned in the introduction, there are different methods to improve the FRT capability of the DFIG. Advanced control methods are discussed mostly in the research papers. These methods control active and reactive power injections, of the DFIG, to restore the PCC voltage during the fault. Most of these methods are complicated and thus they are not so interesting for the industry [16]. Furthermore, during the zero-voltage sag in the PCC, the terminal voltage, as a reference, will be lost and the control method might be inapplicable [17].

Another method is to change the configuration of the DFIG. In this situation, the changes happen in the back-to-back converters. A fault-tolerant DFIG-based wind turbine has been proposed to enhance the FRT capability [10,18]. The configuration of the proposed DFIG is shown in Figure 4. Instead of utilising a general, six-semiconductor-switch converter, a nine-switch GSC has been used. In normal operation, three lower switches are in the on state. When a fault occurs, three lower switches operate to compensate the voltage on the neutral side of the winding in the stator. Another change in the configuration of the conventional DFIG has been studied in [11,19]. The proposed scheme is shown in Figure 4, in grey. Similar to the conventional DFIG, the stator is directly linked to the grid through the terminal. However, an extra converter, which is connected to the star point of the stator winding, is employed to provide an effective active power transfer to the grid, and a better power system disturbance ride-through during the low voltage. As it is clear from Figure 4, these changes in the configuration of the conventional DFIG are probably not of interest to industry. However, to be practically implemented, they should have economical justification.

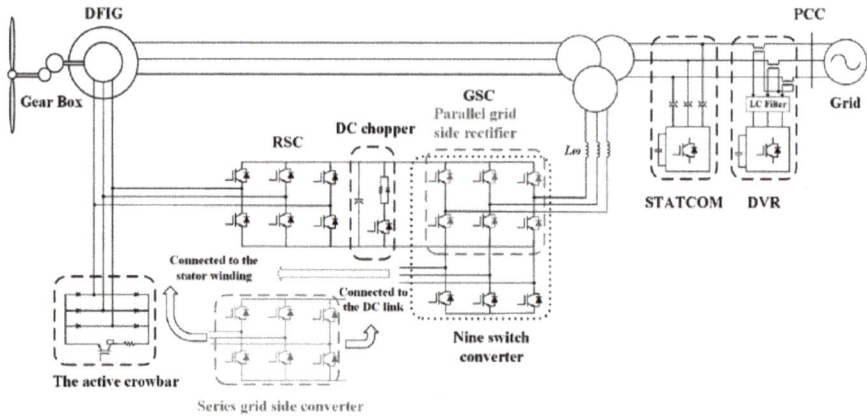

Figure 4. The doubly-fed induction generator (DFIG) configuration with the most well-known fault ride-through (FRT) enhancement approaches. DVR: Dynamic voltage restorers; STATCOM: Static synchronous compensator.

Adding extra hardware is another scheme to improve the FRT capability of the DFIG. Crowbar protection, DC braking chopper, voltage sag compensation devices, and fault current limiting devices are employed as extra hardware in the DFIG to enhance the ride-through capability during the fault condition. The crowbar protection is the most well-known scheme, which is practically employed and utilised in the FRT improvement of the DFIG. The configuration of the crowbar is shown in Figure 4. The crowbar includes a three-phase diode bridge rectifier, a bypass resistance, and a switch (either a semiconductor switch, such as an isolated gate bipolar transistor in the active crowbar, or a switch such as a thyristor in the conventional crowbar). The active crowbar has been introduced to overcome the continuous absorption of the reactive power by the squirrel-cage induction generator in the conventional crowbar [7]. To operate the active crowbar by a semiconductor switch and cease the RSC switching, a threshold current or a threshold voltage is defined regarding the transient over-current in the rotor side or the over-voltage in the DC link, respectively. Whenever the measured current voltage is less than the threshold current voltage, the DFIG converters resume to normal operation [8]. The RSC stop and the reactive power absorption by the active crowbar are the main disadvantages of this method.

DC braking chopper is another scheme to keep the DC-link voltage value in the safe area of operation during the fault condition. The schematic of the DC chopper is depicted in Figure 4. The parallel connected configuration of the DC chopper consists of a resistor in series with a semiconductor switch. An anti-parallel diode is also utilised to protect the chopper switch from voltage spike when the semiconductor switch goes to the off state. However, the DC braking chopper does not effectively restrict the rotor transient over-current. Therefore, high-rated anti-parallel diodes should be employed in the RSC [9]. Similar to the crowbar scheme, when the DC chopper is activated, the RSC switching is ceased.

A static synchronous compensator (STATCOM), as the voltage sag compensation device, has been discussed in [20]. Figure 4 shows the location and simple configuration of the STATCOM. Due to the low capacity of back-to-back converters, in case of the fault in the weak utility, there is a possibility of a voltage instability. Accordingly, voltage sag compensation devices, such as the

STATCOM, have been proposed to support the voltage profile during the fault condition. There are coordinated and independent reactive power controls between the STATCOM and the DFIG [20,21]. In the independent control method, the voltage control is only done by the STATCOM. The STATCOM should be incorporated with a fault current limiting method because it is not solely able to restrict the fault current level in the RSC. Also, the dynamic voltage restorer (DVR) is another voltage sag compensation device, which have been utilised to overcome voltage sag during the fault. In Figure 4, the simple configuration of the DVR is presented. Similar to the STATCOM, auxiliary methods should be incorporated with the DVR, such as changing the control method of the back-to-back converters or the pitch angle control [22]. It should be noted that these mentioned devices require large storage capacity and a high number of semiconductor switches.

From a fault current limiting point of view, the FCLs and the SDBRs are employed to restrict the fault current level in the DFIG. Up until now, many configurations of FCLs have been proposed and studied. The FCL structures can be divided into two categories, with regard to impedance type and components. Considering the impedance type, they are resistive, inductive, resistive-inductive, and resonance type FCLs. Meanwhile, whether a superconductor is utilised or not, there is another category [23]. Recently, a new approach of applying the FCLs has been proposed that has the capability of producing variable resistance [24]. Considering the component type, the FCLs are either a solid state or saturated core transformer [14]. Despite the current limiting characteristics of the FCLs, they have been employed for different applications. From a transient stability point of view, the application of variable and controllable resistance FCLs has been proved [24]. Regarding the location of the FCLs, they are also employed to compensate the voltage sag in the PCC [25].

In order to improve the FRT capability of the DFIG, the FCLs are placed in different locations of the DFIG. Figure 3A shows the studied locations in the literature. In each location, with respect to the FCL impedance type, the impacts on the key parameters of the DFIG are different during the fault condition. From the FCL's location point of view, the FCLs are mostly located at the terminal side and the stator side of the DFIG. Some of the FCLs are placed in the rotor side and a few of the configurations are in the DC link. Meanwhile, from an impedance point of view, the FCLs are mostly the resistive type, as it is clear from Figure 2. In the following section, it will be discussed how the impedance type and the FCL's location could affect the FRT capability of the DFIG based wind turbine. The potential best choice of location and impedance type, in order to obtain the most favourable FRT capability of the DFIG, will be mentioned. In the following section, the FCLs' structures are categorized based on the impedance type.

4. Fault Current Limiting Devices

As mentioned, the FCLs and the SDBRs are employed as the fault current limiting devices to enhance the FRT capability of the DFIGs. At first, different structures of the FCLs and SDBRs are studied regarding the impedance type of each configuration, and then a comparison will be finally done.

4.1. Non-Superconducting FCL

4.1.1. Inductive Type FCL: Non-Controlled FCL

In [26], a cost-effective topology of a non-controlled FCL, to improve the FRT capability of the DFIG, has been studied. The non-controlled FCL is placed in the rotor side, which is shown in Figure 5a. Its configuration is similar to a three-phase bridge type FCL. The limiting inductance is a non-superconductor (NSC), which is cost-effective from an industrial point of view. Regarding the location of the FCL, inserting the NSC either in the stator side or in the rotor side has different impacts.

When the non-controlled FCL is located in the stator side, regarding (9) and (10), the value of back-EMF voltage and the transient inductance of the rotor are changed as follows:

$$\vec{V}_{0r} = \frac{L_m}{L_s + a^2 L_{FCL}} \frac{d\vec{\psi}_s}{dt} \tag{9}$$

$$\sigma_{L1-FCL} = 1 - \frac{L_m^2}{(L_s + a^2 L_{FCL})L_r} \tag{10}$$

Figure 5b shows the variation of the leakage coefficient by changing the FCL's impedance. As it is clear by (9), by increasing the inductance of the FCL in the stator side, the value of rotor back-EMF voltage decreases. Therefore, the RSC has a good controllability in order to counteract the stator flux oscillations. However, the absolute value of the leakage coefficient significantly decreases and damps by increasing the FCL's inductive impedance. In fact, there is the possibility that the rotor transient over-current is not restricted to an acceptable range. It should be mentioned that, by locating the FCL in the stator side, the stator voltage is resumed to some extent. By placing the FCL in the rotor side, the leakage coefficient is expressed as follows:

$$\sigma_{L4-FCL} = 1 - \frac{L_m^2 - a^2 L_{FCL} L_s}{L_s L_r} \tag{11}$$

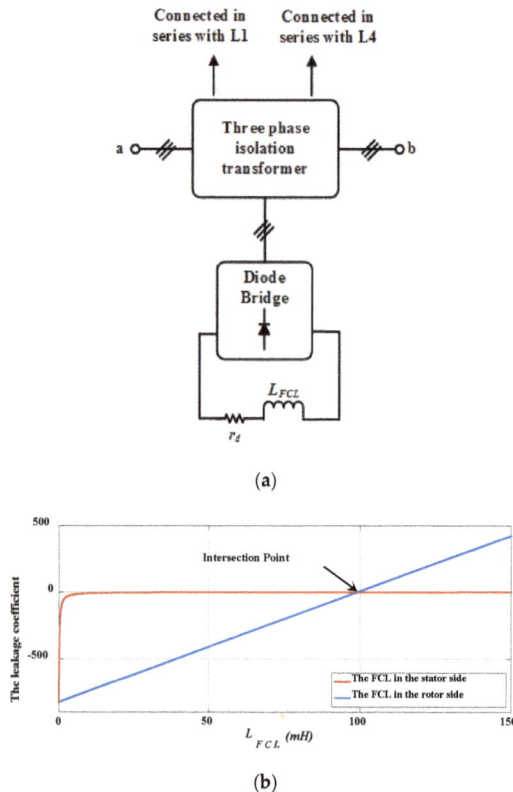

(a)

(b)

Figure 5. (a) The non-controlled inductive fault current limiter [26]; and (b) the leakage coefficient variation, with the FCL inductance variation in the stator and the rotor sides.

When the inductive FCL is located in the rotor side, the variation of the leakage coefficient is also shown in Figure 5b. In order to find out the variation of the leakage coefficient, some points should be considered. For the inductive FCL located in the rotor side, any impedance value lower or higher than the intersection point of curves could result in good rotor transient over-current limitation. Otherwise, the FCL could have an inverse impact on limiting the rotor transient over-current. As a result, if the inductive FCL is going to be in the rotor side, its impedance should be lower or higher and far away than from the intersection point of the curves in order to achieve a good current limitation capability. However, it should be noted that, from limiting the rotor transient over-current perspective, the inductive FCL in the rotor side is more effective than placing the FCL in the stator side. Furthermore, it should be noted that the rotor back-EMF voltage remains unchanged and, consequently, a rather high electromagnetic torque oscillation might occur due to the weak controllability of the RSC on the stator flux linkage. In addition, there is no possibility to restore the stator voltage.

4.1.2. Inductive-Resistive Type FCL: Optimized Located FCL

An optimized located FCL has been proposed in [27], and it is shown in Figure 6. The proposed FCL has been placed in the DC link of the DFIG. However, the RSC switches are turned off during the fault condition, and high-level fault currents pass through the high-rated anti-parallel diodes. The FCL is resistive and is capable of limiting the fault current only in one direction. The FCLs' resistance is bypassed by a semiconductor switch during normal condition. To protect the semiconductor switches in the FCL from over-voltage, a surge arrestor is connected in parallel with each switch of the FCL. It should be noted that the optimized located FCLs can cause over-voltage on the RSC switches during the switching in normal operation. To avoid this destructive voltage, auxiliary circuits should be applied, which result in a high cost and large size.

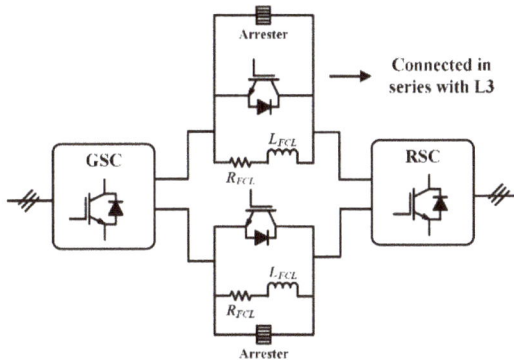

Figure 6. The optimized located FCL [27].

4.1.3. Resistive Type FCL: Thyristor Bridge Type FCL

A thyristor bridge type FCL with bypass resistor (BTFCL-BR) has been presented in [28]. Figure 7a shows the BTFCL-BR. In [28], the stator side location has only been studied. The main advantage of the BFCL-BR is the use of the NSC. Because of employing the NSC, there are high voltage spikes on it, which affect the stator voltage and, consequently, generate severe stator flux and electromagnetic torque oscillations during normal operation. Therefore, the bypass resistor (BR) has been proposed, not only to restrict the high voltage spikes but also to eliminate the current harmonics in normal operation of the power system. During the fault condition, after fault detection, the signal triggering of the thyristors will go to the off state. So, before turning off the thyristors in the zero-current crossing, the bypass resistor and the NSC restrict the stator fault current in the first instances of the fault. After turning off all thyristors, the bypass resistance solely limits the stator fault current. Increasing the resistance in the stator side decreases the time constant of the stator current and, in the meantime, restricts the stator fault current in a good manner. The number of thyristor switches employed can be a disadvantage for this configuration.

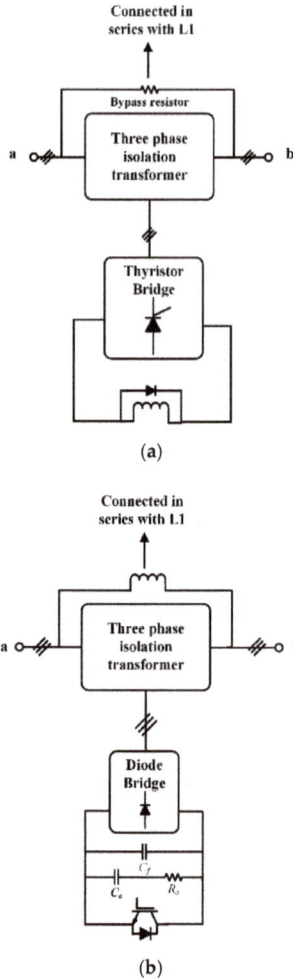

Figure 7. (**a**) The thyristor bridge type fault current limiter with bypass resistor (BTFCL-BR) [28]; and (**b**) the switch type fault current limiter (STFCL) with a snubber circuit [14].

4.1.4. Resistive Type FCL: Switch Type FCL (STFCL)

An STFCL has been proposed in [14], and it is shown in Figure 7b. The STFCL, compared to the BTFCL, uses an inductor as a fault-current-limiting impedance instead of the BR. Furthermore, instead of using a thyristor bridge, a diode bridge has been employed in the STFCL, which decreases the installation cost. A snubber circuit has been used in order to utilize the inductor to suppress the transient over-voltage on the semiconductor switch, when it goes to the off state after the fault detection. After suppressing the first spikes by C_f, which is smaller than C_a, C_a absorbs the excess energy in the stator until its voltage reaches a steady state. Afterwards, the current pass, in the DC side of the diode bridge, is blocked and the inductor does its limiting operation. The STFCL's impact on the DFIG is the same as has been discussed for the non-controlled FCL located in the stator side.

4.1.5. Resistive Type FCL: Variable Resistive Type FCL

In [29], a variable resistive type FCL (VR-FCL) has been placed in the PCC to improve the transient stability of a hybrid power system, including the DFIG, a Photovoltaic (PV) plant, and a synchronous generator. The configuration of the VR-FCL is similar to [30] and is shown in Figure 8a. However, to make better transient stability improvement, the pre-fault conditions, including the pre-fault active power of the hybrid generation units and fault location, have been taken into account. Regarding the pre-fault conditions, to control the VR, three nonlinear controller schemes, including fuzzy logic controller (FLC), static nonlinear controller (SNC), and an adaptive-network-based fuzzy inference system (ANFIS) have been discussed. The operation of the ANFIS-based VR-FCL is better than the FLC, and the FLC is better than the SNC. The factors to evaluate the operation of each controller are the rotor speed deviation of the DFIG, the DC-link voltage deviation of the PV generator, and the load angle deviation of the synchronous generator. The concept of the pre-fault conditions to achieve the maximum transient stability has been previously proposed in [30]. However, it should be noted that the dynamic behaviour of the DFIG is completely different than that of the other generation units [13]. As soon as the fault happens, especially the three-phase fault, the voltage drop on the FCL cannot be enough due to the exponential current variation in the stator of the DFIG, particularly after damping the stator current in the first cycles of the power system frequency.

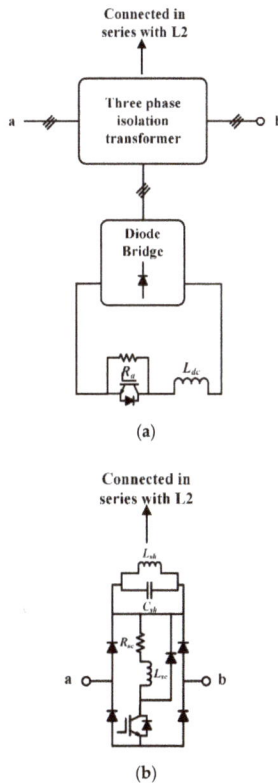

(a)

(b)

Figure 8. (**a**) The variable resistive type fault current limiter (VR-FCL) [29]; and (**b**) the parallel resonance type fault current limiter [31].

4.1.6. Resonance Type FCL: Parallel Resonance Type FCL

In [31], as shown in Figure 8b, a parallel resonance type FCL has been investigated and compared to the bridge type FCL with parallel resistive-inductive shunt impedance. The idea of parallel resonance type FCL has been proposed in [32]. In normal operation, the semiconductor switch is in the on state and the parallel resonance is bypassed. After a fault occurrence, the semiconductor switch goes to the off state. So, the fault current passes through the parallel impedance and is limited by the resonance type FCL, which helps to enhance the FRT capability of the DFIG. Considering the assessment factors, including the deviation of the wind farm terminal voltage, the active power of the wind farm, and the rotor speed of the DFIG, the parallel resonance type FCL has better operation in comparison with the bridge type FCL with parallel resistive-inductive shunt impedance.

4.2. Superconducting FCL

4.2.1. Inductive Type FCL: Superconducting Fault Current Limiter–Magnetic Energy Storage System (SFCL-MES)

Although, the superconducting FCLs are not of interest for industry from an economical point of view [28], many configurations have been proposed to employ the superconductor for the improvement of the FRT capability of the DFIG. An SFCL-MES has been presented in [33,34]. Its configuration is shown in Figure 9a. In addition to limiting the fault current, the SFCL-MES can smooth the active power fluctuations and stabilize the DC-link voltage. The SFCL has been studied at both the rotor side and the stator side. The SFCL-MES is able to control the superconductor (SC) stored energy and the output power of the DFIG. As previously discussed, placing the inductive FCL either at the stator side or at the rotor side has different impacts. By placing the inductive FCL at the stator side, the stator flux oscillations are limited by decreasing the rotor back-EMF. However, as mentioned before, the rotor transient over-current is not restricted well with the inductive FCL in the stator side, compared to the inductive type FCL placed in the rotor side. In fact, locating the inductive type FCL in the rotor side increases the leakage coefficient more than the inductive type FCL placed in the stator side. However, it may decrease the RSC controllability due to the unchanged rotor back-EMF voltage.

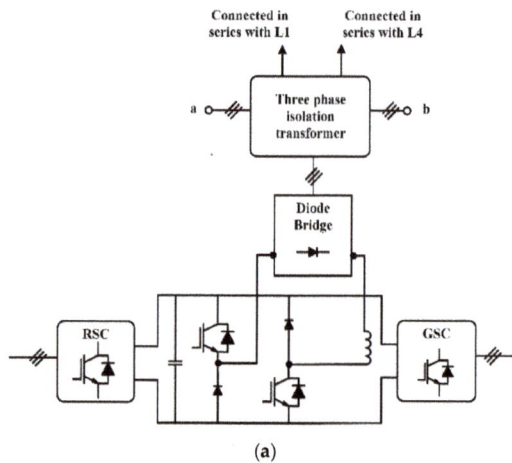

(a)

Figure 9. *Cont.*

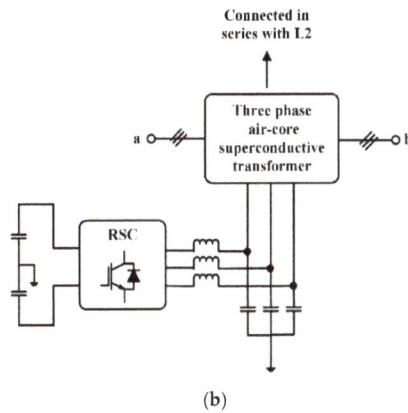

(b)

Figure 9. (a) The superconducting fault current limiter–magnetic energy storage system (SFCL-MES) [33,34]; and (b) the active superconducting fault current limiter [35].

4.2.2. Inductive Type FCL: Active SFCL with Reactive Power Injection

An active SFCL, to compensate the voltage sag in the PCC and restrict the fault current level, has been studied in [35], and is shown in Figure 9b. An air-core superconductive transformer is placed in a series connection in the terminal of the DFIG. A three-phase converter with split DC-link capacitors has been connected to the secondary side of the transformer. The converter injects a controlled value of the current to compensate for the voltage drop on the PCC during the fault condition, and to surpass the current level. However, in this configuration, the cost of the superconductor and converter should be taken into account.

4.2.3. Resistive Type FCL: DC-Resistive SFCL

In [36,37], a DC-resistive SFCL, shown in Figure 10a, is placed in the PCC to limit the fault current level. There is no novelty, both in the configuration and in the control of the FCL. Furthermore, as mentioned, using the SC increases the installation cost of the structure. In fact, the commercial application of the SFCL might be unavailable due to its high cost.

4.2.4. Resistive Type FCL: Resistive-Flux-Coupling Type SFCL

A resistive-flux-coupling type SFCL has been proposed in [38], and is shown in Figure 10b. The SFCL consists of a coupling transformer, an SC, and a semiconductor switch (S_1). Furthermore, an arrestor is employed to overcome the overvoltage in switching instances. In normal operation, S_1 is closed and if the coupling coefficient is supposed to be one, then the FCL's impedance is almost zero. When a fault happens, S_1 goes to the off state and the overvoltage is restricted by the arrestor. The SC enters into the fault current pass to limit the current level. It should be noted that placing a resistive impedance in the stator side can be much more effective than in the rotor side, due to the voltage sag compensation and the greater active power absorption. The resistive type FCL in the stator side limits both the stator and rotor currents effectively.

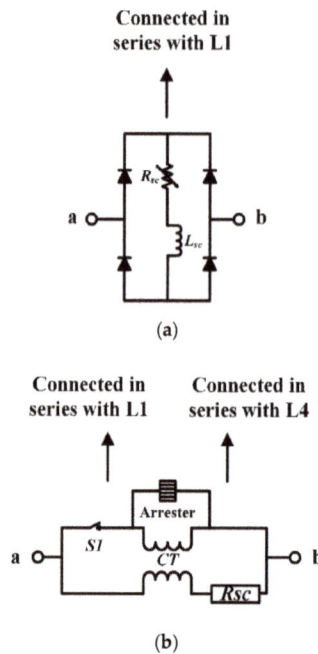

Figure 10. (a) The DC-resistive superconducting fault current limiter [36]; and (b) the resistive-flux-coupling type superconducting fault current limiter (SFCL) [38].

4.2.5. Resistive Type FCL: Resistive Type SFCL with Transient Voltage Control (TVC)

Different configurations of the resistive type SFCL have been placed at the terminal of the DFIG. A resistive type SFCL, cooperated with TVC in the DFIG converters, has been studied in [39]. It has been mentioned that the voltage sag compensation by TVC or the fault current limiting characteristics of the SFCL cannot solely improve the PCC voltage during the fault condition. Post-fault voltage sag compensation in the PCC, by employing the SFCL as a passive voltage compensator, results in a much more reactive power injection by the converters, which act as an active voltage compensator.

4.2.6. Resistive Type FCL: Superconducting Magnetic Energy Storage (SMES) with the SFCL

In [40], instead of employing the TVC in the converters, an SMES has been applied together with the SFCL, as shown in Figure 11a. The main application of the SMES is to smooth the active power fluctuation after the fault current limiting by the SFCL. However, in [39], it has been noted that the SFCL is not solely able to effectively improve the PCC voltage.

In [41], the authors modified the configuration of the SFCL-SMES studied in [40]. Instead of utilising two separate SCs (one in the SFCL and another in the SMES), one common SC has been employed in two DC choppers, as shown in Figure 11b. During fault condition, the DC chopper of the FCL enters into the fault current path. After fault removal, the DC chopper of the SMES operates, and the remaining active power fluctuations are restricted as discussed in [40].

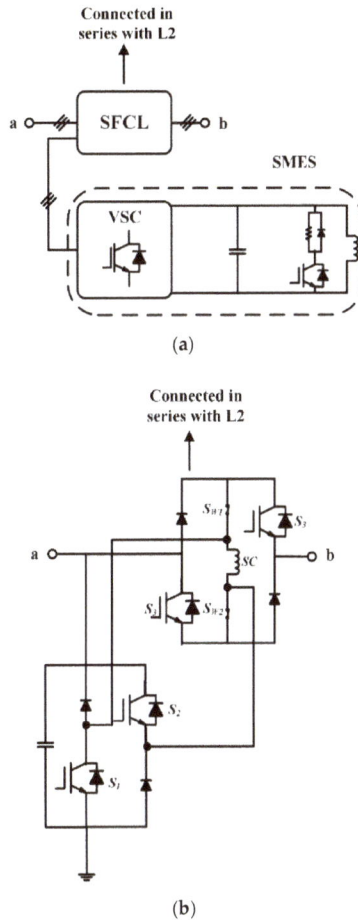

Figure 11. (a) The superconducting magnetic energy storage (SMES) together the SFCL [40]; and (b) the superconducting fault current limiter with superconducting magnetic energy storage, with common superconductor [41].

4.2.7. Resistive Type FCL: Resistive Type SFCL in the Rotor Side

In addition to placing the resistive type SFCL in the terminal of the DFIG, different configurations have been located in the rotor side. In [42,43], the proposed resistive type SFCL in the rotor side operates effectively to limit the rotor transient over-current. Meanwhile, due to active power consumption in the rotor side, the DC-link over-voltage is avoided. However, the RSC controllability gets worse in practice, as mentioned before. Although, the voltage dip in the PCC has been improved by the SFCL in the rotor side, the SFCL in the stator side, either inductive or resistive, has a better performance for voltage sag compensation compared to the rotor side.

In [44], a resistive solid state FCL has been studied in the rotor side. The FCL is composed of antiparallel semiconductor switches, which are employed to bypass or insert the resistance. The FCL operates as a passive compensator to aid the converters to act as an active compensator, as previously discussed in [40].

4.3. Series Dynamic Braking Resistor

An SDBR can be used to improve the FRT capability of the DFIG. The SDBR is placed in different locations and its impact evaluated. In [45], the SDBR is connected in series with the rotor. The rotor currents have been employed to trigger the SDBR and the active crowbar. The important advantage of the SDBR is to decrease the number of switching operations of the crowbar, which avoids the DFIG operating as a squirrel-cage induction generator during the fault condition. Furthermore, due to limiting the rotor over-current, the frequent performance of the DC braking chopper decreases. This means that the charging current of the DC-link capacitor is reduced by the operation of the SDBR. It should be mentioned that the RSC switching stopped with the SDBR operation. Meanwhile, the SDBR in the rotor side reduces the rotor controllability.

In [46,47], the SDBR is placed at the stator side of the DFIG. Its operation has been compared to the DC braking chopper. As discussed, the current control limiting devices in the stator side are effective for the voltage sag compensation. A summary of advantages and disadvantages of the fault current limiting devices are presented in Table 1.

Table 1. A summary of characteristics of the fault current limiting devices.

FCL Type	Location	Impedance Type	Rotor Transient Over-Current Limitation	Stator Fault Current Limitation	Excess Active Power Evacuation	Terminal Voltage Sag Compensation	Rotor Switching State during the Fault	Operation on Different Types of Fault	Components	SC.	Cost
SRCL-MES	L1 and L4	Inductive	L1: Yes (G.) L4: Yes (E.)	L1: Yes (E.) L4: Yes (G.)	No	L1: Yes (G.) L4: No	Continuous with different control, good controllability of the RSC in L1	Not effective in asymmetrical faults	I-T*1 Diodes*6(CSC)*8(VSC) Inductor*1 S-S*0(CSC)*2(VSC)	Yes	High
Switch Type FCL	L1	Inductive	Yes (G.)	Yes (E.)	No	Yes (G.)	Continuous with good RSC controllability	Not effective in asymmetrical faults	I-T*1 Diode*6 Inductor*1 S-S*1 A snubber circuit	No	Low
Active SRCL with Reactive Power Injection	L2	Inductive	Yes (G.)	Yes (E.)	No	Yes (G.)	Continuous with good RSC controllability	Effective for all fault types	I-T*1(superconductive) A low pass filter S-S*6 A split DC-link capacitors	Yes	High
DC Resistive FCL	L2	Resistive	Yes (G.)	Yes (E.)	Yes	Yes (G.)	Continuous with good RSC controllability	Effective for all fault types	Diodes*12 SC*3	Yes	High
Resistive Flux Coupling Type SRCL	L1 and L4	Resistive	L1: Yes (E.) L4: Yes (G.)	L1: Yes (E.) L4: Yes (G.)	Yes	L1: Yes (G.) L4: No	Continuous with different control, good controllability of the RSC in L1	Effective for all fault types	Coupling Transformer*3 SC*3 S-S*3 Arrestor*3	Yes	High
Resistive Type SRCL with SMES	L2	Resistive	Yes (G.)	Yes (E.)	Yes	Yes (E.)	Continuous with good RSC controllability	Effective for all fault types	Parallel Transformer*1 S-S*7 SC.*4 Diode*1 Capacitor*1	Yes	High
Resistive Type SRCL with Transient Voltage Control	L2	Resistive	Yes (G.)	Yes (E.)	Yes	Yes (E.)	Continuous with TVC, good controllability of the RSC	Effective for all fault types	SC.*3	Yes	High

I–T: Isolation Transformer; S–S: Semiconductor Switch; Li: Location no.; E.: Excellent; G.: Good; SC.: Superconductor.

Table 1. Cont.

FCL Type	Location	Impedance Type	Rotor Transient Over-Current Limitation	Stator Fault Current Limitation	Excess Active Power Evacuation	Terminal Voltage Sag Compensation	Rotor Switching State during the Fault	Operation on Different Types of Fault	Components	SC.	Cost
Resistive Type SFCL, SMES with Common SC	L2	Resistive	Yes (G.)	Yes (E.)	Yes	Yes (E.)	Continuous with good RSC controllability	Effective for all fault types	S-S*18 SC.*3 Diode*12 Capacitor*3	Yes	High
Thyristor Bridge Type FCL with Bypass Resistor	L1	Resistive	Yes (G.)	Yes (E.)	Yes	Yes (E.)	Continuous with good RSC controllability	Not effective in asymmetrical faults	I-T*1 Diode*1 Thyristor*6 Bypass resistor*3 Inductance*1	No	Low
Variable Resistive Type FCL	L2	Variable resistance	Yes (G.)	Yes (E.)	Yes, controlled active power absorption	Yes (E.)	Continuous	Not effective in asymmetrical faults	I-T*1 Diode*6 Inductor*1 S-S*1 Resistance*1	No	Low
Optimized Located FCL	L3	Resistive-inductive	Yes (E.)	Yes (G.)	Yes, the rotor active power	No	Blocked	Effective for all fault types	S-S*2 Resistance*2 Inductance*2 Arrestor*2	No	Low
Non-controlled FCL	L1 and L4	Inductive	L1: Yes (G.) L4: Yes (E.)	L1: Yes (E.) L4: Yes (G.)	No	L1: Yes (G.) L4: No	Continuous	Effective for all fault types	I-T*1 Diode*6 Inductor*1	No	Low
Parallel Resonance Type FCL	L2	Resonance	Yes (G.)	Yes (E.)	Yes	Yes (E.)	Continuous	Effective for all fault types	S-S*3 Diode*15 Inductance*6 Capacitance*3	No	Low
SDBR	L1 and L4	Resistive	L1: Yes (G.) L4: Yes (E.)	L1: Yes (E.) L4: Yes (G.)	Yes	L1: Yes (G.) L4: No	Blocked	Effective for all fault types	Resistance*3 Thyristor*6	No	Low

I-T: Isolation Transformer; S-S: Semiconductor Switch; Li: Location no.; E.: Excellent; G.: Good; SC.: Superconductor.

5. Simulation Results

In this section, the impact of the different locations and the type of impedance for the FCLs are studied. From the FCL's operation point of view, they should have a very low impedance during normal condition, fast response after fault detection, high impedance value during the fault, and also fast recovery time after fault removal. However, various configurations of the FCLs are slightly different in the characteristics mentioned above. For instance, regarding the configuration of the FCLs, some types of FCLs are superconductive, but then have almost zero impedance in normal operation. Whilst for the other types, there is a small impedance. Meanwhile, the FCLs may have different response and recovery times considering their configurations.

However, it should be mentioned that after fault occurrence and during the fault steady state, all FCLs should have a high impedance value and the only difference should be related to the impedance type. Therefore, the type of impedance and its impact on the operation of the DFIG and the FRT enhancement were considered at three different locations, including the stator side, the rotor side, and the terminal side. The DC-link location has the same effect as the rotor side location. So, in order to avoid proliferation, location three was not mentioned. The DFIG was connected to the grid through a three-phase transformer and a transmission line. The parameters of the simulation are presented in Table 2. To have a reasonable comparison, the impedance values for the resistive and inductive type FCLs were the same. Meanwhile, to demonstrate how the FCL's impedance, either as an inductive or a resistive, could affect the RSC controllability, the impedance value was chosen in regards to the FCL impact on the stator side. As discussed, the resistive type FCL in the stator side has a very effective operation because of voltage sag compensation, excess active power consumption, and also increasing the RSC controllability. Therefore, the resistive type FCL in the stator side was selected as a reference to calculate the impedance value.

Table 2. Simulated DFIG Specifications.

The DFIG and Transformer	
Rated power	2 MW
Three-phase transformer	0.69/34.5 kV, 60Hz, 5MVA
Rated stator voltage	690 V
Rated frequency	60 Hz
Stator leakage inductance	0.12 p.u.
Rotor leakage inductance	0.12 p.u.
Magnetising inductance	3.45 p.u.
Stator to rotor turns ratio	0.35
Stator resistance	0.011 p.u.
Stator inductance	0.012 p.u.
Nominal wind speed	13 m/s
The DC Chopper	
Rated DC-link voltage	1200 V
DC chopper resistor	0.5 Ω
DC bus capacitor	50 mF
DC-link activation threshold voltage	1.1 p.u.
Transmission Lines	
Length	30 km
Line impedance	0.01 + j0.1 Ω/km
Resistance of grid side filter	0.3 p.u.
Reactance of grid side filter	0.003 p.u.

As mentioned, the different locations have different impacts on the FRT improvement of the DFIG. In this section, the rotor and stator currents are shown in Figure 12.

Inductive Type Fault Limiter

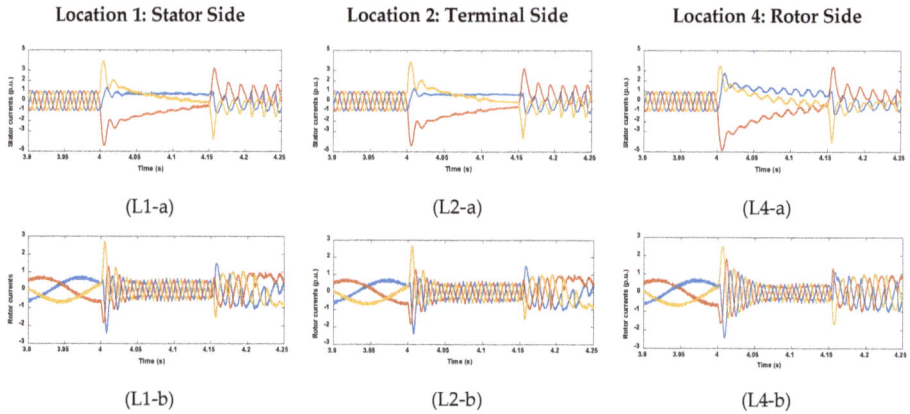

| Location 1: Stator Side | Location 2: Terminal Side | Location 4: Rotor Side |

(L1-a)   (L2-a)   (L4-a)

(L1-b)   (L2-b)   (L4-b)

Figure 12. With the inductive type fault current limiter in the stator side (L1), in the terminal side (L2), and in the rotor side (L4) during a three-phase fault: (**a**) The stator currents; and (**b**) the rotor currents.

At first, the inductive type FCL was placed at three locations in the DFIG. The results are presented in Figure 12. The value of the inductance was less than the intersection point inductance value of the leakage coefficient curves presented in Figure 5b. Therefore, it can be concluded that the absolute value of the leakage coefficient for the inductive FCL in the rotor side was higher than for the inductive FCL in the stator side. As a result, considering Figure 12 and the discussion in Section 4.1, it is expected that the inductive FCL in the stator side, the rotor side, or the terminal side has almost the same impact on the stator current. Regarding the rotor current shown in Figure 12, for the inductive FCL placed in the rotor side, the rotor fault current level in the steady state of the fault was less than the other FCL locations, due to the high leakage coefficient value shown in Figure 5b. This point has been discussed in Section 4.1.

The impacts of the resistive type FCL for the three locations are shown in Figure 13. As mentioned, the operation of the FCL with resistive impedance in the stator side was the most effective one in limiting both the stator and rotor currents level. The resistive impedance in the stator side was capable of compensating for the voltage sag in the DFIG terminal, and was also able to absorb excess active power during the fault condition. For the FCL at the terminal side, its impact was almost the same as the FCL placed at the stator side. Regarding Figure 12, due to decreasing the RSC controllability by employing the FCL at the rotor side, there was a high current level in both the stator and rotor sides. However, in order to achieve a good fault current limitation in the rotor side by the FCL, a high impedance value should be employed in the FCL, either in the inductive type FCL or the resistive type FCL.

Considering the results provided for the rotor and the stator currents, the inductive FCL almost had the same impact on the stator current and the rotor current regardless of its location. However, it should be noted that the FCL located in either the stator side or the terminal side had the capability of compensating for the voltage sag in the terminal of the DFIG as well as increasing the RSC controllability. The resistive FCL was most effective in enhancing the FRT capability of the DFIG if it was located in the stator side or the terminal side. Like the inductive FCL, located in the stator or the terminal sides, the resistive FCL could compensate the voltage sag in the terminal and increase the RSC controllability. But, the resistive FCL, located in the stator side or the terminal side, compared to the inductive FCL, located in the stator side or the terminal side, was able to consume excess active power during the fault condition. In fact, from the FRT capability enhancement point of view, the resistive FCL placed in the stator or the terminal sides had the best operation.

Resistive Type Fault Limiter

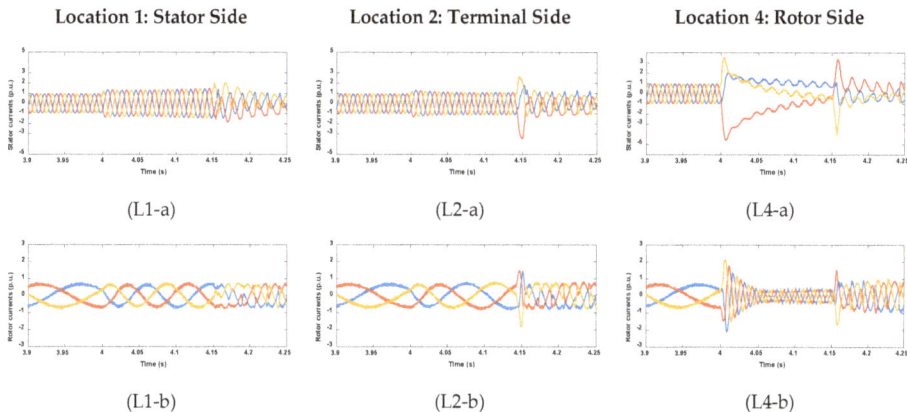

Location 1: Stator Side	Location 2: Terminal Side	Location 4: Rotor Side

(L1-a) (L2-a) (L4-a)

(L1-b) (L2-b) (L4-b)

Figure 13. With the resistive type fault current limiter in the stator side (L1), in the terminal side (L2), and in the rotor side (L4) during a three-phase fault: (**a**) The stator currents; and (**b**) the rotor currents.

6. Conclusions

In this paper, a review was carried out on the improvement of the fault ride-through capability of the doubly-fed induction generator by use of fault current limiting devices. Almost all of the types of fault current limiters, employed in the fault ride-through of the doubly-fed induction generator, have been discussed. For all prototypes, their cost and the impact of their operation on rotor transient over-current limitation, stator current limitation, excess active power evacuation, and terminal voltage sag compensation have been analysed and provided in Table 1. With respect to the comparison, the best operation is granted to the resistive type fault current limiter located in the stator side, due to the voltage sag compensation capability in the doubly-fed induction generator terminal, the excess active power consumption, and the increase in controllability of the rotor side converter. As all fault current limiters have the same performance after fault detection, the operation of the fault current limiters was simulated in regards to their location in the doubly-fed induction generator and the impedance types. Three main locations, including the stator side, the rotor side, and the terminal side, were taken into account. Furthermore, for all three locations, the inductive and resistive type fault current limiters were placed, and their impacts on the rotor and stator current limitations were discussed. Regarding the simulation, it was demonstrated that the resistive type fault current limiter placed in the stator side was the most effective fault current limiter in enhancement of the doubly-fed induction generator. For future works, in order to achieve a good fault current limitation with the fault current limiter at the rotor side, the fault current limiter should utilise a high impedance value compared to the fault current limiter located in the stator side. In addition, for fault current limiters in the stator side, a resistive type FCL should be utilised in order to achieve a good fault ride-through capability enhancement in the doubly-fed induction generator.

Author Contributions: All authors equally contributed in conceptualization, methodology, software, formal analysis, investigation, resources, writing-original draft preparation, writing-review & editing. M.N. and F.B. supervised and administrated the project.

Funding: This research has not received any external funding.

Conflicts of Interest: The authors declare no conflicts of interest.

Abbreviations

The list of abbreviations used in the paper is provided as follows:

FRT	Fault Ride-Through
DFIG	Doubly-Fed Induction Generator
RSC	Rotor Side Converter
PCC	Point of the Common Coupling
FCLs	Fault Current Limiters
SDBRs	Series Dynamic Braking Resistors
GSC	Grid Side Converter
EMF	Electromotive Force
SFCL-MES	Superconducting Fault Current Limiter–Magnetic Energy Storage System
TVC	Transient Voltage Control
SMES	Superconducting Magnetic Energy Storage
STATCOM	Static Synchronous Compensator
DVR	Dynamic Voltage Restorers
BTFCL-BR	Thyristor Bridge Type FCL with Bypass Resistor
NSC	Non-Superconductor
STFCL	Switch Type FCL
VR-FCL	Variable Resistive Type FCL
FLC	Fuzzy Logic Controller
SNC	Static Nonlinear Controller
ANFIS	Adaptive-Network-Based Fuzzy Inference System

References

1. Tsili, M.; Papathanassiou, S. A review of grid code technical requirements for wind farms. *IET Renew. Power Gener.* **2009**, *3*, 308–332. [CrossRef]
2. Hansen, A.D.; Michalke, G. Fault ride-through capability of DFIG wind turbines. *Renew. Energy* **2007**, *32*, 1594–1610. [CrossRef]
3. López, J.; Gubía, E.; Sanchis, P.; Roboam, X.; Marroyo, L. Wind turbines based on doubly fed induction generator under asymmetrical voltage dips. *IEEE Trans. Energy Convers.* **2008**, *23*, 321–330. [CrossRef]
4. Ma, K. *Power Electronics For The Next Generation Wind Turbine System*; Springer: Berlin/Heidelberg, Germany, 2015.
5. Netz, E. *Grid Code; High and Extra High Voltage*; E. ON Netz GmbH: Bayreuth, Germany, 2006; Available online: https://www.eon.com/en.html (accessed on 1 August 2018).
6. Justo, J.J.; Mwasilu, F.; Jung, J.-W. Doubly-fed induction generator based wind turbines: A comprehensive review of fault ride-through strategies. *Renew. Sustain. Energy Rev.* **2015**, *24*, 447–467. [CrossRef]
7. Duong, M.Q.; Sava, G.N.; Grimaccia, F.; Leva, S.; Mussetta, M.; Costinas, S.; Golovanov, N. Improved LVRT based on coordination control of active crowbar and reactive power for doubly fed induction generators. In Proceedings of the ATEE 2015 9th International Symposium on Advanced Topics in Electrical Engineering, Bucharest, Rome, 7–9 May 2015; pp. 650–655.
8. Huchel, L.; Moursi, M.S.E.; Zeineldin, H.H. A parallel capacitor control strategy for enhanced frt capability of DFIG. *IEEE Trans. Sustain. Energy* **2015**, *6*, 303–312. [CrossRef]
9. Pannell, G.; Zahawi, B.; Atkinson, D.J.; Missailidis, P. Evaluation of the performance of a dc-link brake chopper as a dfig low-voltage fault-ride-through device. *IEEE Trans. Energy Convers.* **2013**, *28*, 535–542. [CrossRef]
10. Ambati, B.B.; Kanjiya, P.; Khadkikar, V. A low component count series voltage compensation scheme for dfig wts to enhance fault ride-through capability. *IEEE Trans. Energy Convers.* **2015**, *30*, 208–217. [CrossRef]
11. Flannery, P.S.; Venkataramanan, G. Unbalanced voltage sag ride-through of a doubly fed induction generator wind turbine with series grid-side converter. *IEEE Trans. Ind. Appl.* **2009**, *45*, 1879–1887. [CrossRef]
12. Fathabadi, H. Control of a DFIG-based wind energy conversion system operating under harmonically distorted unbalanced grid voltage along with nonsinusoidal rotor injection conditions. *Energy Convers. Manag.* **2014**, *84*, 60–72. [CrossRef]

13. Lopez, J.; Sanchis, P.; Roboam, X.; Marroyo, L. Dynamic behavior of the doubly fed induction generator during three-phase voltage dips. *IEEE Trans. Energy Convers.* **2007**, *22*, 709–717. [CrossRef]

14. Guo, W.Y.; Xiao, L.; Dai, S.T.; Li, Y.H.; Xu, X.; Zhou, W.W.; Li, L. LVRT capability enhancement of DFIG with switch-type fault current limiter. *IEEE Trans. Ind. Electron.* **2015**, *62*, 332–342. [CrossRef]

15. Naderi, S.B.; Negnevitsky, M.; Jalilian, A.; Tarafdar Hagh, M.; Muttaqi, K.M. Low voltage ride-through enhancement of DFIG-based wind turbine using DC link switchable resistive type fault current limiter. *Int. J. Electr. Power Energy Syst.* **2017**, *86*, 104–119. [CrossRef]

16. Xie, D.L.; Xu, Z.; Yang, L.H.; Østergaard, J.; Xue, Y.H.; Wong, K.P. A comprehensive LVRT control strategy for DFIG wind turbines with enhanced reactive power support. *IEEE Trans. Power Syst.* **2013**, *28*, 3302–3310. [CrossRef]

17. Rahimi, M.; Parniani, M. Efficient control scheme of wind turbines with doubly fed induction generators for low-voltage ride-through capability enhancement. *IET Renew. Power Gener.* **2010**, *4*, 242–252. [CrossRef]

18. Kanjiya, P.; Ambati, B.B.; Khadkikar, V. A novel fault-tolerant DFIG-based wind energy conversion system for seamless operation during grid faults. *IEEE Trans. Power Syst.* **2014**, *29*, 1296–1305. [CrossRef]

19. Flannery, P.S.; Venkataramanan, G. A fault tolerant doubly fed induction generator wind turbine using a parallel grid side rectifier and series grid side converter. *IEEE Trans. Power Electron.* **2008**, *23*, 1126–1135. [CrossRef]

20. Qiao, W.; Harley, R.G.; Venayagamoorthy, G.K. Coordinated reactive power control of a large wind farm and a STATCOM using heuristic dynamic programming. *IEEE Trans. Energy Convers.* **2009**, *24*, 493–503. [CrossRef]

21. Qiao, W.; Harley, R.G.; Venayagamoorthy, G.K. Effects of FACTS devices on a power system which includes a large wind farm. In Proceedings of the 2006 IEEE PES Power Systems Conference and Exposition, Atlanta, GA, USA, 29 October–1 Novermber 2006; pp. 2070–2076.

22. Ibrahim, A.O.; Nguyen, T.H.; Lee, D.C.; Kim, S.C. A fault ride-through technique of DFIG wind turbine systems using dynamic voltage restorers. *IEEE Trans. Energy Convers.* **2011**, *26*, 871–882. [CrossRef]

23. Hagh, M.T.; Naderi, S.B.; Jafari, M. Application of non-superconducting fault current limiter to improve transient stability. In Proceedings of the 2010 IEEE International Conference on Power and Energy, Kuala Lumpur, Malaysia, 29 Novermber–1 December 2010; pp. 646–650.

24. Naderi, S.B.; Negnevitsky, M.; Jalilian, A.; Hagh, M.T.; Muttaqi, K.M. Optimum resistive type fault current limiter: An efficient solution to achieve maximum fault ride-through capability of fixed-speed wind turbines during symmetrical and asymmetrical grid faults. *IEEE Trans. Ind. Appl.* **2017**, *53*, 538–548. [CrossRef]

25. Jafari, M.; Naderi, S.B.; Hagh, M.T.; Abapour, M.; Hosseini, S.H. Voltage sag compensation of point of common coupling (PCC) using fault current limiter. *IEEE Trans. Power Deliv.* **2011**, *26*, 2638–2646. [CrossRef]

26. Naderi, S.B.; Negnevistky, M.; Jalilian, A.; Hagh, M.T. Non-controlled fault current limiter to improve fault ride through capability of DFIG-based wind turbine. In Proceedings of the 2016 IEEE Power and Energy Society General Meeting (PESGM), Boston, MA, USA, 17–21 July 2016; pp. 1–5.

27. Mardani, M.; Fathi, S.H. Fault current limiting in a wind power plant equipped with a DFIG using the interface converter and an optimized located FCL. In Proceedings of the 6th Power Electronics, Drive Systems & Technologies Conference (PEDSTC2015), Tehran, Iran, 3–4 February 2015; pp. 328–333.

28. Guo, W.; Xiao, L.; Dai, S.; Xu, X.; Li, Y.; Wang, Y. Evaluation of the performance of BTFCLS for enhancing lvrt capability of DFIG. *IEEE Trans. Power Electron.* **2015**, *30*, 3623–3637. [CrossRef]

29. Hossain, M.K.; Ali, M.H. Transient stability augmentation of PV/DFIG/SG-based hybrid power system by nonlinear control-based variable resistive FCL. *IEEE Trans. Sustain. Energy* **2015**, *6*, 1638–1649. [CrossRef]

30. Naderi, S.B.; Jafari, M.; Tarafdar Hagh, M. Controllable resistive type fault current limiter (CR-FCL) with frequency and pulse duty-cycle. *Int. J. Electr. Power Energy Syst.* **2014**, *61*, 11–19. [CrossRef]

31. Rashid, G.; Ali, M.H. Application of parallel resonance fault current limiter for fault ride through capability augmentation of DFIG based wind farm. In Proceedings of the 2016 IEEE/PES Transmission and Distribution Conference and Exposition (T & D), Dallas, TX, USA, 2–5 May 2016; pp. 1–5.

32. Naderi, S.B.; Jafari, M.; Hagh, M.T. Parallel-resonance-type fault current limiter. *IEEE Trans. Ind. Electron.* **2013**, *60*, 2538–2546. [CrossRef]

33. Guo, W.; Xiao, L.; Dai, S. Enhancing low-voltage ride-through capability and smoothing output power of DFIG with a superconducting fault-current limiter magnetic energy storage system. *IEEE Trans. Energy Convers.* **2012**, *27*, 277–295. [CrossRef]

34. Guo, W.; Xiao, L.; Dai, S. Fault current limiter-battery energy storage system for the doubly-fed induction generator: Analysis and experimental verification. *IET Gener. Trans. Distrib.* **2016**, *10*, 653–660. [CrossRef]
35. Chen, L.; Zheng, F.; Deng, C.; Li, Z.; Guo, F. Fault ride-through capability improvement of DFIG-based wind turbine by employing a voltage-compensation-type active SFCL. *Can. J. Elect. Comput. Eng.* **2015**, *38*, 132–142. [CrossRef]
36. Hossain, M.M.; Ali, M.H. Transient stability improvement of doubly fed induction generator based variable speed wind generator using DC resistive fault current limiter. *IET Renew. Power Gener.* **2016**, *10*, 150–157. [CrossRef]
37. Hossain, M.; Ali, H. Asymmetric fault ride through capability enhancement of DFIG based variable speed wind generator by DC resistive fault current limiter. In Proceedings of the 2016 IEEE/PES Transmission and Distribution Conference and Exposition (T & D), Dallas, TX, USA, 2–5 May 2016.
38. Chen, L.; Deng, C.; Zheng, F.; Li, S.; Liu, Y.; Liao, Y. Fault ride-through capability enhancement of DFIG-based wind turbine with a flux-coupling-type SFCL employed at different locations. *IEEE Trans. Appl. Supercond.* **2015**, *25*, 15. [CrossRef]
39. Ou, R.; Xiao, X.Y.; Zou, Z.C.; Zhang, Y.; Wang, Y.H. Cooperative control of SFCL and reactive power for improving the transient voltage stability of grid-connected wind farm with DFIGs. *IEEE Trans. Appl. Supercond.* **2016**, *26*, 1–6. [CrossRef]
40. Ngamroo, I.; Karaipoom, T. Cooperative control of SFCL and SMES for enhancing fault ride through capability and smoothing power fluctuation of dfig wind farm. *IEEE Trans. Appl. Supercond.* **2014**, *24*, 1–4. [CrossRef]
41. Ngamroo, I.; Karaipoom, T. Improving low-voltage ride-through performance and alleviating power fluctuation of DFIG wind turbine in dc microgrid by optimal smes with fault current limiting function. *IEEE Trans. Appl. Supercond.* **2014**, *24*, 1–5. [CrossRef]
42. Zou, Z.C.; Chen, X.Y.; Li, C.S.; Xiao, X.Y.; Zhang, Y. Conceptual design and evaluation of a resistive-type SFCL for efficient fault ride through in a DFIG. *IEEE Trans. Appl. Supercond.* **2016**, *26*, 1–9. [CrossRef]
43. Zou, Z.C.; Xiao, X.Y.; Liu, Y.F.; Zhang, Y.; Wang, Y.H. Integrated protection of DFIG-based wind turbine with a resistive-type SFCL under symmetrical and asymmetrical faults. *IEEE Trans. Appl. Supercond.* **2016**, *26*, 1–5. [CrossRef]
44. Mohammadi, J.; Afsharnia, S.; Vaez-Zadeh, S.; Farhangi, S. Improved fault ride through strategy for doubly fed induction generator based wind turbines under both symmetrical and asymmetrical grid faults. *IET Renew. Power Gener.* **2016**, *10*, 1114–1122. [CrossRef]
45. Yang, J.; Fletcher, J.E.; O'Reilly, J. A series-dynamic-resistor-based converter protection scheme for doubly-fed induction generator during various fault conditions. *IEEE Trans. Energy Conver.* **2010**, *25*, 422–432. [CrossRef]
46. Okedu, K.E. Enhancing DFIG wind turbine during three-phase fault using parallel interleaved converters and dynamic resistor. *IEEE Trans. Energy Conver.* **2016**, *10*, 1211–1219. [CrossRef]
47. Okedu, K.E.; Muyeen, S.M.; Takahashi, R.; Tamura, J. Wind farms fault ride through using DFIG with new protection scheme. *IEEE Trans. Sustain. Energy* **2012**, *3*, 242–254. [CrossRef]

applied
sciences

MDPI

Article

Wind Power Forecasting Using Multi-Objective Evolutionary Algorithms for Wavelet Neural Network-Optimized Prediction Intervals

Yanxia Shen *, Xu Wang and Jie Chen

Key Laboratory of Advanced Process Control for Light Industry, Jiangnan University, Wuxi 214122, China; wangxu_0626@163.com (X.W.); chenjjiangnan@163.com (J.C.)
* Correspondence: shenyx@jiangnan.edu.cn; Tel.: +86-138-6186-7517

Received: 12 December 2017; Accepted: 25 January 2018; Published: 26 January 2018

Abstract: The intermittency of renewable energy will increase the uncertainty of the power system, so it is necessary to predict the short-term wind power, after which the electrical power system can operate reliably and safely. Unlike the traditional point forecasting, the purpose of this study is to quantify the potential uncertainties of wind power and to construct prediction intervals (PIs) and prediction models using wavelet neural network (WNN). Lower upper bound estimation (LUBE) of the PIs is achieved by minimizing a multi-objective function covering both interval width and coverage probabilities. Considering the influence of the points out of the PIs to shorten the width of PIs without compromising coverage probability, a new, improved, multi-objective artificial bee colony (MOABC) algorithm combining multi-objective evolutionary knowledge, called EKMOABC, is proposed for the optimization of the forecasting model. In this paper, some comparative simulations are carried out and the results show that the proposed model and algorithm can achieve higher quality PIs for wind power forecasting. Taking into account the intermittency of renewable energy, such a type of wind power forecast can actually provide a more reliable reference for dispatching of the power system.

Keywords: wind power forecasting; wavelet neural network; multi-objective artificial bee colony algorithm; prediction intervals

1. Introduction

In recent years, wind power has grown rapidly in many countries as a type of clean and renewable energy source. However, the uncertainty and intermittency of wind power have brought great challenges to large-scale electrical power systems. Accurate short-term wind power forecasting is a necessary condition to dispatch the electrical power resources in time, reduce the operating costs, and then ensure the electrical power systems operate reliably and safely [1,2].

Many studies on wind power forecasting have been reported which can be mainly divided into two categories: one is based on statistical models, including regression models [3,4], the Kalman filter [5,6], and time series [7,8]. The other is based on artificial intelligence models, such as fuzzy systems [9], neural networks (NN) [10,11], and so on. Compared with the statistical model, artificial intelligence models are more flexible. More and more commercial prediction software used by utility companies have been developed based on artificial intelligent models [12], especially the NN model. These forecasting models are often called point prediction, mainly aiming to forecast values in the future, with less care about the prediction reliability, which is of limited value for uncertainties in the data or variability in the underlying system [13].

Recently there have been some reports about prediction intervals (PIs), which are considered as an excellent tool for the quantification of the wind power uncertainties that have never been considered in point predictions. Typical PIs are composed of a lower bound, an upper bound, and a confidence level indicating prediction reliability. PIs provide not only an interval covering the target value, but also a coverage probability indicating the prediction accuracy. In the literature, several models have been proposed for the construction of PIs and assessments of the uncertainties of prediction results. The PIs' models [14,15], based on neural networks, require special assumptions about the data distribution and the computational complexity is massive. Then, a model called the lower upper bound estimation (LUBE) model was proposed in [16] with more reliability and simplicity.

From the perspective of making decisions, the larger the coverage probability and smaller width of PIs, the more accurate the wind power forecasting is. However, in fact, in order to achieve a larger coverage probability, the width of the intervals will increase simultaneously, and vice versa. Thus, this is a typical multi-objective optimization problem. Previously, the literature solved this problem by translating the multi-objective problem into a single-objective problem with penalty parameters [13,16,17], such as [13]; the coverage probability is transformed to a hard constraint and then the problem becomes to the minimization of the width of the PIs.

In this paper, this two-objective optimization problem is directly considered to use the multi-objective artificial bee colony algorithm with no penalty parameters for optimization. Then, a new formulation to calculate the width of the PIs is proposed. In this formulation, the prediction value of the PIs is considered as a misleading result, especially when the deviation is large, so the width can be closer to that of the primary without any negative effects of the value out of PIs. Considering that the wind power data is nonlinear, highly dimensional, and strongly coupled, the Wavelet Neural Network (WNN) is a more suitable and flexible neural network that can deal with the data of wind power efficiently, because it can instigate a superior system model for complex and seismic applications in comparison to the NN with a sigmoidal activation function [18,19], so it is used to construct the PI model.

In addition, a new multi-objective artificial bee colony (MOABC) algorithm is proposed to optimize the parameters of WNN. The basic MOABC algorithm is extended from the artificial bee colony (ABC) algorithm, inheriting the structure of the multi-objective evolution algorithm (MOEA). Although the ABC algorithm is simpler and more efficient for parameter optimization, the pure inheritance of MOEA cannot exploit the advantages of the ABC algorithm. Thus, a new MOABC based on integrated evolution knowledge (EKMOABC) is proposed. The elite population knowledge and the other population knowledge are integrated to guide the evolution of the employed bees and maintain the diversity of the population. A strategy of combining the individual dominance relationship with the population distribution relationship is introduced into the probability selection of onlooker bees. Finally, a more strict strategy for updating the archive is put forward to reduce the cost of the proposed method.

The rest of this paper is organized as follows: Section 2 provides a brief review about the evaluation indices of the PIs. The multi-objective optimization model for the PIs' construction is explained in Section 3. Section 4 introduces a new MOABC based on evolution knowledge and the main steps of the PIs' construction. Experimental results and analysis are demonstrated in Section 5. Finally, Section 6 concludes this paper and discusses the future work.

2. PI Assessments

2.1. PI Coverage Probability

The PI coverage probability (PICP) is the most important characteristic for the reliability, as it indicates the probability that the targets lie in the constructed PIs. PICP can be calculated as follows:

$$\text{PICP} = \frac{1}{N_p} \sum_{i=1}^{N_p} \rho_i \tag{1}$$

where N_p is the number of the sample data. ρ_i is defined as follows:

$$\rho_i = \begin{cases} 0, & if \quad y_i \notin [L_i, U_i] \\ 1, & if \quad y_i \in [L_i, U_i] \end{cases} \tag{2}$$

where y_i is the target value. L_i and U_i are, separately, the lower bound and the upper bound of the PIs. The value of ρ_i depends on whether the target value is covered by the PIs. Therefore, the more target values are covered by the PIs, the higher the PICP and the more reliable the PIs are. The ideal value of the PICP is 100%, which means that all the target values are covered.

2.2. PIs' Normalized Average Width

If a proper PICP is chosen, a width (the maximum upper bound and minimum lower bound) of the PIs as large as possible will be the only thing to be decided so that all the target values are covered. However, too large a width is useless for making the decision in the electrical power system. Thus, another significant evaluation index of the PIs is defined as the normalized average width (PINAW), which is calculated as follows:

$$PINAW = \frac{1}{N_p R} \sum_{i=1}^{N_p} (U_i - L_i) \tag{3}$$

where R is the range of the target value. It is the normalized parameter for calculating PINAW in percentage regardless of the magnitudes of the target values. If the value of PICP is under control (a fixed value), the PIs are more accurate when PINAW becomes smaller.

PINAW is always used to assess whether the target value is covered by PIs or not. When the target value is out of the PIs, it will bring a negative effect to the width of the PIs. Thus, inspired by this, a new evaluation index for the width, called the PIs covered-normalized average width (PICAW), is developed as follows:

$$PICAW = \frac{1}{R} \left(\frac{1}{N_{p+}} \sum_{i=1}^{N_{p+}} (U_i - L_i) + \lambda \frac{1}{N_{p-}} \sum_{i=1}^{N_{p-}} (U_j - L_j) \right) \tag{4}$$

where N_{p+} and N_{p-} represent the number of target values covered by the PIs or not, separately. λ is the control parameter which can magnify the difference between the target values and PIs. In practice $\lambda > 1$. When $\lambda = 1$, PICAW turns into PINAW. Using both the target values and the PIs, more accuracy can be obtained for PI construction by PICAW, especially when the target value is far away from the PIs.

3. Multi-Objective Optimization Model for WNN-Based PI Construction

3.1. PI Multi-Objective Optimization Criteria

PICP or PICAW can evaluate the quality of PIs from different aspects. In order to obtain the high quality of PIs, a large PICP and small PICAW are both required. Obviously, this is a multi-objective optimization problem for maximizing the coverage probability and minimizing the width of PIs simultaneously. The most famous method for this problem is to transform two primary PIs' assessments into a single one by some hyperparameters. A most common index, called the coverage width-based criterion (CWC), is defined as follows:

$$CWC = PINAW(1 + \gamma(PICP)e^{-\eta(PICP-\mu)}) \tag{5}$$

where μ is an expected coverage probability, η is a control parameter that magnifies the differences between PICP and μ when the coverage probability hardly achieves the expected value. $\gamma(PICP) = 0$ when $PICP \geq \mu$ and $\gamma(PICP) = 1$ when $PICP < \mu$. In CWC, the most critical parameter is η. If η is small, it will be insufficient to obtain the expected coverage probability. In contrast, if η is too large, there will be too much of a penalty to obtain the optimized solution. Thus, for the CWC, how to choose a suitable η becomes a key problem. η is often set empirically in most of the literature. If an unreasonable η is chosen, the multi-objective optimization problem cannot be transformed into a single optimization problem completely and the quality of PIs cannot be guaranteed.

In this paper, the multi-objective optimization problem is solved directly by an improved MOABC algorithm. A PIs' multi-objective optimization criteria (PIMOC) is proposed as follows:

$$
\begin{cases}
\min & \alpha = 1 - \text{PICP} \\
\min & \text{PICAW} \\
\text{s.t.} & 0 \le \alpha \le 1, \quad 0 \le \text{PICAW} \le 1
\end{cases}
\tag{6}
$$

where α is transformed from the confidence level to satisfy the need of multi-objective optimization problem. Taking into account the real requirement of the electrical power system, the range of constraint conditions will be reduced. Thus, the confident level α is restricted in $[0, 20\%]$ and the width of the PIs is restricted in $[0, 25\%]$ after normalization.

3.2. WNN-Based PI Construction

WNN is a neural network based on the wavelet transform (WT). It is powerful for frequency component analysis and suitable for signals which are composed of high-frequency components with short duration and low-frequency components with long duration [20,21]. A brief review of WT is described as follows:

Assuming that $f(t)$ is a square integral function, it can be expressed as:

$$
f(t) = \iint W(a,b)\psi(\frac{t-b}{a})dbda
\tag{7}
$$

where $W(a,b)$ is the continuous wavelet transformation of $f(t)$ and defined as:

$$
W(a,b) = \int f(t)\psi_{a,b}^{*}(t)dt
\tag{8}
$$

with:

$$
\psi_{a,b}^{*}(t) = \frac{1}{\sqrt{a}}\psi(\frac{t-b}{a})
\tag{9}
$$

where a and b are the scaling parameter and shifting parameter, respectively.

$X_i = \{x_1, \cdots, x_i, \cdots, x_N\}$ is the N input sample of WNN, ω_{is} is the input parameter between the input layer node and the hidden layer node, and β_{sj} is the output parameter that connects the hidden layer and the output layer. The calculation equation of the output of the hidden layer is as follows:

$$
g(s) = \psi\left((\sum_{i=1}^{N}\omega_{is}x_i - b_s)/a_s\right), \quad s = 1,2,\cdots k
\tag{10}
$$

where $g(s)$ is the output of the s node of the hidden layer, ψ is the small wave basis function, and k is the number of hidden layer nodes of the WNN. The calculation equation of the output layer is as follows:

$$
Y_j = \sum_{s=1}^{k}\beta_{sj}g(s), s = 1,2,\cdots k
\tag{11}
$$

Then, the M output $Y_j = \{y_1, \cdots, y_j, \cdots, y_M\}$ of the WNN is designed as:

$$
y_j = \sum_{s=1}^{k}\beta_{sj}\psi\left((\sum_{i=1}^{N}\omega_{is}x_i - b_s)/a_s\right)
\tag{12}
$$

The upper and lower bounds of the prediction interval of wind power are directly given by the dual-output WNN, which effectively avoids the complex process of the traditional interval prediction method for the probability error analysis. Thus, a symbolic WNN-based on PIs' construction model is shown in Figure 1.

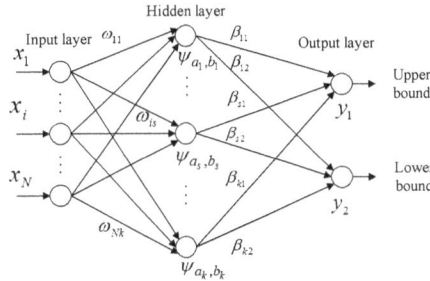

Figure 1. WNN-based on the PI construction model. WNN: wavelet neural network; PI: prediction interval.

In order to avoid the irrational effect of the number of input and hidden layer nodes on the prediction results, the C-C phase space reconstruction method is used to determine the number of input layer nodes N. According to the Kolmogorov theorem, the number of hidden layer nodes K is determined, and the structure of the PI model for the WNN is N–K–2. The Morlet wavelet with strong adaptive ability is selected as the wavelet basis function, and the equation is as follows:

$$\psi(t) = \cos(1.75t) \cdot e^{-t^2/2} \tag{13}$$

By choosing suitable input parameters, output parameters, scaling parameters, and shifting parameters of the WNN, effective and reliable PIs of wind power will be constructed.

4. Evolution Knowledge MOABC Algorithm (EKMOABC) for WNN-Optimized Construction of PIs

To choose the parameters of WNN reasonably and solve the multi-objective optimization problem for PI construction, an improved MOABC algorithm, called Evolution Knowledge MOABC (EKMOABC), is proposed. The main steps of EKMOABC are summarized as follows:

4.1. Initialization

The input parameters, the output parameters, the scaling parameters and shifting parameters of WNN are initialized at this step. The wind power data are divided into a training set, a validation set, and a test set, and then they are normalized to $[0, 1]$. The parameters of EKMOABC, including the number of food sources N, the maximum iterations T_m, the maximum number of solutions in archive I_m, the maximum obsolete number of scout bees D_m, the probability' parameters η_1 and η_2, are also initialized.

4.2. Preliminary Iteration

Preliminary iteration is crucial for the repeatability of the model. Firstly, each food source is initialized as:

$$x_{ij} = L + rand(0,1) \times (U - L) \tag{14}$$

where x_{ij} is one of the food sources. $i \in 1, 2, \cdots, N/2, j \in 1, 2, \cdots, D$ and D is the number of parameters to be optimized. L and U are the minimum lower bound and maximum upper bound of the solutions, respectively. Then the assessment criteria PICP and PICAW are calculated.

4.3. Pareto Dominance

Compared with single-objective optimization, the solution of the multi-objective optimization problem is a trade-off of the performance, which is called the Pareto optimal set based on Pareto dominance.

For the decision vector $x_a, x_b \in \Omega$, x_a dominates x_b (denoted as $x_a \succ x_b$) if $\forall f_i(x_a) \leq f_i(x_b)$ and $\exists f_i(x_a) < f_i(x_b)$. Therefore, a set of decision vectors $v \in \Omega$ is called a non-dominate solution when $\neg \exists x \in \Omega : x \prec v$. Furthermore, a Pareto optimal front set is defined as a group of non-dominate solutions.

In this paper, α and PICAW are the optimization objectives. Thus, when the objectives satisfy the equation below:

$$\alpha_i \leq \alpha_j, \text{PICAW}_i \leq \text{PICAW}_j \tag{15}$$

where α and PICAW cannot be met simultaneously, it is said that the solution i dominates solution j, which can be denoted as $i \succ j$. When the dominant relationship of different food sources is achieved, the non-dominate solution is chosen into the archive.

4.4. Employed Bees Evolution Based on Guidance of Elite Population Knowledge

Elite population represents the optimal information of population and it is beneficial for the population to converge rapidly [20]. Thus, a more effective strategy is adopted to choose an elite population and guide the employed bees' evolution. In the first iteration, the crowding distance of solutions in the archive is calculated and the maximum crowding distance is chosen as the elite solution. Additionally, objective functions should be ordered first according to Equation (16) when the crowding distance is calculated. The crowding distance is defined as:

$$d_i = \begin{cases} \sqrt{\frac{1}{2}\sum_{k=1}^{m} \left((f_i^k - f_i^{k-1})^2 + (f_i^k - f_i^{k+1})^2 \right)}, & 1 < i < s \\ \inf, & else \end{cases} \tag{16}$$

where f_i^k is the target value (the assessment criteria of PIs). m is the number of the target value. s is the number of solutions in the archive. At the next iteration, an elite solution will be chosen and compared with the former one. When the new elite solution dominates the old one, it will be saved and guide the employed bees' evolution instead. Otherwise, the old one will be saved. However, when neither elite solution dominates the other one, the roulette method will be adopted for making the decision.

In this paper, both dominant and non-dominant solutions of employed bees evolve in different ways. The evolution method is as follows:

$$\begin{cases} v_{tj} = x_{tj} + r_1(x_{tj} - x_{kj}) + r_2(x_{tj} - x_{bj}), & s_i = 0 \\ v_f = x_f + r_3(x_f - x_b), & s_j = 1 \end{cases} \tag{17}$$

where s_i is a flag. $s_i = 0$ represents the dominant solution, while $s_j = 1$ represents the non-dominant solution. x_b is the elite solution. x_t and x_f are the non-dominant solution, and $t \neq f$. $r_1, r_2, r_3 \in [-1, 1]$. This measure is applied in both the dominated front and non-dominated front. When the non-dominant solutions crossover with the elite solution, it is easier to produce excellent solutions and keep the elite population alive. On the other hand, when dominant solutions crossover with both the elite solution and the non-dominant solution, it is beneficial to maintain the diversity of the population.

4.5. Probability Choice Equation Combining the Dominance and Distribution Relationships

Probability choice equation is used to choose the solutions to evolve deeply and balance the capacity between exploit and explore. Different from the single-objective optimization, there may be a large number of solutions for multi-objective optimization which cannot dominate other solutions at the same time. The quality of the Pareto optimal set is related to both the dominance relationship and distribution relationship. If only one of them is considered in the probability choice equation, it will hardly obtain the high-quality of PIs. Therefore, a typical probability choice equation combining the dominance and distribution relationships is proposed as follows:

$$P_i = \begin{cases} \frac{1}{1+(1-s_i)e^{\eta_1 l_i}}, & s_i = 0 \\ \frac{1}{1+s_i e^{-\eta_2 d_i}}, & s_i = 1 \end{cases} \tag{18}$$

where P_i is the probability, d_i is the crowding distance. η_1 and η_2 are the probability choice parameters and l_i is defined as:

$$l_i = \min_{j=1}^{s} \sqrt{\sum_{k=1}^{m} (f_{ik} - f_{jk})^2} \tag{19}$$

where f_i is the dominant solution and f_j is the non-dominant solution. s is the number of solutions in the archive.

From Equations (18) and (19), we can see that the proposed equation of probability choice combines the dominance relationship and the distribution relationship. When the solution is dominated by other solutions, it will be punished by the distance from the Pareto optimal set. From Equation (18), when the solution is dominated, the probability will be limited to $[0, 0.5]$ by the parameter η_1. However, when the solution is non-dominated, the probability will be chosen in the range of $[0.5, 1]$ due to the superiority of non-dominant solutions. Thus, even though the dominant solution is closer to the Pareto optimal set, its probability will be never be greater than that of the non-dominant solution, which will be of great benefit for population evolution both in convergence and distribution.

4.6. Onlooker Bees Evolution

Onlooker bees evolve through the roulette method according to the probability calculated above. The evolution method of onlooker bees is the same as the employed bees.

4.7. Strategy to Update the Archive

The archive is used to save the best solutions ever found and the final solutions in the archive are called the Pareto optimal set. Thus, the strategy for updating the archive is crucial for the quality of the results. Traditionally, the non-dominant solutions are chosen for the archive and weeded out when it is dominated or the crowding distance is too small. In the preliminary iteration, there are few non-dominant solutions and it is reasonable to update the archive. However, in the later iterations, the amount of non-dominant solutions increases quickly, which means some better solutions should be chosen for the archive and some would be deleted. Then the cost of using the traditional strategy will increase. Thus, a stricter strategy is chosen to update the archive. When the number of solutions in the archive reaches the maximum value I_m, the solutions with the smallest crowding distance will be deleted, and the maximum crowding distance of these deleted solutions are calculated and recorded as d_t^{max}. When a solution satisfies the condition for being saved into the archive, one stricter step will be checked. Firstly, the distance between this solution and the other solutions in the archive is calculated. Then the minimum distance will be recorded as d_t^{min} and only when $d_t^{min} > d_t^{max}$, the solution can be chosen to be saved into the archive. The new strategy for archive updating reduces the cost of the algorithm and improves the distribution performance of the Pareto optimal set.

4.8. Scout Bees Evolution

When the evolution of the employed bees and onlooker bees are completed and the solution is not improved, the number of obsolete scout bees will be added by 1. Once it reaches its maximum value D_m, the employed bees will be transformed into scout bees to produce a new solution randomly.

4.9. Termination

The condition for finishing the algorithm is that the iterations reach its maximum value T_m.

5. Experiments and Results

To validate the multi-objective optimization model for PI construction, the wind power data sampled every 10 min from the Alberta interconnected electric system in 2015 are applied. A total of 6000 datasets from 1 January are chosen as the research data with 80% for training and the other 20% for validating and testing, respectively. Firstly, the datasets are normalized in $[0, 1]$ to adjust the parameters of WNN and avoid the influence from different magnitudes of the datasets. Then the structure of WNN using the Morlet wavelet as the basic wavelet is determined. It is necessary to balance the complexity and the learning capacity of WNN. The C-C phase space reconstruction and Kolmogorov theorem are adopted to choose the optimal WNN structure as 4-8-2. Finally, the parameters of WNN and EKMOABC will be set as mentioned in Section 4.

The CWC for PI construction is a single-objective optimization problem, while PIMOC is a multi-objective optimization problem. To valid the experiment results, the basic ABC algorithm in [21] is used to optimize the CWC, and the MOABC algorithm in [22] to optimize the WNN for PIMOC. The food source number of ABC and MOABC are both set to 40, the maximum iteration time is set to 200, the maximum number of solutions in the archive is 20, and the maximum number of obsolete scout bees is 50.

5.1. Performance Comparisons between CWC with Different Parameters and PIMOC

To validate the influences of parameters on CWC, the control parameter η in Equation (5) is chosen as $\eta = 10$, 50, 100 and μ is set as 80%, 85%, 90%, 95%, and 99%. Each experiment is repeated 10 times with different η and μ. The average results are shown in Table 1.

Table 1. Comparative results with different η and μ of CWC. CWC: coverage width-based criterion; PICP: prediction interval coverage probability; PICAW: prediction intervals covered-normalized average width.

η \ μ	Name	$\mu = 0.80$	$\mu = 0.85$	$\mu = 0.90$	$\mu = 0.95$	$\mu = 0.99$
$\eta = 10$	PICP	0.785	0.836	0.892	0.944	0.985
	PICAW	0.137	0.152	0.203	0.235	0.256
$\eta = 50$	PICP	0.822	0.864	0.923	0.963	0.995
	PICAW	0.152	0.198	0.232	0.251	0.284
$\eta = 100$	PICP	0.844	0.872	0.947	0.981	1.000
	PICAW	0.175	0.218	0.259	0.288	0.323

From Table 1, it is obvious that PICP increases with the increase of μ when η is fixed, and PICAW also becomes larger, which makes the accuracy of the PIs worse. On the other hand, η has a great effect on the quality of PIs with the fixed μ. A small η is helpful to improve the accuracy of PIs with a small width of PIs, but it is difficult to ensure that PICP reaches its expected μ and meets the reliability requirements of the PIs. A large η can enhance the reliability of the PIs, and the PICP is always higher than its expected value. However, this will lead to a large width of the PIs, which then makes the solving process into a local optimum. Thus, the controlling parameter η is an uncertain factor in PI construction with CWC, which is unfavorable for solving this optimization problem in high quality.

In order to compare the prediction results between CWC (with the ABC-WNN as the prediction model) and PIMOC (with the MOABC-WNN as the model), the experiments with a group of μ (from 80–99%) in different η are performed and the results are shown in Figure 2.

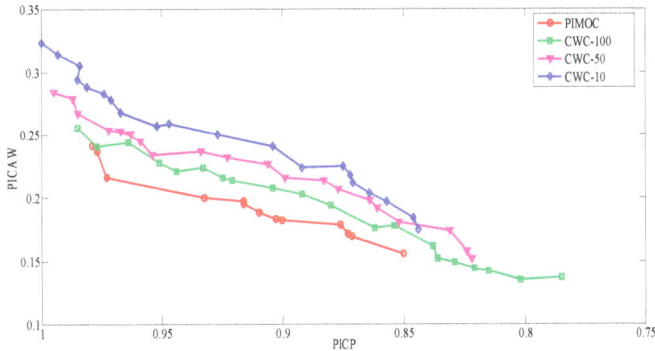

Figure 2. Comparative results between CWC and PIMOC. CWC: coverage width-based criterion; PIMOC: PIs' multi-objective optimization criteria; PICP: PI coverage probability.

As is shown in Figure 2, the prediction results of PIMOC are better than CWC both in the criteria of PICP and PINAW. With different η, the PICP and PICAW of CWC change greatly. During the optimization of CWC, the solutions that cannot satisfy the given confidence level μ will be penalized by η and determined whether to be saved or not according to η. Perhaps a large η can ensure that the requirement of the confidence level μ should be met, but some good solutions which have a PICP slightly less than its expected value, but a good PI width, will be weeded out. However, if a small η is chosen and then the solutions with a slightly worse confidence level, but a small width of the PIs is saved, those solutions dissatisfy the requirement of the confidence level and will not be penalized sufficiently. As a result, the confidence level of all the solutions will be lowered. Thus, a suitable η is of great significance to CWC, but it is always chosen empirically. In PIMOC, this multi-objective optimization problem is solved by adopting a multi-objective evolution algorithm directly instead of transforming it into a single-objective optimization problem, which avoids the choosing of η and it is a benefit for improving the quality of PIs both in accuracy and reliability.

For a better explanation, the PIs of CWC are shown in Figures 3–5 with three different η (the value of η is same as it in Table 1) and the PIs of PIMOC are shown in Figure 6, where Pareto optimal sets are sorted in ascending order according to PICP and the first, fifth, tenth, fifteenth, and twentieth PIs are plotted to make a comparison with the PIs of CWC.

As is shown in Figures 3–5, with the control parameter η increasing, the prediction accuracy of the PIs is improved, while the width is larger and the reliability becomes worse, which is just the same as in Table 1. An unreasonable choice of η may easily cause some excellent solutions unreserved in the next iteration. From Equation (5), it can be seen that, in CWC, the quality of the PIs is assessed impartially only based on a single synthetical criterion. While in PIMOC, some excellent solutions are reserved by the Pareto dominance strategy avoiding the choice of η, and the assessment is carried out according to both PIMOC, both in accuracy and reliability. Then we can see that the performances of PIMOC in Figure 6 are better than all those of CWC in Figures 3–5.

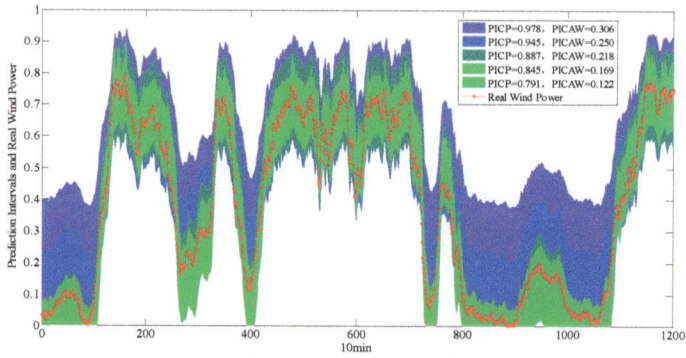

Figure 3. Prediction intervals with CWC ABC-based when $\eta = 10$. CWC: coverage width-based criterion; ABC: artificial bee colony; PICAW: PIs covered-normalized average width; PICP: PI coverage probability.

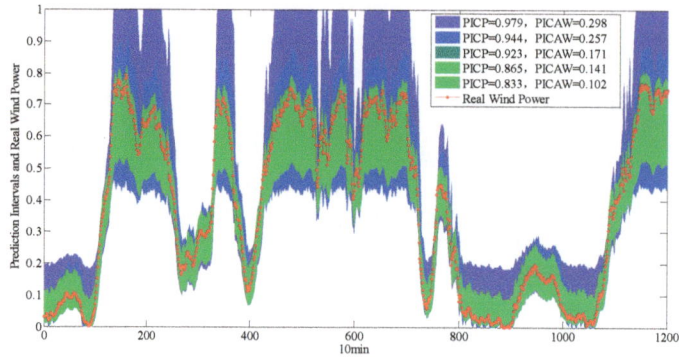

Figure 4. Prediction intervals with CWC ABC-based when $\eta = 50$. CWC: coverage width-based criterion; ABC: artificial bee colony; PICAW: PIs covered-normalized average width; PICP: PI coverage probability.

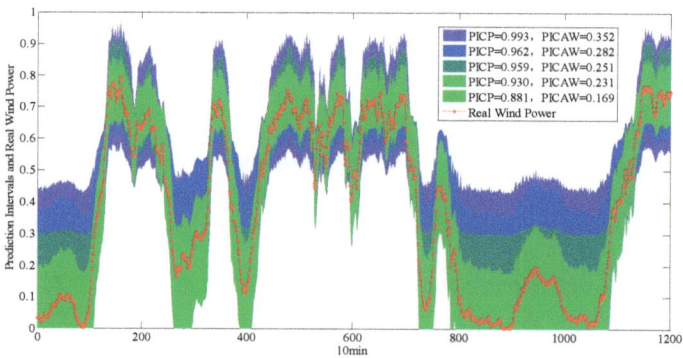

Figure 5. Prediction intervals with ABC-based CWC when $\eta = 100$. CWC: coverage width-based criterion; ABC: artificial bee colony; PICAW: PIs covered-normalized average width; PICP: PI coverage probability.

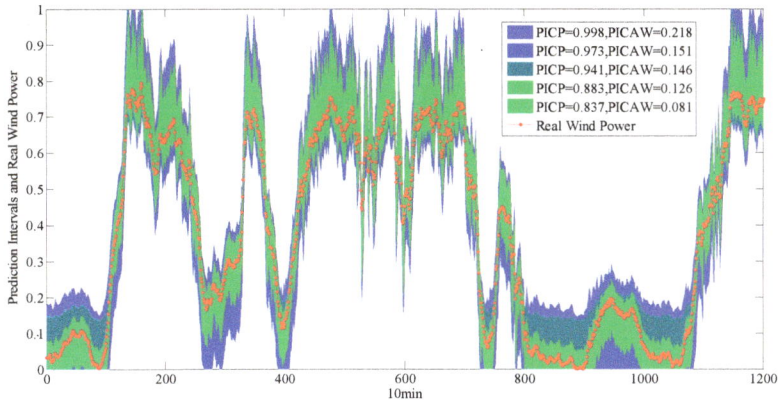

Figure 6. Prediction intervals with PIMOC based on MOABC. PIMOC: PIs' multi-objective optimization criteria; MOABC: multi-objective artificial bee colony; PICAW: PIs covered-normalized average width; PICP: PI coverage probability.

5.2. Performances of the Multi-Objective Evolution Algorithm in PI Construction

To evaluate the proposed EKMOABC for solving the multi-objective optimization problem with PIMOC, three classic multi-objective optimization algorithms, including the basic MOABC, NSGAII (non-dominated sorting genetic algorithm II) [23], and MOPSO (multi-objective particle swarm optimization) [24], are used to optimize the parameters of WNN for contrast experiments. The population size, the maximum number of iterations, and the maximum number of solutions in the archive are set to 40, 200, 20, respectively, for all algorithms. The crossover probability and the mutation probability of NSGAII are set to 0.8 and 0.2, respectively. The inertia weight and the learning factor of MOPSO are set to 0.8 and 2, respectively. The probability choices parameter η_1 and η_2 of EKMOABC are set to 50 and 10. The maximum obsolete number of scout bees in MOABC and EKMOABC are all set to 50. The comparative results for PI construction with four different multi-objective evolutionary optimization algorithms are shown in Figure 7.

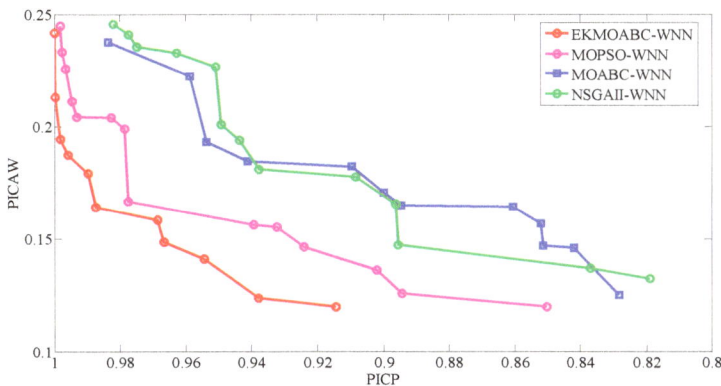

Figure 7. Comparative results with different multi-objective evolutionary algorithms. EKMOABC: evolutionary knowledge multi-objective artificial bee colony; MOPSO: multi-objective particle swarm optimization; MOABC: multi-objective artificial bee colony; NSGAII: non-dominated sorting genetic algorithm II; PICAW: PIs covered-normalized average width; PICP: PI coverage probability.

As shown in Figure 7, Compared with the other three evolutionary optimization algorithms (NSGAII, MOPSO, and MOABC) in WNN optimization, the proposed EKMOABC-WNN can ensure a higher confidence level and narrower width of PIs, especially when the PICP is above 97%, where there is no suitable width of PIs that can be chosen by the other three algorithms. The Pareto optimal set with EKMOABC-WNN has not only better convergent performance, but also distributing performance. One of the reasons is that NSGAII, MOPSO, and MOABC may be easily trapped into a local optimum and their searching capacities will become so weak that the better solutions cannot be searched in the constraint range. Another reason is that the distribution of the Pareto set is not considered sufficiently in the process of iteration in NSGAII, MOPSO, and MOABC, which results in the Pareto set non-uniform distribution. In addition, the EKMOABC is based on the relationship between Pareto domination and distribution. The advanced probability choice equation of EKMOABC can avoid the solutions being trapped into local optimum effectively and it is a benefit for the distribution of the Pareto optimal set. The strategy of the guidance with elite population knowledge plays an important role in searching for better solutions and improving the convergent performance of the Pareto optimal set.

6. Conclusions

Wind power forecasting is of great importance for electrical power systems because of the uncertainties of climate, especially when the renewable energies are merged into the grid. Compared with point forecasting, PI forecasting is an effective way for assessing the uncertainty of wind power. However, the traditional method for prediction interval construction is carried out under some special assumptions and suffers from computational complexity. In this paper, an improved MOABC method was proposed to optimize the parameters of WNN for PI construction in wind power. Instead of transforming into a single-objective optimization problem, the primary multi-objective optimization problem is solved directly by the improved MOABC, called EKMOABC. In addition, a new multi-objective criterion, called PIMOC, was proposed, with which the effect of the prediction value of the PIs is considered. Comparative results between CWC and PIMOC in PI construction demonstrate that the CWC is affected by the controlling parameter η greatly while the PIMOC can avoid the choice of this parameter, so the quality of the PIs based on PIMOC is more accurate and reliable. On the other hand, a multi-objective algorithm, called EKMOABC, was also proposed for the solutions of PIMOC. Some simulation experiments comparing with NSGAII, MOPSO, and MOABC showed that the EKMOABC-based WNN for wind power PI forecasting has a narrower width and higher confidence, which means higher accuracy and reliability.

Acknowledgments: This work was supported in part by the National Nature Science Foundation under grant 61573167 and grant 61572237, in part by the Fundamental Research Funds for the Central Universities under grant JUSRP31106 and grant JUSRP51510, and in part by the Postgraduate Research and Practice Innovation Program of Jiangsu Province under grant KYCX17_1488.

Author Contributions: Yanxia Shen conceived the experiment and wrote the paper; Xu Wang helped in the experiment and writing; and Jie Chen performed the experiments.

Conflicts of Interest: The authors declare no conflict of interest.

References

1. Wang, Q.; Martinez-Anido, C.B.; Wu, H.; Florita, A.R.; Hodge, B.M. Quantifying the economic and grid reliability impacts of improved wind power forecasting. *IEEE Trans. Sustain. Energy* **2016**, *7*, 1525–1537. [CrossRef]
2. Tascikaraoglu, A.; Uzunoglu, M. A review of combined approaches for prediction of short-term wind speed and power. *Renew. Sustain. Energy Rev.* **2014**, *34*, 243–254. [CrossRef]
3. Haque, A.U.; Nehrir, M.H.; Mandal, P. A hybrid intelligent model for deterministic and quantile regression approach for probabilistic wind power forecasting. *IEEE Trans. Power Syst.* **2014**, *29*, 1663–1672. [CrossRef]

4. Hu, J.; Wang, J. Short-term wind speed prediction using empirical wavelet transform and Gaussian process regression. *Energy* **2015**, *93*, 1456–1466. [CrossRef]

5. Che, Y.; Peng, X.; Monache, L.D.; Kawaguchi, T.; Xiao, F. A wind power forecasting system based on the weather research and forecasting model and Kalman filtering over a wind-farm in Japan. *J. Renew. Sustain. Energy* **2016**, *8*, 319–329. [CrossRef]

6. Zuluaga, C.D.; Álvarez, M.A.; Giraldo, E. Short-term wind speed prediction based on robust Kalman filtering: An experimental comparison. *Appl. Energy* **2015**, *156*, 321–330. [CrossRef]

7. Yan, J.; Li, K.; Bai, E.; Yang, Z.; Foley, A. Time series wind power forecasting based on variant Gaussian Process and TLBO. *Neurocomputing* **2016**, *189*, 135–144. [CrossRef]

8. Zhao, Y.; Ye, L.; Li, Z.; Song, X.; Lang, Y.; Su, J. A novel bidirectional mechanism based on time series model for wind power forecasting. *Appl. Energy* **2016**, *177*, 793–803. [CrossRef]

9. Osório, G.J.; Matias, J.C.O.; Catalão, J.P.S. Short-term wind power forecasting using adaptive neuro-fuzzy inference system combined with evolutionary particle swarm optimization, wavelet transform and mutual information. *Renew. Energy* **2015**, *75*, 301–307. [CrossRef]

10. Ata, R. Artificial neural networks applications in wind energy systems: A review. *Renew. Sustain. Energy Rev.* **2015**, *49*, 534–562. [CrossRef]

11. Li, S.; Wang, P.; Goel, L. Wind power forecasting using neural network ensembles with feature selection. *IEEE Trans. Sustain. Energy* **2017**, *6*, 1447–1456. [CrossRef]

12. Tewari, S.; Geyer, C.J.; Mohan, N. A statistical model for wind power forecast error and its application to the estimation of penalties in liberalized markets. *IEEE Trans. Power Syst.* **2011**, *26*, 2031–2039. [CrossRef]

13. Quan, H.; Srinivasan, D.; Khosravi, A. Short-term load and wind power forecasting using neural network-based prediction intervals. *IEEE Trans. Neural Netw. Learn. Syst.* **2014**, *25*, 303. [CrossRef] [PubMed]

14. Quan, H.; Srinivasan, D.; Khosravi, A. Incorporating wind power forecast uncertainties into stochastic unit commitment using neural network-based prediction intervals. *IEEE Trans. Neural Netw. Learn. Syst.* **2015**, *26*, 2123–2135. [CrossRef] [PubMed]

15. Quan, H.; Srinivasan, D.; Khosravi, A. Particle swarm optimization for construction of neural network-based prediction intervals. *Neurocomputing* **2014**, *127*, 172–180. [CrossRef]

16. Khosravi, A.; Nahavandi, S.; Creighton, D.; Atiya, A.F. Lower upper bound estimation method for construction of neural network-based prediction intervals. *IEEE Trans. Neural Netw.* **2011**, *22*, 337–346. [CrossRef] [PubMed]

17. Khosravi, A.; Nahavandi, S.; Creighton, D. Construction of optimal prediction intervals for load forecasting problems. *IEEE Trans. Power Syst.* **2010**, *25*, 1496–1503. [CrossRef]

18. Catalão, J.P.S.; Pousinho, H.M.I.; Mendes, V.M.F. Short-term wind power forecasting in Portugal by neural networks and wavelet transform. *Renew. Energy* **2011**, *36*, 1245–1251. [CrossRef]

19. Rafiei, M.; Niknam, T.; Khooban, M.H. Probabilistic forecasting of hourly electricity price by generalization of elm for usage in improved wavelet neural network. *IEEE Trans. Ind. Inform.* **2016**, *13*, 71–79. [CrossRef]

20. Ibrahim, A.; Rahnamayan, S.; Martin, M.V.; Deb, K. Elite NSGA-III: An improved evolutionary many-objective optimization algorithm. In Proceedings of the 2016 IEEE Congress on Evolutionary Computation (CEC), Vancouver, BC, Canada, 24–29 July 2016; pp. 973–982. [CrossRef]

21. Karaboga, D.; Gorkemli, B. A quick artificial bee colony (qABC) algorithm and its performance on optimization problems. *Appl. Soft Comput. J.* **2014**, *23*, 227–238. [CrossRef]

22. Akbari, R.; Hedayatzadeh, R.; Ziarati, K.; Hassanizadeh, B. A multi-objective artificial bee colony algorithm. *Swarm Evol. Comput.* **2012**, *2*, 39–52. [CrossRef]

23. Deb, K.; Pratap, A.; Agarwal, S.; Meyarivan, T. A fast and elitist multi-objective genetic algorithm: NSGA-II. *IEEE Trans. Evol. Comput.* **2002**, *6*, 182–197. [CrossRef]

24. Coello, C.A.C.; Pulido, G.T.; Lechuga, M.S. Handling multiple objectives with particle swarm optimization. *IEEE Trans. Evol. Comput.* **2004**, *8*, 256–279. [CrossRef]

applied
sciences

MDPI

Article

Output Power Smoothing Control for a Wind Farm Based on the Allocation of Wind Turbines

Ying Zhu [1],*, Haixiang Zang [1], Lexiang Cheng [2] and Shengyu Gao [2]

[1]   College of Energy and Electrical Engineering, Hohai University, Nanjing 211100, China;
      zanghaixiang@hhu.edu.cn
[2]   State Grid Nanjing Power Supply Company, State Grid Jiangsu Electric Power CO. LTD.,
      Nanjing 210008, China; chenglx@js.sgcc.com.cn (L.C.); gaosy@js.sgcc.com.cn (S.G.)
*   Correspondence: yingzhu@hhu.edu.cn; Tel.: +86-139-2142-6216

Received: 18 May 2018; Accepted: 13 June 2018; Published: 15 June 2018

Abstract: This paper presents a new output power smoothing control strategy for a wind farm based on the allocation of wind turbines. The wind turbines in the wind farm are divided into control wind turbines (CWT) and power wind turbines (PWT), separately. The PWTs are expected to output as much power as possible and a maximum power point tracking (MPPT) control strategy combining the rotor inertia based power smoothing method is adopted. The CWTs are in charge of the output power smoothing for the whole wind farm by giving the calculated appropriate power. The battery energy storage system (BESS) with small capacity is installed to be the support and its charge and discharge times are greatly reduced comparing with the traditional ESSs based power smoothing strategies. The simulation model of the permanent magnet synchronous generators (PMSG) based wind farm by considering the wake effect is built in Matlab/Simulink to test the proposed power smoothing method. Three different working modes of the wind farm are given in the simulation and the simulation results verify the effectiveness of the proposed power smoothing control strategy.

Keywords: power smoothing; wind farm; wind turbine allocation; rotor inertia; power wind turbine; control wind turbine; battery energy storage system; wake effect

---

1. Introduction

Wind energy is the most mature and promising renewable energy source in the world at present. The doubly-fed induction generator (DFIG) based and direct-drive permanent magnet synchronous generator (PMSG) based systems are the most popular wind energy conversion systems (WECS), which are widely used in wind farms [1]. Though both of the two WECSs have their own advantages and disadvantages, the direct-drive PMSG based WECSs are preferred in high power applications owing to the simple structure, high efficiency, and high reliability due to the absence of gearboxes [1–3]. However, due to the stochastic nature of wind, the output power of the wind farm is fluctuating, which brings negative influences to the stability and economy of the grid operation. Specifically, the problems caused by the wind power fluctuations are listed as: frequency deviation, voltage flicker, low power quality, and harm to sensitive loads [3–5]. To solve these problems, several methods are proposed to smooth the output power of the wind farm, which can be basically divided into two major categories: direct power control and indirect power control.

The direct power control usually depends on using the wind turbine (WT) itself without energy storage devices, such as through the DC bus voltage control, rotor inertia control, and the pitch angle control [3,6–11]. However, the DC bus voltage control may cause the excessive voltage ripple, thus affecting normal operation of the WECS [5]. For the rotor inertia control, the rotor speed is always controlled to exceed the rated value to store the kinetic energy. According to the research [5], the efficiency of the kinetic energy control relying on the generator inertia is close to the maximum power point tracking (MPPT) control. The trade-off between the maximum power tracking and power smoothing is explored through the optimal control of WTs based on the rotating kinetic energy reserves in references [7,8]. The pitch angle control has to operate at a fast rate, thus mechanical stress accrues in the wind turbine blade. Besides, the direct power control methods always focus on the output power smoothing of a single WT, which may lead to the poor effect for the entire wind farm. In addition, all the smoothing capability of the direct power control method is constrained by the rated power of the WTs.

The indirect power control is realized by using energy storage systems (ESS), such as the battery, the supercapacitor, the superconductor magnetic energy storage (SMES), the flywheel, and so on [12–19]. But all existing ESSs have certain problems in practical application. The charge-discharge cycles of the battery are limited, although many types of batteries are quite mature at present [12,13]. The supercapacitor has the much longer life of charge-discharge processing than that of batteries, but it cost too much to be used on a large scale [14]. SMES is a large superconducting coil that can store electric energy in the magnetic field produced by the flow of a DC current through it. The effectiveness of the SMES is demonstrated by many studies, but it is still too costly [15,16]. The flywheel ESS utilizes the kinetic energy of a rotating disc, which depends on the square of the rotational speed. It has the advantages of long cycling life, large energy storage capacity, and high reliability [17,18]. However, the drawbacks of immature technology, large size, and high standing losses cannot be ignored in practical use at present. Then hybrid ESSs for smoothing out wind power fluctuations are proposed in literatures to overcome the disadvantages of the single ESS, but the control strategies are more complicated [19].

As above, all the existing power smoothing control strategies have both advantages and disadvantages, which result in difficulties for practical application. In order to solve such problems, a power smoothing method based on the allocation of the wind turbines and the rotor inertia is proposed in this paper, which is quite different from the above-mentioned power smoothing methods. In this paper, the direct-drive PMSG based WECS is chosen in the wind farm. The WTs of the wind farm are divided into two classes: control wind turbines (CWT) and power wind turbines (PWT). The PWTs are expected to output as much power as possible and then the MPPT control strategies are adopted by combing the power smoothing strategy using rotor inertia. The increment of wind power can be stored in the generator as a form of kinetic energy and released when the wind power decreases. The CWTs are in charge of the output power smoothing for the entire wind farm by giving the appropriate power according to the output power need of the wind farm. The battery energy storage system (BESS) with small capacity is installed as the back support of the CWTs to satisfy the high power demanding of the grid. The simulation model of the wind farm based on the proposed control strategies is built in Matlab/Simulink and the efficiency of the proposed strategy is verified by the simulation results.

2. Wind Energy Conversion System Description

2.1. Wind Turbine Model

The mechanical power generated by a wind turbine is [20]:

$$P_w = \rho \pi R^2 C_p(\lambda, \beta) v^3 / 2, \tag{1}$$

where ρ is air density, R is turbine blade radius, v is wind speed and C_p is power coefficient which is a function of the tip speed ratio (TSR) λ and blade pitch angle β. λ is defined as:

$$\lambda = \omega_o R / v, \tag{2}$$

where ω_o indicates the rotational speed in rad/s.

The output torque of the wind turbine is:

$$T = \rho \pi R^3 C_p(\lambda, \beta) v^2 / 2\lambda \tag{3}$$

According to Equation (2), the rotor speed can be adjusted to keep λ at the optimum value λ_{opt} and then maximize the power coefficient as C_{pmax}. The wind turbine aerodynamic efficiency $C_p (\lambda, \beta)$ is given by the following Equation [21]:

$$\begin{cases} C_p(\lambda, \beta) = 0.5176(\frac{116}{\alpha} - 0.4\beta - 5)e^{-21/\alpha} + 0.0068\lambda \\ \frac{1}{\alpha} = \frac{1}{\lambda + 0.08\beta} - \frac{0.035}{\beta^3 + 1} \end{cases} \tag{4}$$

where α is the intermediate variable to obtain the C_p by using λ and β. The curves of the power coefficient C_p and the TSR λ with different blade pitch angles are shown in Figure 1. It can be seen that the maximum power coefficient is decreasing by increasing the pitch angle. Therefore, the output power of the wind turbine can be limited to the rated value through the variable pitch angle control.

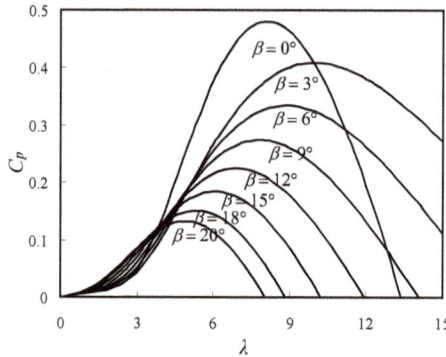

Figure 1. Curve of C_p (λ, β).

According to (1) and (2), the maximum power output of the wind turbine is obtained as:

$$P_{wopt} = \rho \pi R^5 C_{pmax} \omega^3 / 2\lambda_{opt}^3 = k_{opt}\omega^3. \tag{5}$$

Then the optimum torque is expressed as

$$T_{opt} = \rho \pi R^5 C_{pmax} \omega^2 / 2\lambda_{opt}^3 = k_{opt}\omega^2, \tag{6}$$

where k_{opt} is the coefficient associated with the wind turbine characteristics.

2.2. PMSG Model

The voltage equations of the PMSG in decoupled dq-axes rotating reference frame are expressed as [21]:

$$\begin{cases} u_{ds} = R_s i_{ds} + p\psi_{ds} - \omega_e\psi_{qs} \\ u_{qs} = R_s i_{qs} + p\psi_{qs} + \omega_e\psi_{ds} \end{cases} \tag{7}$$

where u_{ds}, u_{qs} are the stator winding voltages in dq-axes. i_{ds}, i_{qs} are the stator winding currents in dq-axes. ψ_{ds}, ψ_{qs} are the stator winding magnet fluxes in dq-axes. ω_e is the rotor speed, R_s is the resistance of the stator windings.

The magnet flux equations of PMSG in dq-axes rotating reference frame are expressed as

$$\begin{cases} \psi_{ds} = L_d i_{ds} + \psi_f \\ \psi_{qs} = L_q i_{qs} \end{cases},\qquad (8)$$

where ψ_f is the permanent magnet flux. L_d, L_q are the stator winding inductances in dq-axes.
The electromagnetic torque of PMSG is given as:

$$T_e = n_p[\psi_f i_{qs} + (L_d - L_q)i_{ds}i_{qs}],\qquad (9)$$

where n_p is the pole pairs of the PMSG.
The motion equation of the PMSG is expressed as:

$$T_L - F\omega - T_e = J\frac{d\omega}{dt},\qquad (10)$$

where T_L is the input load torque, F is friction coefficient, J is the moment of inertia, ω is the mechanical rotor speed.

2.3. Configuration of the Wind Farm

The layout of the wind farm model built in this paper is shown in Figure 2. The wind farm comprised nine PMSG based wind turbines arranged in three rows and three columns. The traditional grid-connection mode of the PMSG based WECS, which was composed of the PMSG and the back-to-back converter, was not adopted in the wind farm model of this paper. Here, the alternating voltage generated by the PMSG was converted into the direct voltage by the generator-side converter and then connected to the DC bus. Apart from the high power transfer capability and cost-effectiveness in a longer transmission system, HVDC is also considered as an environmentally friendly technology owing to its low corona losses and ozone generation rates [22]. Thus, the power is delivered and converged in the cables through the mode of high voltage direct current (HVDC) in this paper. Afterwards, the direct current power was inverted to the alternating current power and transmitted into the grid through the shared grid-connected inverter. The number of grid-connected inverters was decreased, as compared to the traditional grid-connection mode of the PMSG based WECS and then the cost was reduced. The nine wind turbines were divided into three groups according to the topographic location, turbines 1–3 is group one, turbines 4–6 is group two and turbines 7–9 is group three. The BESS was connected to the HVDC DC link to be the support of the CWTs as shown in Figure 2.

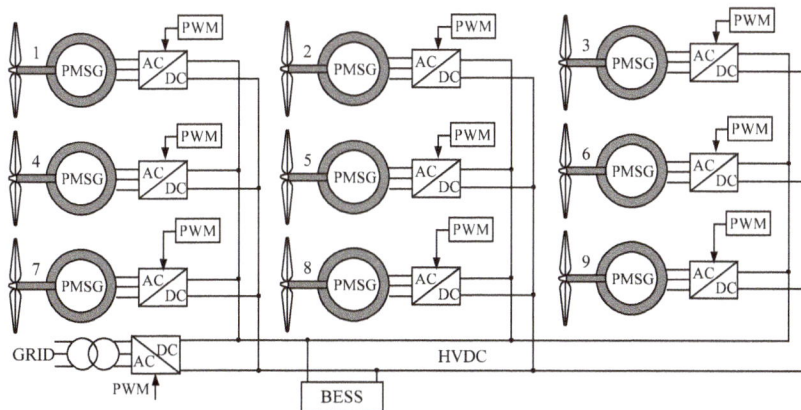

Figure 2. The layout of the wind farm model.

2.4. Wake Effect Model of the Wind Farm

In a wind farm, the wind speed of downstream wind turbines was affected by the upstream turbines due to the air flow and the power loss that happened is called wake effect. The Jensen wake model is used commonly in the wind farm simulation due to its simplicity and validity for the far field [23,24]. In this paper, the Jensen model was adopted to generate the wind speed of the downstream WTs as shown in Figure 3a. The basic Jensen wake effect model is shown in Figure 3 and the wind speed of downstream WTs is expressed as [23]:

$$V_x = V_{in}\left[1 - \left(1 - \sqrt{1 - C_t}\right)\left(\frac{d}{d_x}\right)^2\right],$$

(11)

where V_x is the downstream wind speed, V_{in} is the free stream wind speed, d is the diameter of the WT rotor, x is the distance between the WTs, k is the wake expansion coefficient which represents the effects of atmospheric stability, and $d_x = d + 2kx$. In general, k is set as 0.075 for onshore wind farms and 0.05 for offshore wind farms. In this paper, k was set as 0.075. C_t is the thrust coefficient which is usually given by the wind turbine manufacturer or can be calculated according to the simulation software by giving the required data. In this paper, C_t was set as 0.8 when the wind speed was between 3 and 12 m/s by consulting the Vestas V80 type [24].

The leeside speed can be obtained according to Equation (11) by assuming the x is zero as:

$$V_o = V_{in}\sqrt{1 - C_t}$$

(12)

In practical wind farms, free wind speed can be affected by both obstacles and other WTs around which is called the wind shade effect [23]. Then the Equation (11) can be modified as by considering this effect:

$$V_x = V_{in}\left[1 - \left(1 - \sqrt{1 - C_t}\right)\left(\frac{d}{d_x}\right)^2\left(\frac{A_s}{A}\right)\right],$$

(13)

where A_s is the shade area caused by the other upstream WTs, and A is area of the downstream WT. Most wind farms consist of a large number of WTs, thus the cumulative shade effect of each WT should be considered. Thus, the Equation (13) is modified as

$$V_x = V_{in}\left[1 - \sum_i^n\left(1 - \sqrt{1 - C_t}\right)\left(\frac{d}{d_{xi}}\right)^2\left(\frac{A_{si}}{A}\right)\right]:$$

(14)

where n is the number of influencing upstream WTs, A_{si} is the shade area caused by upstream WT i. The method for calculating the shade area is introduced in [24].

The height of WTs and the wind direction are also important factors in the consideration of wake effect and the details can be seen in [23]. In this paper, the height of the nine WTs was the same and the wind direction was set as from west to east. The layout of the wind turbines was chosen according to the industry standards, as shown in Figure 3b, while the distance between the WTs in the west-east direction was 6d and the distance between the WTs in the north-south direction was 5d [24]. The WTs of the second row was affected by the first row WTs and the wind speed of the third row was affected by both the first and second row WTs. The wind speed deficit of WTs at the second row and third row can be calculated by the equations in [25,26]. The Jensen's wake effect model adopted in this paper was used to generate the wind speed for all WTs in the wind farm by giving the nature stochastic wind speed to the first row WTs.

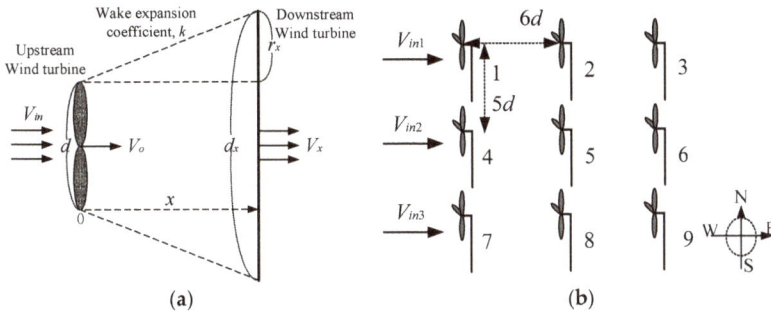

Figure 3. The wake effect model. (**a**) Jensen model principle; (**b**) wind turbine distribution structure.

3. Power Smoothing Control Strategy

3.1. Power Allocation of the PWTs, CWTs and the BESS

The wind turbines of the wind farm were divided into CWTs and PWTs with different control targets in the proposed power smoothing control strategy of this paper. The PWTs adopted the MPPT control strategies by combing the rotor inertia based power smoothing strategy. The CWTs were in charge of the output power smoothing for the wind farm by giving the appropriate power according to the output power need and the right calculations. The BESS with small capacity was configured as the back support of the CWTs to satisfy the high power demanding of the grid. This paper focuses on the wind turbine allocation based power smoothing control strategy itself and then the quantity configuration of the CWTs and PWTs was not discussed due to the limited space. Here, in the wind farm shown in Figure 2, the wind turbines 1–6 were chosen to be the PWTs while the turbines 7–9 were the CWTs.

The power allocation method of the WTs and the BESS for the proposed power smoothing control strategy is shown in Figure 4. All the WTs of the wind farm were controlled to output the maximum power if the grid demand power of the wind farm P_{given} exceeded the maximum capability of the WTs P_{Σ_max}. At this time, the demand power cannot be reached and the BESS will work in the discharging state as shown in the gray box of Figure 4. The discharge power of the BESS was obtained by P_{given} subtracting the output power of the all WTs. When the maximum output power of the wind farm was more than the demand power, the judgment will be continued.

Then when the demand power was more than the sum of the maximum output power of the PWTs and the minimum output power of the CWTs, the BESS will be on the rest as shown in the yellow box of Figure 4 and the PWTs will be controlled to output the maximum power. The reference power of the CWTs can be obtained by P_{given} subtracting the output power of PWTs. Otherwise, the CWTs were controlled to output the minimum power and the process will continue by comparing the state of charge (SOC) of the BESS with the specified maximum value SOC_{max}.

The BESS was on the rest when the real-time SOC exceeded the SOC_{max}, as shown in the yellow box of Figure 4. At this time the control target of the PWTs should be changed as the reference power of the PWTs were given by P_{given} subtracting the minimum output power of CWTs. Otherwise, the BESS will be in the charging state, as shown in the blue box of Figure 4 and the main judgment process will continue. The calculated charging power should be compared with the maximum charging power capability of the BESS P_{ess_max}.

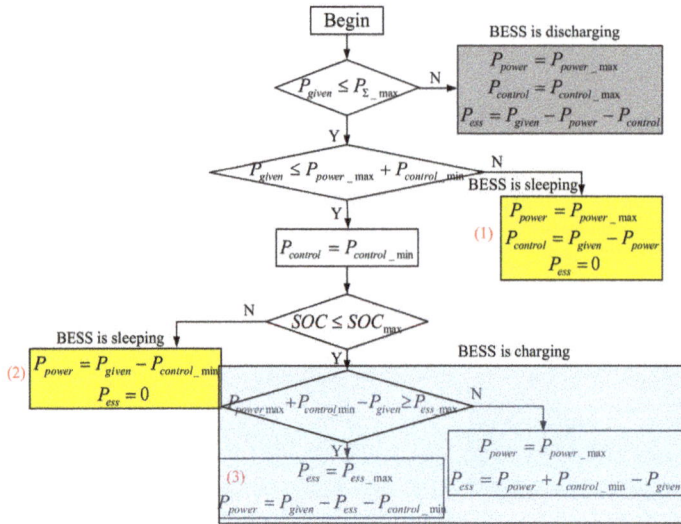

Figure 4. Block diagram of the power allocation.

If the calculated charging power was beyond P_{ess_max}, the reference charging power was set to the value P_{ess_max}. Then the reference power of PWTs will be obtained according to P_{given}, P_{ess} and $P_{control\_min}$. If the calculated charging power was less than P_{ess_max}, the PWTs will be controlled to output the maximum power and the extra power was stored in the BESS. The red digital markers (1)–(3) in Figure 4 are referring to the general power allocation of CWTs and PWTs. The detailed reference power for each CWT and PWT can be obtained according to the wind speed relationship between each WT. Though the Jensen wake effect model is widely used in engineering due to its simplicity, it is not accurate enough in prediction over short-term period, as well as for wind farm application. Thus, the wind speed of each WT needed in the control process was not calculated according to the wake effect model. It can be achieved through the anemoscope simply. Alternatively, short-term wind speed forecasting can be utilized to get the wind speed, which will be studied by authors next.

In Figure 4, P_{power_max} and $P_{control\_max}$ indicate the maximum output power of the power wind turbines and control wind turbines, respectively. The maximum output power was based on the equation $P_w = \rho \pi R^2 C_p(\lambda, \beta) v^3 / 2$ while C_p is the maximum value (0.48 in the manuscript). In the practical application, the efficiency of the wind turbine η should be considered and then $P_{power\_max} = \eta \rho \pi R^2 C_{pmax} v_{power}^3 / 2$, $P_{control\_max} = \eta \rho \pi R^2 C_{pmax} v_{control}^3 / 2$. In the simulation, many losses, such as iron loss and mechanical loss, are not considered. Then the efficiency of the wind turbines were relatively high and the efficiency was set as 0.95 in this paper. The minimum output power of CWTs $P_{control\_min}$ can be changed according to the wind speed condition and wind farm or grid demand.

3.2. Control Strategy for the PWTs

The control purpose of the PWTs was to output the relatively smoother maximum power. Then the power given MPPT control with the rotor inertia based power smoothing control was adopted. The rotor inertia based power smoothing method can reduce the rotor torque ripple and then reduce the mechanical stress of the generator. The idea of the rotor inertia control was to use the kinetic energy in the rotor inertia of the wind turbine to smooth the output power of the system, especially when the inertia was significant, it can be regarded as an energy storage device [27]. The kinetic energy of the wind turbine will increase due to the acceleration of the rotor speed when the wind speed increases fast. Otherwise, the kinetic energy will release due to the decline of the rotor speed when the wind speed decreases.

The average value of the maximum wind power in Equation (5) can be calculated as [5]:

$$P_{avg} = \frac{1}{T} \int_{t-T}^{t} P_{wopt} dt,$$ (15)

where t is the real time and T is the sampling period. The power difference of the optimum value and the average value determines the storing and restoring power in inertia of the wind turbine. Then the integral of the difference power is the redundant kinetic energy. The kinetic energy of the wind turbine can is calculated as [5]:

$$E = \frac{1}{2} J \omega^2.$$ (16)

Thus, the reference kinetic energy can be given as:

$$E^* = \frac{1}{2} J \omega^2 + \int (P_{wopt} - P_{avg}).$$ (17)

Then the reference rotor speed is

$$\omega^* = \sqrt{2E^*/J}.$$ (18)

Finally, the output power smoothing by using the kinetic energy can be realized by giving the reference wind turbine rotor speed shown in Equation (18). The power smoothing effect relies on parameters, such as the wind turbine rotor inertia J and the sampling period T. The greater the rotor inertia is, the better the smoothing effect is. However, the big rotor inertia will bring a lot of difficulties for the system control, such as reducing the control response speed.

As shown in Equation (17), the traditional kinetic energy control compares the optimum power and the average value of the optimum power in a period to obtain the needed smoothing power. The principle of taking an average is simple, but the sampling period and system sampling precision are very important in the process. The low-pass filter is used widely in the wind power smoothing filed combing with the energy storage devices. In this paper, the low-pass filter is used to replace the averaging step to simplify the control process and then the Equation (17) will be changed as:

$$E^* = \frac{1}{2} J \omega^2 + \int (P_{wopt} - P_{filter}),$$ (19)

where P_{filter} is the maximum power after low-pass filtering. The rotor inertial power smoothing based MPPT control by adopting the low-pass filter is shown in Figure 5. The cut-off frequency of the low-pass filter is set according to the power smoothing demand of the grid. The lower the cut-off frequency is, the smoother the output power is. T_s is the sampling time of the kinetic energy.

After calculating the reference rotor speed ω^*, the actual speed ω is compared with ω^* and the difference was imported to a Proportional Integral (PI) controller. For realizing the vector control of the PMSG, the real three phase currents (i_a, i_b, i_c) were transformed into direct axis (d-axis) current i_{ds} and quadrature axis (q-axis) current i_{qs} through the combination of the Clark transformation and Park transformation (abc/dq) using the rotor position θ. The reference q-axis current i^*_{qs} was obtained from the speed PI loop. $i_d = 0$ control was adopted widely due to its simplicity and practicality [28]. Thus, the reference d-axis current i^*_{ds} was set to zero to simplify the control. The reference current was compared with the actual current and the difference is input to the current PI loop, as shown in Figure 6. Then the reference d-axis voltage U^*_{ds} and q-axis voltage U^*_{qs} were achieved from the current PI loops. The reference α-axis voltage U^*_{as} and β-axis voltage $U^*_{\beta s}$ in the two-phase static frame were obtained according to the inverse Park transformation. The space vector pulse width modulation (SVPWM) method [29] was adopted here to generate the PWM signal of the generator-side converter (S_a, S_b, S_c). The complete control diagram of the PWTs is presented in Figure 6.

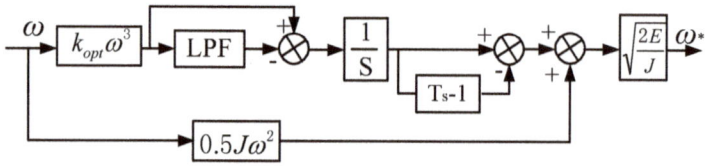

Figure 5. Rotor inertial power smoothing control by adopting the low-pass filter.

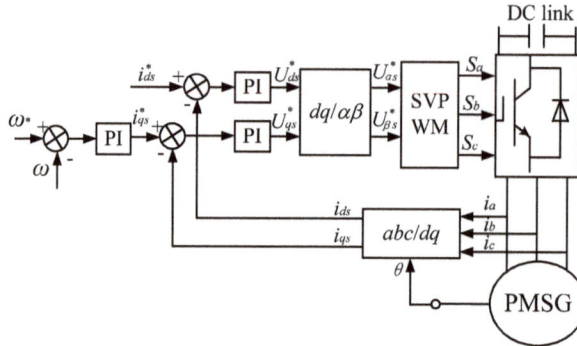

Figure 6. Control diagram of the power wind turbines.

3.3. Control Strategy for the CWTs

The reference power given control method was adopted for the CWTs and its control diagram is shown in Figure 6. When the control target of the CWTs was the MPPT control, the reference power will be set to the maximum power according to Equation (5). When the target is changed to smooth output power, the reference power will be obtained according to the block diagram shown in Figure 4. The given power and the actual power were compared and the difference was adjusted by the PI module to obtain the *q*-axis current. The reference *d*-axis current was set to zero to simplify the control. Similar to the control of PWTs, the SVPWM method was applied in the CWTs control, as shown in Figure 7.

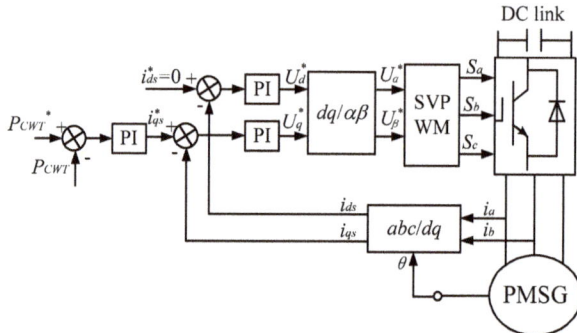

Figure 7. Control diagram of the control wind turbines.

3.4. Control Strategy for the BESS

The BESS adopted in this paper was the backup support of CWTs to satisfy the high power requirements of the grid. As analyzed in Section 3.1, the charge and discharge times of the BESS were less than those of the BESS used in the traditional BESS based power smoothing control strategy.

The BESS was connected to the HVDC DC link with a two-quadrant DC/DC converter, as shown in Figure 8. The DC/DC converter can work in both buck and boost modes according to the switch conditions of the two Insulated Gate Bipolar Transistors (IGBTs) S_1 and S_2. The duty ratios of switches S_1 and S_2 were controlled to regulate the power flow between output power of the nine WTs and grid demand power. The reference power of the BESS was obtained according to Figure 4.

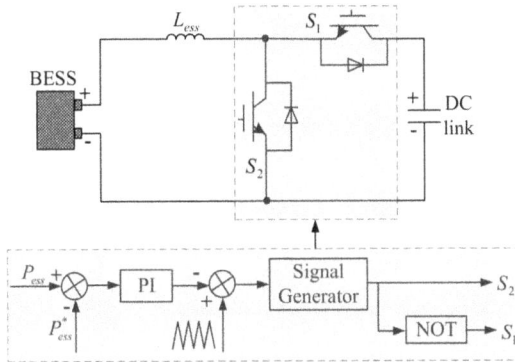

Figure 8. Structure and control of the energy storage system based on DC/DC converter.

When output power of the nine WTs was higher than the grid demand power, the redundant power will be absorbed in the BESS. Then S_1 was controlled to be activated and the BESS was charged, while the DC/DC converter operated in the buck mode. On the contrary, when the output power of the nine WTs was lower than the grid demand power, S_2 will be in the turn-on operation and the BESS was discharged while the DC/DC converter operated in the boost mode.

3.5. Evaluation of the Power Efficiency

The efficiency of the power smoothing control method can be evaluated through the power smoothing function P^*_{level} and maximum energy function P^*_{max}, which are expressed as [5]:

$$P^*_{level} = \int_0^t \left| \frac{dP_o(t)}{dt} \right| dt; \tag{20}$$

$$P^*_{max} = \int_0^t P_o(t)dt, \tag{21}$$

where $P_o(t)$ is the output power at the time t.

If the power smoothing function P^*_{level} is small, the fluctuation of the output power is small. In other words, the performance of the power smoothing is great. The output power efficiency is also an important factor to power smoothing control method. The efficiency cannot be sacrificed too much for realizing the power smoothing. The larger the maximum energy function P^*_{max} is, the higher the output power efficiency is. Thus, the two factors will be calculated in the simulation to evaluate the effectiveness of proposed power smoothing control strategy.

4. Simulation Results

The simulation model of the wind farm was built in the MATLAB/Simulink, which was based on the above-mentioned models and control strategies. The DC link voltage and the output reactive power were controlled through the grid-side converter. The variable pitch angle control was also applied in the simulation. The detailed control strategies of the grid-side converter and the pitch angle control were not presented in this paper and the detailed control strategy can be seen in [30,31]. The phase-locked-loop method was used to detect the grid voltage and phase information [32]. The parameters of the wind turbine and the PMSG are listed in Table 1. The size of the grid-side DC/AC inverter was chosen as 4.16 kV/20 MW which was based on the whole output power capacity of the wind farm (18 MW) by adding some margin. The reference voltage of the HVDC was set to 3500 V and the reference reactive power of the grid-connected inverter was set to zero. The minimum output power of CWTs was set as 200 kW. The size of the BESS was chosen according to the grid demand and the output power of wind turbines, which can be estimated from the wind speed. The small capacity BESS with the parameter 4 MW/3 MWh was configured in the simulation.

Table 1. Parameters of the simulation.

Wind Turbine		PMSG	
Blade radius (m)	35	Stator resistant (Ω)	0.01
Air density (kg/m^3)	1.225	Inductance (mH)	0.835
Optimum TSR	8.1	Rotor inertia (kg·m^2)	8500
Optimum power coefficient	0.48	PM linkage (Wb)	8.76
Rated wind speed (m/s)	12	Pole pairs	32
Rated power (MW)	2	Rated power (MW)	2

Three working modes of the wind farm were studied in the simulation, which were the MPPT mode, the power smoothing mode, and the specified power output mode, respectively. The first working mode of the wind farm was to output the maximum power then the optimum power given MPPT control was adopted for all the nine WTs, while using the control diagram shown in Figure 7. The BESS was adopted to smooth the output power of the nine WTs based on the traditional method. The reference power of the BESS was given as the difference of the output power of WTs and the output power through a low-pass filter. The simulation results of the WTs under the MPPT control are shown in Figures 9 and 10.

The stochastic nature of the wind speed generated by the software TurbSim was given in the simulation. The wind speed waveforms of the nine turbines are shown in Figure 9a–c, while those of 1–3 WTs are given in Figure 9a, 4–6 WTs are given in Figure 9b and 7–9 WTs are given in Figure 9c. The wind speed of each wind turbine was given different values by considering the wake effect to close to the real wind farm in the simulation. The wind speeds of three group wind turbines were totally different, while the wind speed of each wind turbine in one group was decreasing from the upstream to downstream with the similar shape. Figure 9d–f show the rotor speed of the nine WTs respectively. It can be seen that the rotor speeds of nine WTs varied with the wind speeds to realize the MPPT control.

The torque of nine WTs is shown in Figure 10a,c,e and the output power is shown in Figure 10b,d,f, respectively. The MPPT control was realized while the torque and output power of the WTs varied with the wind speed. The total output power is shown in Figure 10g and it is clearly observed that the output power was consistent with the ideal MPPT power by ignoring the system losses in the simulation. The output power of the wind farm was smoothed by the BESS, as shown in Figure 10g. The power of the BESS was presented in Figure 10i, where the BESS was charging when the value was greater than 0 and discharging when the value was less than 0. It can be seen that the BESS was charging and discharging all the time under the traditional power smoothing control, which would cause the life of BESS to decrease seriously. Figure 10h demonstrates the DC link voltage of the wind farm, from which it can be seen that the voltage remained at the reference value 3500 V only, with little fluctuation which was about 5 V in the steady state, as shown in the extended waveform. Then the control strategy of the grid-side converter was verified.

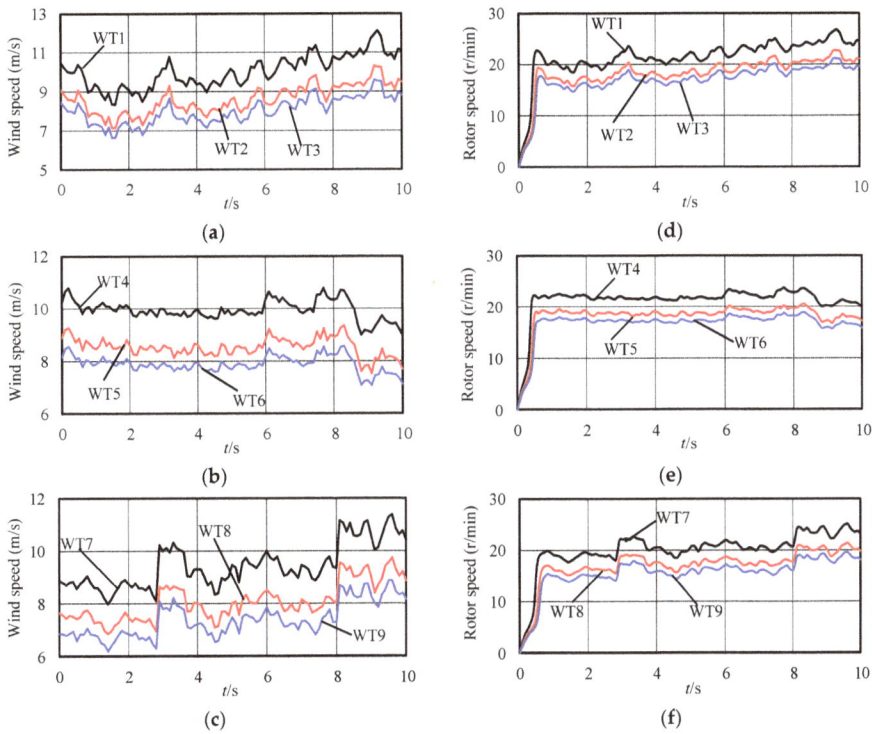

Figure 9. Time histories of the wind speed and rotor speed in the wind farm under the maximum power point tracking (MPPT) control. (**a**) Wind speed of the first group WTs (1–3). (**b**) Wind speed of the second group WTs (4–6). (**c**) Wind speed of the third group WTs (7–9). (**d**) Rotor speed of the first group WTs (1–3). (**e**) Rotor speed of the second group WTs (4–6). (**f**) Rotor speed of the third group WTs (7–9).

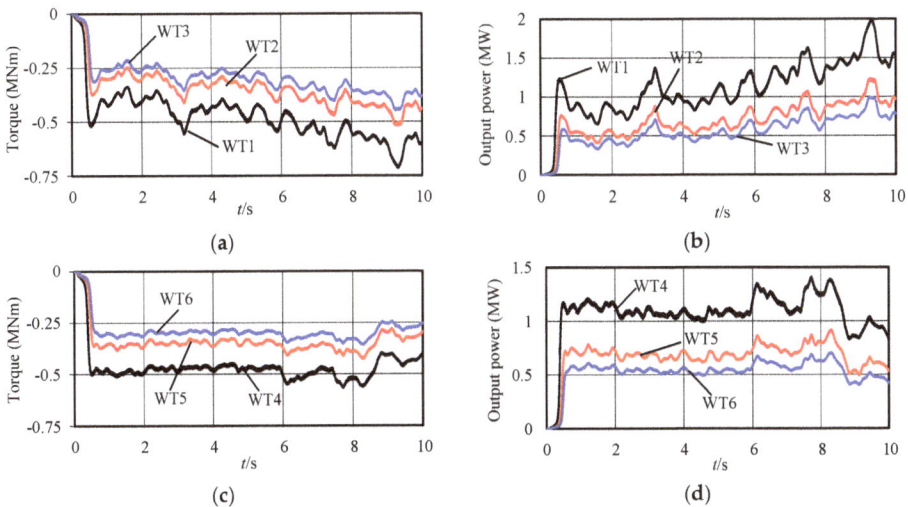

Figure 10. *Cont.*

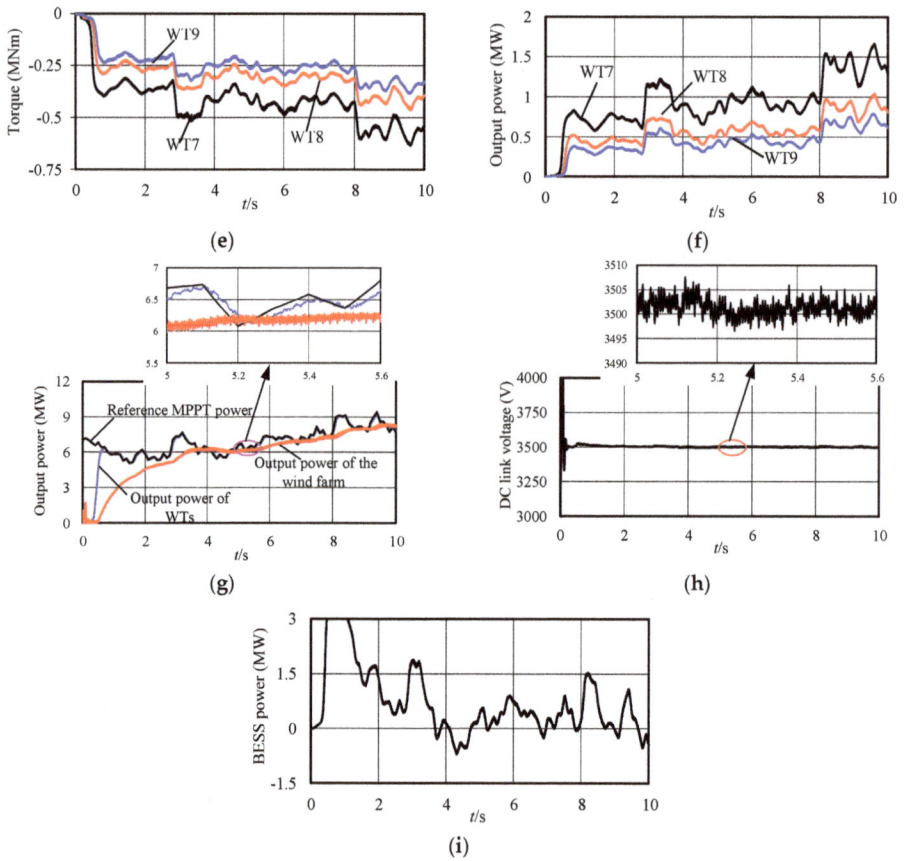

Figure 10. Time histories of the torque, power of WTs in the wind farm under the maximum power point tracking (MPPT) operation mode. (**a**) Torque of the first group WTs (1–3). (**b**) Torque of the second group WTs (4–6). (**c**) Torque of the third group WTs (7–9). (**d**) Output power of the first group WTs (1–3). (**e**) Output power of the second group WTs (4–6). (**f**) Output power of the third group WTs (7–9). (**g**) Total output power. (**h**) DC link voltage. (**i**) Power of the BESS.

The second working mode of the wind farm was to output the smoothing power and the simulation results are shown in Figures 11 and 12. At this time, the proposed power smoothing control strategy based on the allocation and rotor inertia of wind turbines was adopted. The MPPT method combing the rotor inertia control was applied to the PWTs 1–6, while the given power control was adopted for the CWTs 7–9. The sampling period of the kinetic energy based power smoothing strategy for PWTs was 0.001. The cut off frequency of the low-pass filter was 10 Hz. The simulation waveforms of the PWTs are shown in Figure 11 including the rotor speed, torque, and output power. The rotor speed of PWTs shown in Figure 11a,d did not follow the optimum MPPT speed, which are shown in Figure 9d,e. It can be seen from Figure 11b,e that the torque of the PWTs was almost straight with very tiny fluctuations. Then, the mechanical stress of WTs was reduced effectively. By comparing the output power Figure 11c,f with Figure 10d,e, it can be observed that the output power of PWTs under the proposed rotor inertia based method was smoother than the power under MPPT control. Thus, the effectiveness of power smoothing method for PWTs based on the rotor inertia control was verified.

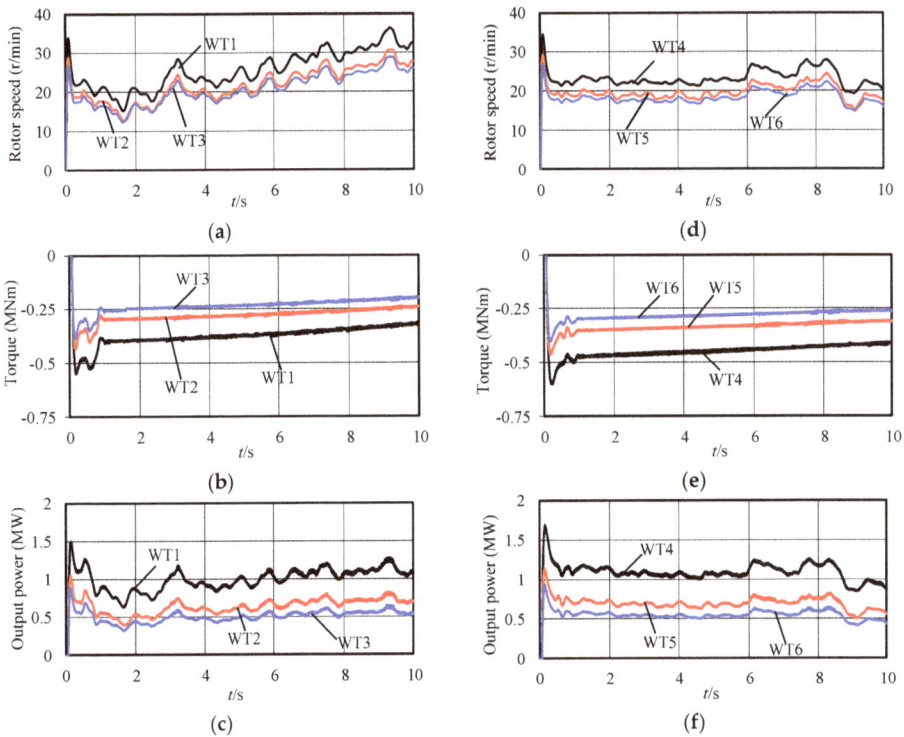

Figure 11. Time histories of the PWTs in the wind farm under the power smoothing operation mode. (a) Rotor speed of PWTs (1–3). (b) Torque of PWTs (1–3). (c) Output power of PWTs (1–3). (d) Rotor speed of PWTs (4–6). (e) Torque of PWTs (4–6). (f) Output power of PWTs (4–6).

The output power of the three CWTs is shown in Figure 12a–c respectively. The reference power calculated according to the block diagram in Figure 4 was given to CWTs after 1s, while the maximum power was given to the PWTs before 1s. The power of each CWT was allocated according to the wind speed relationship which was obtained by calculating the wake effect. As shown in Figure 12a–c, the output power of CWTs was different from that under the MPPT control mode which is shown in Figure 10f and the actual output power of the three CWT can match the reference given power well. It can be seen from Figure 12e that the output power of WTs was smoothed through the proposed control strategy by comparing with the MPPT power. However, the output power of WTs could not be smoothed when the needed smoothing power exceeded the maximum output power of CWTs. In this case, the BESS was joined in to smooth the rest power and the power of BESS is shown in Figure 12d. Thus, the output power of the wind farm was smoother than the output power of WTs. Figure 12d shows that the BESS was not working all the time and then the charge and discharge times was reduced significantly compared to that using the traditional power smoothing method (shown in Figure 10i). Thus, the battery lift would be extended greatly. The DC link voltage can remain constant as shown in Figure 12f with little fluctuation around 5 V in the steady state.

The maximum energy function of the MPPT power, the output power of WTs and output power of the wind farm is shown in Figure 12g. It reflects that the maximum energy function of the output power of wind farm was slightly less than the MPPT output power. The energy loss mainly happened in the PWTs by adopting the kinetic energy based power smoothing control. The power smoothing function is shown in Figure 12h, from which it can be seen that the value of output power by WTs was obviously smaller than that of MPPT control. Then, it can be summarized that the proposed method can realize the good performance of power smoothing. The power smoothing effect was better when the BESS is joined in as shown in Figure 12h.

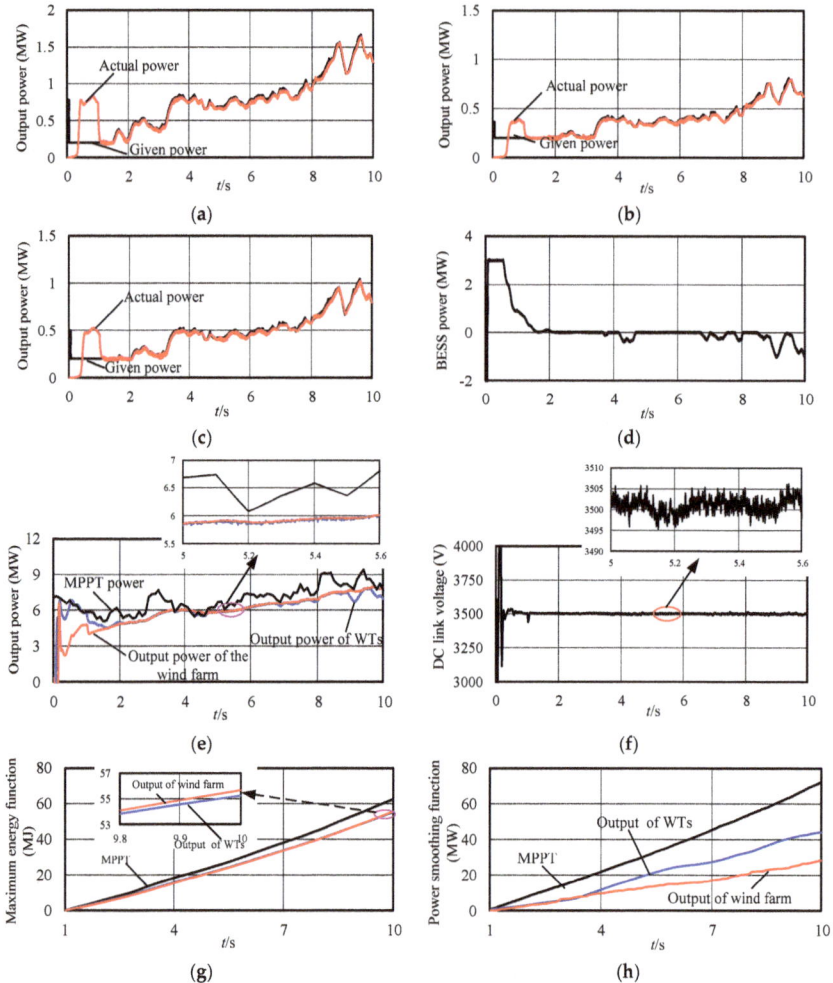

Figure 12. Time histories of the CWTs and the wind farm under the power smoothing operation mode. (**a**) Output power of CWT (WT7). (**b**) Output power of CWT (WT8). (**c**) Output power of CWT (WT9). (**d**) Power of the BESS. (**e**) Total output power of the wind farm. (**f**) DC link voltage. (**g**) Maximum energy function. (**h**) Power smoothing function.

In some situations, the grid needs the wind farm to have the ability to output the specified power to meet the balance of the grid and loads. Thus, the third working mode of the wind farm was to output the determined power. The step-changed power demand (6 MW, 4 MW, 8 MW, 6 MW, 7 MW) for the wind farm was given in the simulation. The reference demand power was given to test the effectiveness of the proposed method in both cases of power increasing and power decreasing. Besides, the maximum reference demand power is limited by the sum of the maximum output power of wind turbines and the BESS. The simulation results of this case are shown in Figure 13. Being the same as that in the second working mode, the CWTs were given the reference power calculated according to block diagram Figure 4 after 1 s. The simulation results of PWTs were the same as that shown in Figure 11 and not given here. The output power of CWTs is demonstrated in Figure 13a–c, from which it can be seen that the CWTs outputted constant minimum power 200 kW between 2 s to 4 s due to the low grid demand power. The power of the BESS is shown in Figure 13, and it can be found that the BESS was working in the charge, discharge, and rest working modes according to the different power grid demands. The output power of the wind farm can follow the given step-changed demand power as shown in Figure 13e. As displayed in Figure 13f, the DC link voltage basically remained at the reference value, but with acceptable mutation due to the sudden power changes.

The new wind speed condition of CWTs was given in the simulation to increase persuasion of the proposed power smoothing strategy, while the wind speed of the PWTs was the same as that in Figure 9. The wind speed of CWTs is shown in Figure 14a, from which it can be seen that the shape of downstream WT was different from the upstream WT. Then, the rotor torque of each CWT had different shapes, as demonstrated in Figure 14b. The actual output power of the three CWT can match the reference given power well as shown in Figure 14c–e. Figure 14f shows the output power of the wind farm and it can be observed that the output power was smoothed by comparison with the reference MPPT power.

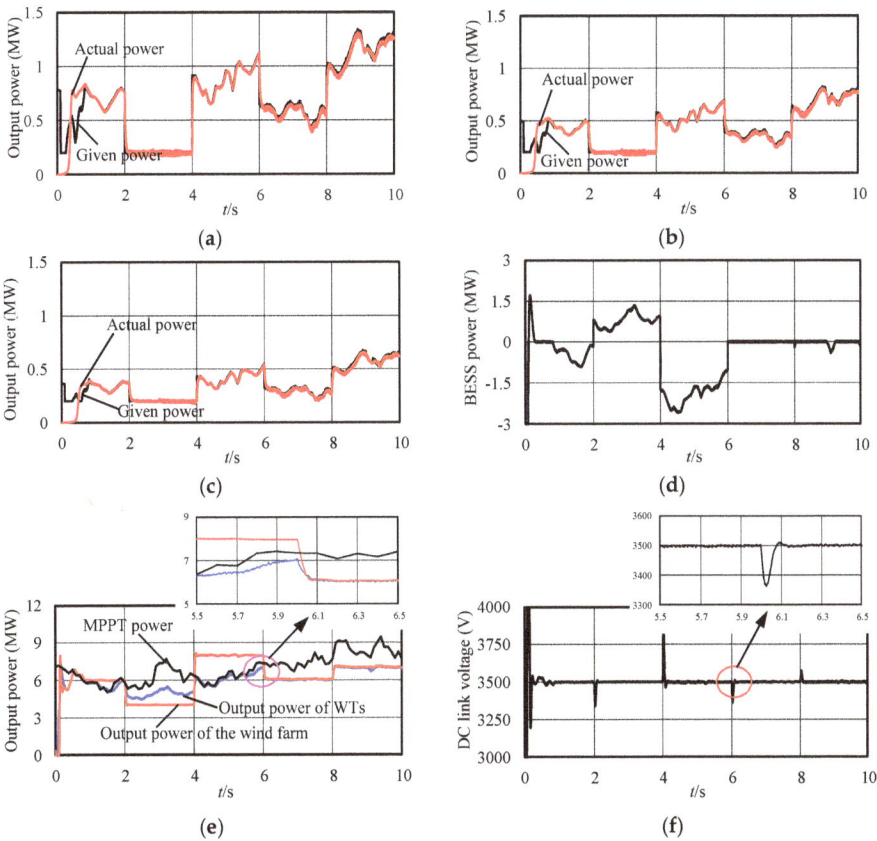

Figure 13. Time histories of the wind farm under the specified power given operation mode. (**a**) Output power of CWT (WT7). (**b**) Output power of CWT (WT8). (**c**) Output power of CWT (WT9). (**d**) Total output power of the wind farm. (**e**) Power of the BESS; (**f**) DC link voltage.

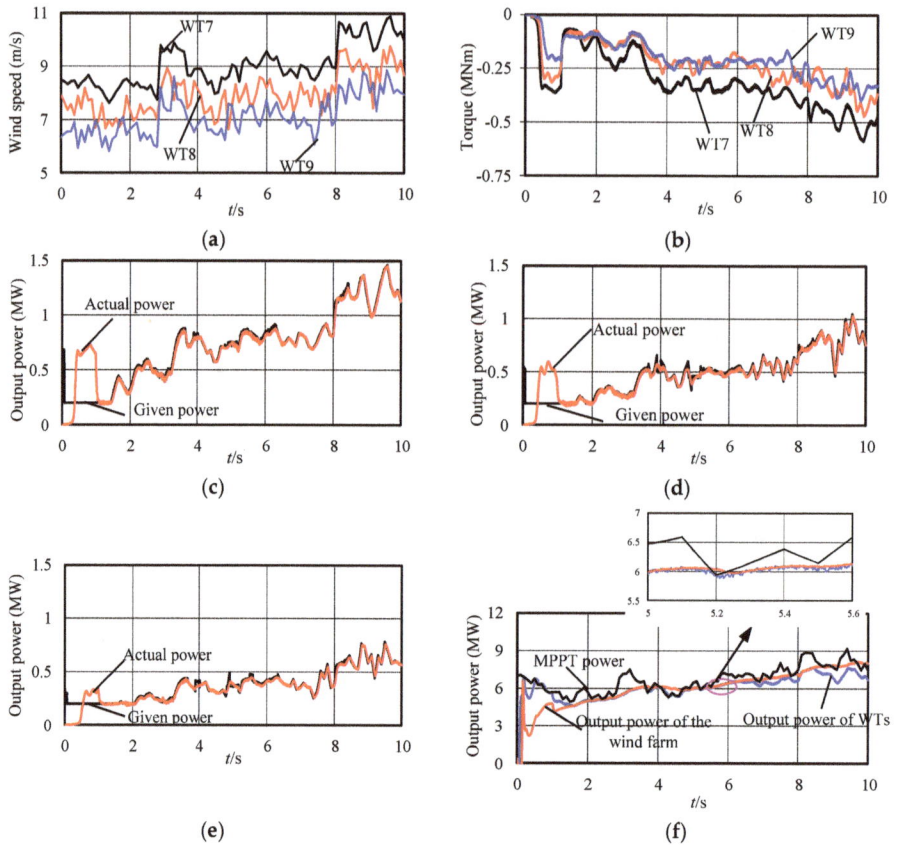

Figure 14. Time histories of the wind farm when the wind speed condition of CWTs is changed. (a) wind speed of CWTs. (b) Rotor torque of CWTs. (c) Output power of CWT (WT7). (d) Output power of CWT (WT8). (e) Output power of CWT (WT9). (f) Total output power of the wind farm.

Thus, the proposed power smoothing control strategy through the reasonable allocation of the WTs in the wind farm and the rotor inertia was verified by the simulation results. The rotor inertia based power smoothing method for PWTs can reduce the torque ripple and mechanical stress significantly. The CWTs can output the appropriate power according to the reference value. The configured BESS were not working all the time to extend the life. The control for the grid-side converter was also verified through the constant DC link voltage. The proposed control strategy can respond to different needs of the grid, which was verified by the simulation results under different working modes.

5. Conclusions

This paper proposed a novel power smoothing control strategy for the wind farm by using the allocation of wind turbines. The WTs of the wind farm are divided into two classes: CWTs and PWTs with different control strategies. The PWTs were controlled to output the maximum power by combining with the rotor inertia based power smoothing method. Power given control was applied to the CWTs. The CWTs and small capacity BESS worked together to take charge of the total power output of the wind farm. The BESS was controlled to participate only when the power limit of CWTs was reached to reduce the charge and discharge times.

The wind farm model was built based on Matlab/Simulink to verify the proposed power smoothing control strategy by considering the wake effect. The grid-connected converter control was also adopted to fulfill the complete functionality of the wind farm. Three working modes of the wind farm as the MPPT control, power smoothing control, power given control are given in the simulation. It has been shown that the output power can meet the different needs of the grid. The effectiveness of the proposed power smoothing strategy was verified by the simulation results.

Author Contributions: Y.Z. was in charge of the work listed as: conceiving and proposing the control strategies, designing the simulations, making charts and writing the paper; H.Z. analyzed the simulation data and drew some of the charts. L.C. and S.G. analyzed and confirmed the simulation parameters, diagrams and results.

Acknowledgments: This work was supported in part by the National Natural Science foundation of China (51507050), the Natural Science foundation of Jiangsu Province (BK20150822), the Fundamental Research Funds for the Central Universities (Project No. 2018B15314) and Science and Technology Project of SGCC (J2017071).

Conflicts of Interest: The authors declare no conflicts of interest. The founding sponsors had no role in the design of the study; in the collection, analyses, or interpretation of data; in the writing of the manuscript, and in the decision to publish the results.

References

1. Cheng, M.; Zhu, Y. The state of the art of wind energy conversion systems and technologies: A review. *Energy Convers. Manag.* **2014**, *88*, 332–347. [CrossRef]
2. Muyeen, S.M.; Takahashi, R.; Murata, T.; Tamura, J. A variable speed wind turbine control strategy to meet wind farm grid code requirements. *IEEE Trans. Power Syst.* **2010**, *25*, 331–340. [CrossRef]
3. Uehara, A.; Pratap, A.; Goya, T.; Senjyu, T.; Yona, A.; Urasaki, N.; Funabashi, T. A coordinated control method to smooth wind power fluctuations of a PMSG-based WECS. *IEEE Trans. Energy Convers.* **2011**, *26*, 550–558. [CrossRef]
4. Aigner, T.; Jaehnert, S.; Doormain, G.L.; Gjengedal, T. The effect of large-scale wind power on system balancing in Northern Europe. *IEEE Trans. Sustain. Energy* **2012**, *3*, 751–759. [CrossRef]
5. Howlader, A.M.; Senjyu, T.; Saber, A.Y. An integrated power smoothing control for a grid-interactive wind farm considering wake effects. *IEEE Syst. J.* **2015**, *9*, 954–965. [CrossRef]
6. Howlader, A.M.; Urasaki, N.; Yona, A.; Senjyu, T.; Saber, A.Y. A review of output power smoothing methods for wind energy conversion systems. *Renew. Sustain. Energy Rev.* **2013**, *26*, 135–146. [CrossRef]
7. Simon, D.R.; Johan, D.; Johan, M. Power smoothing in large wind farms using optimal control of rotating kinetic energy reserves. *Wind Energy* **2015**, *18*, 1777–1791.
8. Johan, M.; Simon, D.R.; Johan, D. Smoothing turbulence-induced power fluctuations in large wind farms by optimal control of the rotating kinetic energy of the turbines. *J. Phys. Conf. Ser.* **2014**, *524*, 012187.
9. Rawn, B.G.; Lehn, P.W.; Maggiore, M. Disturbance margin for quantifying limits on power smoothing by wind turbines. *IEEE Trans. Control Syst. Technol.* **2013**, *21*, 1795–1807. [CrossRef]
10. Varzaneh, S.G.; Gharehpetian, G.B.; Abedi, M. Output power smoothing of variable speed wind farms using rotor-inertia. *Electr. Power Syst. Res.* **2014**, *116*, 208–217. [CrossRef]
11. Rijcke, S.D.; Tielens, P.; Rawn, B.; Hertem, D.V.; Driesen, J. Trading energy yield for frequency regulation: Optimal control of kinetic energy in wind farms. *IEEE Trans. Power Syst.* **2015**, *30*, 2469–2478. [CrossRef]
12. Zou, J.; Peng, C.; Shi, J.; Xin, X.; Zhang, Z. State-of-charge optimizing control approach of battery energy storage system for wind farm. *IET Renew. Power Gener.* **2015**, *9*, 647–652. [CrossRef]
13. Zhang, F.; Xu, Z.; Meng, K. Optimal sizing of substation-scale energy storage station considering seasonal variations in wind energy. *IET Gener. Transm. Distrib.* **2016**, *10*, 3241–3250. [CrossRef]
14. Jayasinghe, S.D.G.; Vilathgamuwa, D.M. Flying supercapacitors as power smoothing elements in wind generation. *IEEE Trans. Ind. Electron.* **2013**, *60*, 2909–2918. [CrossRef]
15. Muyeen, S.M.; Hasanien, H.M.; Al-Durra, A. Transient stability enhancement of wind farms connected to a multi-machine power system by using an adaptive ANN-controlled SMES. *Energy Convers. Manag.* **2014**, *78*, 412–420. [CrossRef]
16. Hasanien, H.M. A set-membership affine projection algorithm-based adaptive-controlled SMES units for wind farms output power smoothing. *IEEE Trans. Sustain. Energy* **2014**, *5*, 1226–1233. [CrossRef]

17. Diaz-Gonzalez, F.; Bianchi, F.D.; Sunper, A.; Gomis-Bellmunt, O. Control of a flywheel energy storage system for power smoothing in wind power plants. *IEEE Trans. Energy Convers.* **2014**, *29*, 204–214. [CrossRef]
18. Islam, F.; Al-Durra, A.; Muyeen, S.M. Smoothing of wind farm output by prediction and supervisory-control-unit-based FESS. *IEEE Trans. Sustain. Energy* **2013**, *4*, 925–933. [CrossRef]
19. Jiang, Q.; Hong, H. Wavelet-based capacity configuration and coordinated control of hybrid energy storage system for smoothing out wind power fluctuations. *IEEE Trans. Power Syst.* **2013**, *28*, 1363–1372. [CrossRef]
20. Pucci, M.; Cirrincione, M. Neural MPPT control of wind generators with induction machines without speed sensors. *IEEE Trans. Ind. Electron.* **2011**, *58*, 37–47. [CrossRef]
21. Rosyadi, M.; Muyeen, S.M.; Takahashi, R.; Tamura, J. A Design Fuzzy Logic Controller for a Permanent Magnet Wind Generator to Enhance the Dynamic Stability of Wind Farms. *Appl. Sci.* **2012**, *2*, 780–800. [CrossRef]
22. Muyeen, S.M.; Hasanien, H.M. Operation and control of HVDC stations using continuous mixed p-norm-based adaptive fuzzy technique. *IET Gener. Transm. Distrib.* **2017**, *11*, 2275–2282. [CrossRef]
23. Kim, H.; Singh, C.; Sprintson, A. Simulation and estimation of reliability in a wind farm considering the wake effect. *IEEE Trans. Sustain. Energy.* **2012**, *3*, 274–282. [CrossRef]
24. Gil, M.P.; Gomis-Bellmunt, O.; Sumper, A.; Bergas-Jane, J. Power generation efficiency analysis of offshore wind farms connected to a SLPC (single large power converter) operated with variable frequencies considering wake effects. *Energy* **2012**, *37*, 455–468.
25. Tian, J.; Zhou, D.; Su, C.; Soltani, M.; Chen, Z.; Blaabjerg, F. Wind turbine power curve design for optimal power generation in wind farms considering wake effect. *Energies* **2017**, *10*, 395. [CrossRef]
26. Tian, J.; Zhou, D.; Su, C.; Blaabjerg, F.; Chen, Z. Optimal Control to Increase Energy Production of Wind Farm Considering Wake Effect and Lifetime Estimation. *Appl. Sci.* **2017**, *7*, 65. [CrossRef]
27. Lee, J.; Muljadi, E.; Sorensen, P.; Kang, Y.C. Releasable kinetic energy-based inertial control of a DFIG wind power plant. *IEEE Trans. Sustain. Energy* **2016**, *7*, 279–288. [CrossRef]
28. Hang, J.; Zhang, J.; Cheng, M.; Ding, S. Detection and discrimination of open phase fault in permanent magnet synchronous motor drive system. *IEEE Trans. Power Electron.* **2016**, *31*, 4697–4709. [CrossRef]
29. Gupta, A.K.; Khambadkone, A.M. A general space vector PWM algorithm for multilevel inverters, including operation in overmodulation range. *IEEE Trans. Power Electron.* **2007**, *22*, 517–526. [CrossRef]
30. Geng, H.; Yang, G. Output power control for variable-speed variable-pitch wind generation systems. *IEEE Trans. Energy Convers.* **2010**, *25*, 494–503. [CrossRef]
31. Zhu, Y.; Zang, H.; Fu, Q. Grid-connected Control Strategies for a Dual Power Flow Wind Energy Conversion System. *Electr. Power Compon. Syst.* **2017**, *45*, 1–14. [CrossRef]
32. Wang, Q.; Cheng, M.; Jiang, Y.; Zuo, W.; Buja, G. A simple active and reactive power control for applications of single-phase electric springs. *IEEE Trans. Ind. Electron.* **2018**, *65*, 6291–6300. [CrossRef]

applied
sciences

MDPI

Article

Optimisation of the Structure of a Wind Farm—Kinetic Energy Storage for Improving the Reliability of Electricity Supplies

Andrzej Tomczewski and Leszek Kasprzyk *

Faculty of Electrical Engineering, Poznań University of Technology, Piotrowo 3A str., 60-965 Poznań, Poland; andrzej.tomczewski@put.poznan.pl
* Correspondence: leszek.kasprzyk@put.poznan.pl; Tel.: +48-061-665-2389

Received: 15 July 2018; Accepted: 20 August 2018; Published: 23 August 2018

Abstract: An important issue in the correct operation of the power system is the reliability of the electricity supply from generation systems. This particular problem especially concerns renewable sources, the output power of which is variable over time and additionally has a stochastic character. The solution used in the work to improve the reliability indicators of wind farm sources is the partial stabilization of their output power achieved through cooperation with the kinetic energy storage. Excessive increase in storage capacity is associated with a large increase in investment and operating costs. It is therefore important to determine the minimum storage capacity required to maintain the accepted criteria for the reliability of energy supply. In this paper, a population meta-heuristics algorithm was used for this purpose. The obtained results confirm the possibility of limiting the energy capacity of the flywheels, they also indicate its non-linear character as a function of selected parameters of the reliability of energy supplies from wind farms.

Keywords: optimization; kinetic energy storage; wind farm; reliability of electricity supplies

---

1. Introduction

The prospect of exhausting fuel resources, air pollution, growing environmental awareness and legal regulations contribute to the increasing popularity of renewable energy sources for the production of electricity (RES-E). This trend can be observed for both low-power sources, which typically supply individual facilities (including, in particular, photovoltaic systems) and high-power facilities, such as water power plantsas well as wind and photovoltaic farms [1,2]. In some countries, the share of renewable energy sources (RES) in the power grid is greater than the share of conventional sources (Norway is the leader in this field; 98% of electricity in Norway is generated by renewable power plants—mainly water power plants) [3–5]. Few EU countries can boast such a high share of renewable sources in power generation. According to the "Renewable energy in Europe—2017 Update" report, on average, about 30% of electricity in member-states was generated from renewable sources [3].

One should pay attention to the fact that a dynamic increase in the number of renewable energy sources has some disadvantages related to a lack of power generation stability, which depends on the current weather conditions and low operating inertia. The problem applies especially to solar and wind sources. Heterogeneity can stabilise the operation of a power grid. It involves a high share of prosumer solutions and integration by means of intelligent systems (e.g., power generation predictions). Another solution is to use energy storage units, which can buffer the generated power (e.g., power generated by photovoltaic systems during the day will be released at night). Unfortunately, in the area of electrical power engineering, the existing energy storage systems are insufficient on a large scale, except for pumped-storage power stations (and thermal storages in southern countries). Moreover, it is expensive to build new energy storage systems (ESS) intended to work with RES-E. The cost can reach thousands of Euro/kWh in the case of electrical-chemical and kinetic storage units [6–8]. This is why scientific centres worldwide carry out research on ESS that can function in conjunction with RES-E, especially photovoltaic cells and wind turbines. References [9–15] focused on the issue of predicting the generating capacity of RES-E to adapt in advance with the power generated by other systems. There are the examples of papers [16–24] concern with the selection of the structure of energy sources and storages according to location needs and conditions. The assessment of the ESS effectiveness and profitability of their use is of paramount importance. The issue was mentioned in papers [2,12,22,23].

However, the literature, does not provide exhaustive information on the reliability of electrical energy generation in the context of the contribution of unstable renewable sources to the power grid. This is why the authors decided to concentrate on the topic of cooperation between wind farms and kinetic energy storage. Minimising the volume of ESS fulfils the set criterion including the reduction of costs related to the RES-E and ESS system installation. Analyses were carried out using historical wind speed data and, hence, the identified indicators are not probabilities or expected values but specific values identified for the specific analysis period.

2. Characteristics of a Wind Farm—Kinetic Energy Storage System

2.1. Description of the System, Power Flow Algorithm

A wind farm is an unstable source with random values of the instantaneous power $P_1(t)$ in the range $[0, P_{WFn}]$, where P_{WFn} stands for the total of the rated power values of all turbines. The annual and multi-annual deterministic components add to the stochastic nature of the wind speed changes, which means that a period of one year should be taken as the minimum period of analysis for such systems.

In relation to the above, even high rated power wind farms, where turbines are located at a certain distance from one another, do not guarantee stable power released to the power grid in either short (minutes) or long (hours, days) time intervals [16]. The stability and predictability of power generation by the sources within the power grid are essential in order to guarantee an adequate level of power supply reliability and determine the energy safety of a country. In the case of wind sources, the best results of the output power stability (power released to the power grid) are achieved by using energy storage devices with a capacity suited to the power plant efficiency, its geographical location and assumed period of continuous operation. The capacity of the storage not only depends on the farm's efficiency and the mean annual wind speed v_{wAvg}, but also on the dynamics of wind speed changes specific for the local conditions, and on the statistical parameters of breaks in power generation: their mean duration, the number of subsequent intervals between the periods when the energy storage can be charged, etc. The author has presented detailed results of such studies in Reference [24].

Figure 1 presents a schematic diagram of a wind farm cooperating with a global (common for all turbines) storage of kinetic type. The power generated in each turbine $P_{1(i)}(t)$ (for $i = 1, 2, \ldots, N_{WT}$, where N_{WT} is the number of turbines in the farm) is supplied to a node and then—depending on the instantaneous power of farm P_1 and the state of charge (SOC) of the ESS—to storage P_2 and/or power grid P_3. The storage can also operate with power P_2 when discharged, and then the stored energy is supplied to the power grid. The (+) sign of power P_2 means that the storage is discharged, while (−) stands for its charged conditions. Power P_3 is supplied to the power grid. The power is the sum of the power of the turbines and the storage, reduced by the ΔP power loss in the transformer and the line supplying power to the grid:

$$P_3(t) = P_1(t) + P_2(t, \text{SOC}) - \Delta P \tag{1}$$

The CFEM (*control flow of energy module*) is responsible for controlling the flow of energy between the wind farm, storage, and power grid.

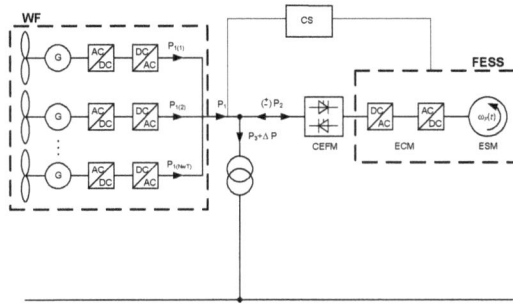

Figure 1. Schematic diagram of a wind farm cooperating with a kinetic magazine of a global type (CS: control system; ECM: Energy Conversion Module; ESM: Energy Storage Module; FESS: flywheel energy storage system; WF: wind farm).

It is possible to improve the reliability of electrical energy supplies to the power grid from a wind farm-kinetic energy storage system when an appropriate control algorithm is implemented that stores the flow of energy between the turbines, storage and power grid.

This assumes that the minimum set power P_{3min} on the system outlet is maintained at all times with the maximum duration T_{max}, even when the wind speed drops below the turbine cut-in speed v_{cut-in} (then $P_1 = 0$). At an appropriately selected energy storage capacity A_{FESSn} [24], it is possible to recover power in the periods when the power generated by the farm is lower than the assumed minimum value P_{3min}. In this way, the system guarantees the maintenance of power supplied to the power grid regardless of weather conditions in a period not longer than T_{max} and with power not lower than P_{3min}. The implementation of energy storage results in an increase in the predictability of the source generation related to the partial stabilisation of its output power P_3.

From the point of view of the acquired functionalities of the system presented in Figure 1, the control system is among its most important elements. The control system responds to the instantaneous values of wind speed $v_w(t)$ and the storage charging level $SOC(t)$. Considering the turbine cut-in speed (v_{cut-in}), reaching the rated power (v_n) and the output power of the farm P_{3min} (v_{3min}), turbine cut-out time ($v_{cut-out}$) and the generated power values corresponding to them, one can identify four operating conditions of a wind farm-kinetic energy storage system: independent operation of the farm, storage charging, independent operation of the storage and simultaneous operation of the storage and farm. The conditions and system diagrams corresponding to them are presented in Figure 2. The arrows mark the energy flow directions that depend on the current wind speed, the acquired power level P_{3min}, the capacity of A_{FESSn} storage and its charging level, as well as the mutual relations between power P_1, P_{2max}, P_3, and P_{3min}, where P_{2max} is the maximum power of storage.

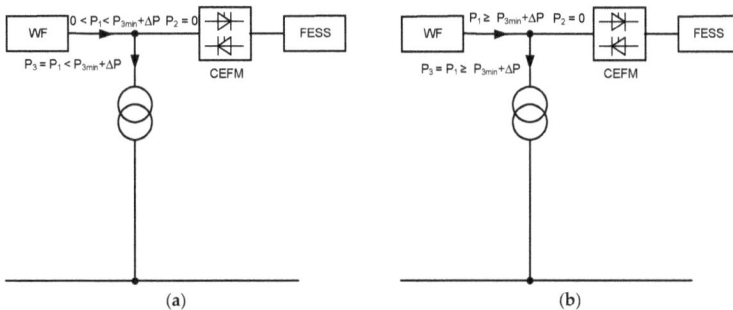

(a)   (b)

Figure 2. *Cont.*

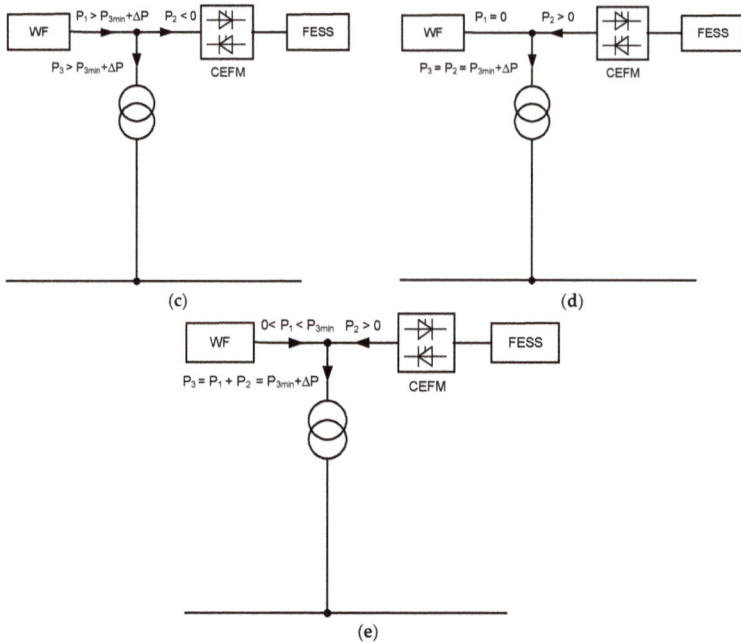

Figure 2. Power flow in the wind turbine (WT)-FESS system for different system operating conditions: (a,b) independent operation of a wind farm; (c) storage charging; (d) independent operation of the storage (discharging); (e) simultaneous operation of a wind farm and a storage system ($P_{3'}$—output power of the system considering power losses ΔP in the transformer and the line).

2.2. Wind Speed Measurement Data

A simulation of the operation of the system described in Section 2.1, establishing the course of instantaneous power, generated power, and the impact of a kinetic energy storage system on the reliability of electrical energy supplies to the power grid, requires the use of data which characterise the wind conditions. Measurements of the wind speed were used for this paper. Each sample represented the mean wind speed in the period $\Delta t_M \approx 47$ s, while the number of samples used for one year was to 670,000. The duration of the kth sample is expressed based on the formula $t_k = k \cdot \Delta t_M$, for $k = 1, 2, \dots$, N, where N stands for the number of samples.

All measurements used in the paper were carried out with an NP-3 (Far Data) anemometer in the radiation transfer station in Strzyżów near Rzeszów—south-eastern Poland at a height of $h_M = 10$ m AGL. The reference measurement data were re-calculated for the actual altitude of the turbine nacelle h_{WT}, for the needs of the study, according to the relationship describing the exponential form of the vertical profile of wind speed changes [25]:

$$v_{WT}(t_k) = v_{wM}(t_k) \left(\frac{h_{WT}}{h_M} \right)^{\alpha} \tag{2}$$

where: k is wind speed sample number, v_{WT} is wind speed at the height of the wind turbine nacelle h_{WT}, v_{wM} is wind speed at the measurement height h_M, and α is coefficient depending on the land roughness.

3. Mathematical Model of the System

3.1. Turbine and Wind Farm

The identification of a turbine output power by means of an analysis, based on the wind speed measurement values requires the use of a mathematical and numerical model of the turbine [25]. In order to examine the reliability of power supplies from the wind farm to the power grid, a simplified model was used for the wind speed sampling period Δt_M. In the simplified model, the wind turbine plays the role of a functional unit modelling the power characteristics $P_1 = f(v_w(t))$. Based on linear interpolation between two points of the turbine power's discrete characteristics, the power generated for the kth sample of the wind speed $v_w(t_k)$ can be identified based on the following formula:

$$
P_1(v_w(t_k)) = \begin{cases} P_1 + \frac{P_2 - P_1}{v_2 - v_1}(v_w(t_k) - v_1) \text{ for } v_{\text{cut}-\text{in}} \leq v_w(t_k) < v_{wn} \\ P_{WTn} \text{ for } v_{wn} \leq v_w(t_k) \leq v_{\text{cut}-\text{out}} \\ 0 \text{ for other } v_w(t_k) \end{cases} \tag{3}
$$

where: v_1, v_2 are wind speed values on the discrete characteristics of the turbine power between which the speed $v_1 \leq v_w(t_k) \leq v_2$ is positioned, P_1 and P_2 correspond to the wind speed, v_1 and v_2 are turbine power values read from discrete power characteristics.

In order to approximate the farm model to its actual operating conditions, differences in the position and, hence, the actual wind speed on the nacelle level were taken into consideration for each N_{WT} turbine. To that end, random changes ($\pm 5\%$) in the wind speed (in the moment t_k) were considered for each turbine as compared to measurement samples. Thus, for the ith turbine and kth sample, the wind speed (in the moment t_k) is:

$$
v_{w(i)}(t_k) = (1 \pm rnd\,(-0.05,\,0.05))v_w(t_k) \tag{4}
$$

where: $rnd\,(-0.05,\,0.05)$ is a random value (normal deposition) from the $[-0.05, 0.05]$ range. Then, the power released to the power grid from the reference farm is:

$$
P_3(t_k) = \sum_{i=1}^{N_{WT}} P_{1(i)}((1 \pm rnd\,(-0.05,\,0.05))v_w(t_k)) - \Delta P \tag{5}
$$

3.2. Kinetic Energy Storage

Kinetic energy storage systems use an indirect method of electrical energy storing in the form of flywheel rotational kinetic energy. Its shape, material and maximum rotational speed determine the storage dimensions, capacity, and weight, and are simultaneously the basic criteria of their division [9,24,26–28]. Figure 3 presents a general construction diagram of a kinetic energy storage system using a three-phase AC machine.

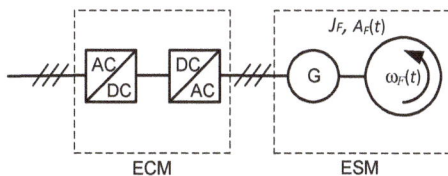

Figure 3. Structure of kinetic energy storage using a three-phase AC machine (A_F: flywheel mechanical energy, G: generator, ω_F: rotational speed of the system, J_F: moment of inertia of the system (mainly the flywheel), AC/DC, DC/AC: power electronics converters.

Analyses carried out to identify the reliability of power supply to the power grid do not require the consideration of all aspects of kinetic storage modelling and, especially, of dynamic states. A change in the storage charging level as a function of time (subsequent wind speed measurement samples) is important for the issue in question. Therefore, storage modelling was limited to the following equation of the rotational movement dynamics:

$$J_F \frac{d\omega_F}{dt} = \frac{P_2}{\omega_F} + \omega_F K_1 + K_2 \tag{6}$$

where: J_F is the moment of inertia of the system, P_2 is the storage charging/discharge power, ω_F is the rotational speed of the rotor, K_1 is the coefficient of friction, K_2 is the Coulomb's friction.

Solving Equation (6), when the storage capacity, dimensions of the rotational mass and scope of changes in the working speed and efficiency of electrical systems, which are part of the storage, are known, one can identify the course of changes in the power stored in the storage system when it is connected to a wind turbine and power grid, and execute the algorithm given in Section 2.1.

3.3. Power Electronics Systems Controlling the Flow of Power

From the point of view of the executed algorithm (point 2.1), power electronics systems related to DC electrical energy conversion into AC and vice versa (in the kinetic storage structure) and control of the energy flow between the turbine, storage and power grid according to Equation (1), are all important elements of the system presented in Figure 1. The efficiency of the abovementioned systems is a parameter that greatly affects the effectiveness of power conversion. A typical relationship of a power electronics converter η_{INV} as a function of load $\eta_{INV} = f (P/P_{INVn})$ is used in the acquired model, where P is converted power, and P_{INVn} is the rated power of the converter assembly. Figure 4 presents the assumed efficiency characteristics of power electronics systems.

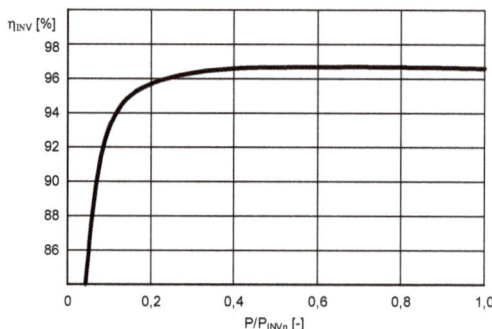

Figure 4. Changes in the efficiency of power electronics systems used in the analysed system as a function of a relative value of converted power (P/P_{INVn}).

4. Indicators of Electrical Energy Supplies from a Wind Farm—Kinetic Energy Storage System to the Power Grid

An assessment of electrical energy generation reliability for the assumed period involves the identification of the extent to which power demand $P_z(t)$ is covered by the capacity of subsystem $P_w(t)$. Information about reliability established for future periods (e.g., following changes in the structure of the interesting part of the grid) includes load connection, additional unstable sources etc. A model of power generation reliability, as a difference in the value of stochastic processes $P_z(t)$ and $P_w(t)$, is also defined as a stochastic process D(t) which is a power deficit. This is described by the general relationship [29]:

$$D(t) = \begin{cases} P_z(t) - P_w(t) \text{ for } P_z(t) > P_w(t) \\ 0 \text{ for } P_z(t) \le P_w(t) \end{cases} \tag{7}$$

Indicators related to the duration of power deficit, power not supplied to customers and frequency and duration of occurrence, loss of energy expectation (LOEE), frequency and duration indices (F&D), expected capacity deficiency (ECD), probability of capacity deficiency (PCD), expected loss of load (XLOL) etc.—are used as quantitative parameters describing a stochastic process of a power deficit D(t) [29–33].

A method based on an analysis of the changes in the level of the output power P_3 for a one-year period was used to compare the reliability of electrical energy supplies to the power grid from a wind farm—kinetic energy storage system with various structures. The application of an energy storage system helps to shape the changes in the power supplied to the system as compared to the power level resulting from changes in the wind energy. The applied indicators are similar to the parameters defined for a power deficit stochastic process D(t) but they still take into consideration the acquired power flow control algorithm (Section 2.1). Furthermore, they are identified based on historical data. The following indicators are used further in this paper:

- total annual time of power generation with a value lower than P_{3min}, covering only the periods lasting up to T_{max} (according to the system operation algorithm, the generated power deficit is replenished by the energy from the storage) at the storage capacity A_{FESS}:

$$T_{sumTmax}(A_{FESS}) = \sum_{i=1}^{M} T_{Tmax(i)} \tag{8}$$

where: $T_{Tmax(i)}$ is the duration of the ith interval which is shorter than or equal to T_{max} in which the power supplied to the grid P_3 is lower than the assumed minimum power P_{3min}, and M is the number of intervals T_{Tmax} in which power deficit occurs,
- percentage coefficient of elimination of periods when the generated power value is lower than P_{3min}, considering only the periods which last up to T_{max}:

$$\Delta T_{sumTmax\%} = \frac{T_{sumTmax}(0) - T_{sumTmax}(A_{FESS})}{T_{sumTmax}(0)} \times 100\% \tag{9}$$

where: $T_{sumTmax}(0)$ is time $T_{sumTmax}$ identified for a system without storage, $T_{sumTmax}(A_{FESSn})$ is the time $T_{sumTmax}$ identified for a system with energy storage with a rated capacity A_{FESS} (the diagram explaining the difference between T_{max}, $T_{sumTmax}$ and $T_{sumTmax}(0)$ was presented in Figure 5),
- power deficit accounting only for periods lasting up to T_{max}:

$$\Delta A = \sum_{i=1}^{M} T_{Tmax(i)}(P_{3min} - P_{3avg(i)}) \tag{10}$$

where: $P_{3avg(i)}$ is the mean power supplied to the grid from the wind farm—kinetic energy storage system in the ith interval shorter than or equal to T_{max}, in which the power released to the grid P_3 is lower than the assumed minimum power P_{3min}.

Equation (8) identifies the total time composed of periods no longer than T_{max}, in which the deficit of power generated by the system ($P_3 < P_{3min}$) was not completely covered by the energy in the storage. Efforts are made to minimise the indicator, while its value close to zero means complete elimination of power outages in the assumed maximum duration T_{max}. High values of Equation (9) mean elimination of the majority of outages with duration up to T_{max}. Efforts are then made to keep the value at 100%. Equation (10) characterises the system operation on the power side and helps to identify the power which was not supplied to the power grid in periods lasting up to T_{max}. Efforts are made to keep the value of Equation (10) as low as possible, and ideally at 0.

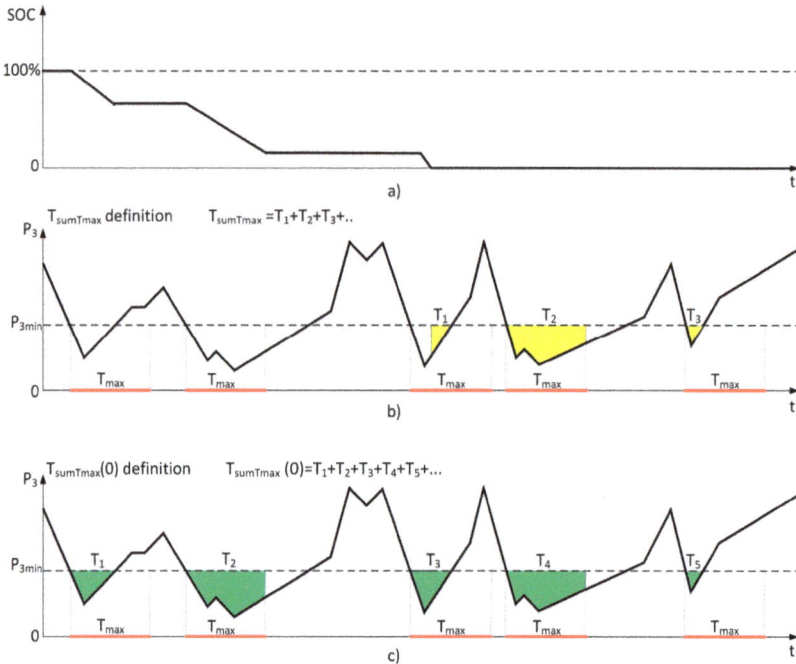

Figure 5. The diagram presenting the difference between T_{max}, $T_{sumTmax}$ and $T_{sumTmax}(0)$: (**a**) FESS's state of charge; (**b**) process of output power changes P_3 with marked periods shorter than T_{max}, in which the system does not provide minimum power P_{3min}; (**c**) process of changes in output power P_3 with marked periods shorter than T_{max}, in which the system does not provide minimum power P_{3min}.

5. Optimisation of a Wind Farm—Kinetic Energy Storage System

5.1. Purpose of Optimisation, Function and Constraints

The purpose of the optimisation task solved in the paper is to identify the structure of a wind farm—kinetic energy storage system, which for the set rated power of the farm P_{WTn} and its geographical location minimises the capacity of A_{FESS} storage limiting the annual time of supplies of power values to the power grid below P_{3min} (considering only the periods which last up to T_{max}) to $T_{sumTmaxY}$. The vector of decisive variables x includes: wind turbine type N_{WT} (x_1), turbine height h_{WT} (x_2), number of turbines n_{WT} (x_3), type of kinetic storage N_{FESS} (x_4), and the number of connected energy storage modules n_{FESS} (x_5). The minimisation task defined in this way takes the following form:

$$\dot{A}_{FESS} = \min\{A_{FESS}(x)\} = f(N_{WT}, n_{WT}, h_{WT}, N_{FESS}, n_{FESS}, w) \tag{11}$$

where: $\dot{A}_{FESS}$ is the minimum capacity value meeting the requirements for the set operating conditions of the system, as w is the collection of algorithm parameters of energy flow between the farm, storage, and power grid.

The type of turbine used (variable x_1) determines its power curve and at the same time the power generated in the moments specified by available wind speed samples. In the case of turbine, several variants are available differing in the installation height of the rotor (variable x_2), which, according to

Equation (2) to the vertical profile of wind speed variations, also affects the level of power generated by the turbine. The number of turbines specified by the variable x_3 allows obtaining the nominal power of the P_{WFn} farm assumed in the algorithm. With the assumed minimum power P_{3min} (definition in Section 2.1), the number and duration of periods of power generation by the farm below the level P_{3min} depend on the variables x_1, x_2 and x_3 in Figure 5. Individual modules of kinetic magazines are characterized by different energy capacities, but the most important from the point of view of the analysed problem are the values of their maximum charging and discharging power (these parameters are determined by the variable x_4). The last of these parameters in various ways adapt the properties of the energy storage in the field of exchange the energy, to the real dynamics of wind speed changes at a specific farm location (the course of changes in wind speed). The last variable x_5 affects the total capacity and power of the used kinetic energy storage system.

Equation (11) can be solved only when considering the collection of constraints, i.e., the relationships (equality and/or inequality) which identify the size of the acceptable solution area X. Structural constraints in the analysed task (related directly to the vector of decisive variables) expressed in a standardised form include:

- True power of the farm $P_{WFr}(\mathbf{x})$:

$$1 - \frac{P_{WFr}(x)}{0.9P_{WFn}} \le 0 \cap \frac{P_{WFr}(x)}{1.1P_{WFn}} - 1 \le 0 \tag{12}$$

- Size of turbine database (number of designs and turbine heights) and kinetic storage units (number of energy storage types).

Constraints in the true power of the farm $P_{WFr}(\mathbf{x})$ are related to the inclusion in the calculation process of a finite collection of turbine designs whose multiple power values are not always equal to the assumed rated power of the farm P_{WFn}.

Functional constraints (not related to vector $\mathbf{x}$) cover the parameters which are a result of a numerical analysis of the system operation (output power P_3) and are related to characteristic indicators of the control algorithm of the energy flow between the farm, storage, and power grid. In the analysed task, the functional constraints in a standardised form cover:

- time $T_{sumTmax}(\mathbf{x})$ — Equation (6):

$$\frac{T_{sumTmax}(x)}{T_{sumTmax}} - 1 \le 0 \tag{13}$$

where: $T_{sumTmaxY}$ is the total annual boundary time of power generation below P_{3min} assumed in the optimisation task, covering only the periods lasting up to T_{max};
- maximum power of kinetic storage $P_{FESSMax}(\mathbf{x})$:

$$1 - \frac{P_{FESSMax}(x)}{P_{3min}} \le 0 \tag{14}$$

Interrelations between the parameters of the selected turbine (power characteristic, height) and a single module of energy storage (rated energy capacity, maximum charging and discharging power) directly affect, for the given location (wind conditions), the $T_{sumTmax}$ parameter in Figure 5. Equation (13) of limiting the indicated parameter to the value of $T_{sumTmaxY}$ therefore requires the use of different storage capacities for various types of turbines and energy storage. Therefore, the parameters determined in Equation (11) determine the value of the energy storage capacity, and the application of the optimization algorithm described in the paper allows the determination of the optimal structure of the wind farm—kinetic energy storage, minimizing the capacity of the storage working in it, while fulfilling all technical requirements for the cooperation of the farm with power system.

5.2. Selection of Optimisation Method, Description of Application Developed

The selection of the optimisation method dedicated to the solution defined in Section 5.1 of the task requires a detailed analysis of several elements, including in particular: the form of the function of the objective, the number of decisive variables and the size of the acceptable solution area [25,34–36]. Additionally, one should consider the possibilities of including in the optimisation complex numerical calculations used for identifying the value of the function of the objective and constraint control. The experience of the authors in the effective use of the method to solve similar tasks is very important [37].

The multi-modal nature of the acquired function of Equation (11) requires the use of stochastic methods or heuristics in order to seek its global extreme. Even though only five variables were defined in the reference task, the task remains complex and time-consuming, as a result of the multiple identification of changes in the power value in the analysed system, the integer nature of decisive variables and the hidden occurrence of variables in the criterion function. In relation to the above, a genetic algorithm population method is used to optimise the structure of a wind farm-kinetic energy storage system, which leads to the minimisation of storage capacity A_{FESS}, according to the system operation algorithm described in Section 2.1. An important advantage of the method is related to its ability to modify the basic parameters in order to improve its effectiveness when performing detailed tasks. A disadvantage in the case of tasks with highly complex calculations is the reduced reproducibility of the results [35,36]. It is also necessary to perform initial tests whose results help to identify the appropriate AG parameters that would improve the reproducibility of the results and reduce calculation time.

The effectiveness of the proposed method in solving tasks of global optimisation in the area of electrical engineering and renewable energy was proven in a number of scientific publications, e.g., [38,39] and the authors used it successfully to optimise the structure of complex electrical light systems, the shape and parameters of high-power lines, and to minimise the cost of power generation in hybrid generation systems [37,40–45].

Based on the mathematical model described in Section 3 and a selected optimisation method in Matlab (Lincence no: 975466, Version 2014b, Poznan University of Technology, Poznan, Poland, 2014) and MS Visual Studio environment, an application was developed intended to optimise the structure of a wind farm—kinetic energy storage system with a view to minimising energy storage capacity. The application uses proprietary structures and classes related to different types of energy storage, wind turbines, PV modules, power electronics systems and indicators of the reliability of power supplies to the power grid mentioned in Section 4, as well as functions from the Global optimisation toolbox of the Matlab environment to implement the modified genetic algorithm method.

The assessment of the quality of solutions obtained through optimisation required an analysis of power: generated by farm P_1, energy storage P_2, and supplied to the power grid P_3 in the analysed system. To that end, measurements of the wind speed in a one-year period and data from wind turbines and PV modules collected in a database developed for the purpose of the application were used.

The Augmented Lagrangian Genetic Algorithm (ALGA) interior penalty function method [46] was employed to consider the constraints of the optimisation tasks (Equations (12)–(14)). The ALGA method was implemented in the Matlab environment.

5.3. Optimisation Calculations

A search for the structure of a wind farm—kinetic energy storage system which minimises the capacity of energy storage A_{FESS} (Equation (11)) for the assumed output parameters of the system, was carried out for wind farms with two power values: $P_{WFn(1)} = 5$ MW and $P_{WFn(2)} = 10$ MW (indices expressed as integers stand for a parameter option). Six time values were taken into account concerning power generation periods with power values below P_{3min}: $T_{max(1)} = 5$ min, $T_{max(2)} = 10$ min, $T_{max(3)} = 15$ min, $T_{max(4)} = 20$ min, $T_{max(5)} = 25$ min and $T_{max(6)} = 30$ min; two minimum power values $P_{3min(1)}$ = 10%P_{WFn}, and $P_{3min(2)}$ = 20%P_{WFn}; and four values of the acceptable total time: $T_{sumTmax(1)}$ = 75 h, $T_{sumTmax(2)}$ = 100 h, $T_{sumTmax(3)}$ = 125 h, and $T_{sumTmax(4)}$ = 150 h.

The calculations were carried out using the developed optimisation algorithm and application described in Section 5.2 in order to conduct a preliminary study involving changes in the AG parameters to solve the test task. The obtained results helped to draw conclusions relating to the value of the algorithm parameters which contribute to the final solution before the maximum number of iterations, and reduce computing time to ca. 3 h. Finally, a genetic algorithm was applied using the remainder selection method, the Gaussian mutation and elite strategy with a transfer of three best individuals. The number of individuals N_p = 50 and the number of generations N_g = 50 were established experimentally.

Figure 6 presents the changes in the value of the best individual's (solution) adaptation as a function of the generation number for five activations of the algorithm (Figure 6b) for a farm with rated power $P_{WFn(2)}$ = 10 MW, power $P_{3min(2)}$ = 20%P_{WFn} = 2 MW, $T_{max(5)}$ = 25 min, and $T_{sumTmax(1)}$ = 75 h.

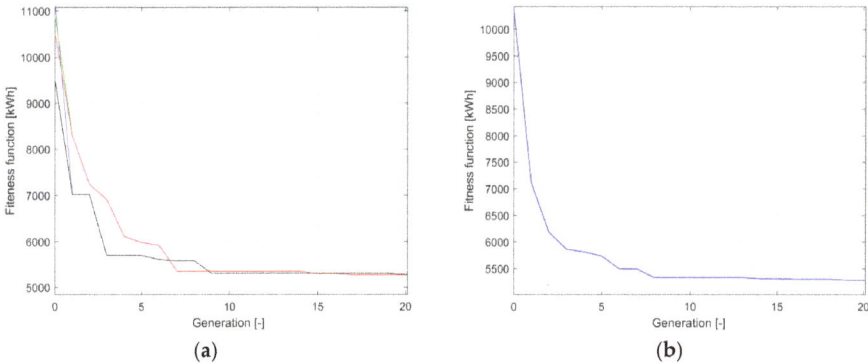

Figure 6. Changes in the value of the function of the best individual's adaptation as a function of the generation number for a farm power $P_{WFn(1)}$ = 10 MW, power $P_{3min(2)}$ = 20%P_{WFn} = 2 MW, $T_{max(3)}$ = 25 min and $T_{sumTmax(1)}$ = 75 h: (**a**) for five activations of the genetic algorithm; (**b**) mean value from ten activations of the genetic algorithm.

Figure 7 presents the results of optimisation calculations as changes in the minimum capacity of a kinetic storagesystem which meets the required parameters of a system controlling the flow of energy between the farm, storage and power grid as a function of time T_{max} for two wind farms with power $P_{WFn(1)}$ = 5 MW (Figure 7a) and power $P_{WFn(2)}$ = 10 MW (Figure 7b). In each case, the calculations were made for two minimum power values $P_{3min(1)}$ = 10%P_{WFn} and $P_{3min(2)}$ = 20%P_{WFn} and $T_{sumTmax(1)}$ = 100 h.

An essential issue for the analysed class of systems is to identify the relationship between the minimum capacity of the storage $A_{FESSmin}$ and the time of eliminated outages T_{max} depending on the maximum total time $T_{sumTmax}$. Optimisation calculations representing the aforementioned task were performed for a farm with power $P_{WFn(2)}$ = 10 MW, two power values $P_{3min(1)}$ = 10%P_{WFn} = 1 MW and $P_{3min(2)}$ = 20%P_{WFn} = 2 MW and four times $T_{sumTmax}$:$T_{sumTmax(1)}$ = 75 h, $T_{sumTmax(2)}$ = 100 h, $T_{sumTmax(3)}$ = 125 h and $T_{sumTmax(4)}$ = 150 h. The results (value of the minimum capacity of energy storage $A_{FESSmin}$ meeting the assumed algorithm of the system operation as a function of time T_{max}) are presented in Figure 8a ($P_{3min(1)}$ = 10%P_{WFn} = 1 MW) and 8b ($P_{3min(2)}$ = 20%P_{WFn} = 2 MW).

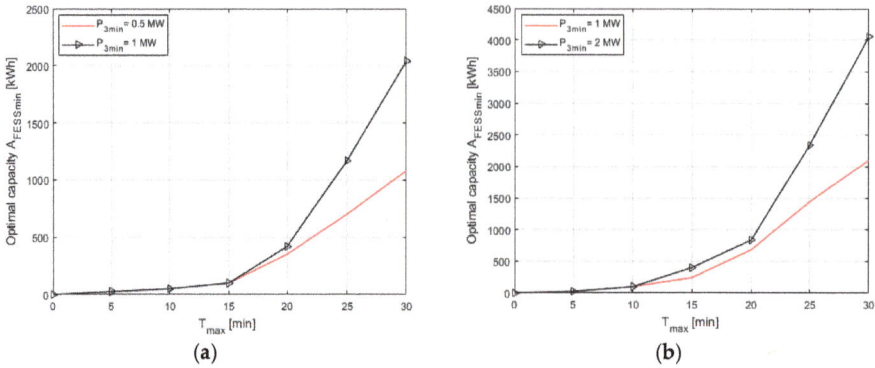

Figure 7. Changes in the optimum (minimum) capacity of a kinetic energy storage system $A_{FESSmin}$ as a function of time T_{max} for two power values $P_{3min(1)} = 10\%P_{WFn}$ and $P_{3min(2)} = 20\%P_{WFn}$ and time $T_{sumTmax(x)} = 100$ h for: (**a**) a farm with rated power $P_{WFn(1)} = 5$ MW; (**b**) a farm with rated power $P_{WFn(2)} = 10$ MW.

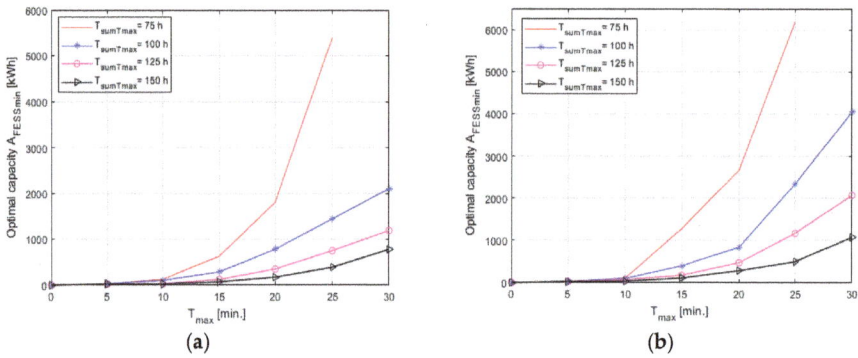

Figure 8. Changes in the minimum (optimum) capacity of a kinetic energy storage system $A_{FESSmin}$ as a function of time T_{max} for a farm power $P_{WFn(2)} = 10$ MW, time: $T_{sumTmax} = 75$ h, $T_{sumTmax} = 100$ h, $T_{sumTmax} = 125$ h, $T_{sumTmax} = 150$ h and the optimised system data: (**a**) $P_{3min(1)} = 10\%P_{WFn} = 1$ MW; (**b**) $P_{3min(2)} = 20\%P_{WFn} = 2$ MW.

Figure 9 presents the relationship between the minimum capacity of the storage unit $A_{FESSmin}$ and the maximum total time $T_{sumTmax}$ for two values of time $T_{max(3)} = 15$ min and $T_{max(5)} = 25$ min.

Details of the optimisation results of the wind farm-kinetic energy storage system for rated power $P_{WFn(2)} = 10$ MW and selected parameters P_{3min}, $T_{sumTmax}$, and time T_{max} are presented in Table 1.

Figure 10 presents the changes in the percentage value of the coefficient of elimination of power generation periods with power values below P_{3min} for solutions optimum as a function of time $T_{sumTmax}$ for two power values $P_{3min(1)} = 10\%P_{WFn}$ and $P_{3min(2)} = 20\%P_{WFn}$, and two time values $T_{max(3)} = 15$ min (Figure 9a) and $T_{max(4)} = 20$ min (Figure 9b). The data were developed based on information given in Table 1.

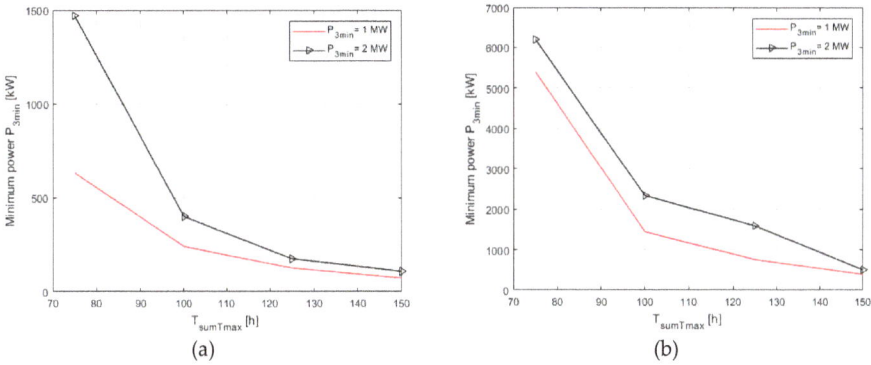

Figure 9. Changes in the minimum (optimum) capacity of the kinetic storage unit $A_{FESSmin}$ in the function of time $T_{sumTmax}$ for: (**a**) $T_{max(3)}$ = 15 min; (**b**) $T_{max(5)}$ = 25 min.

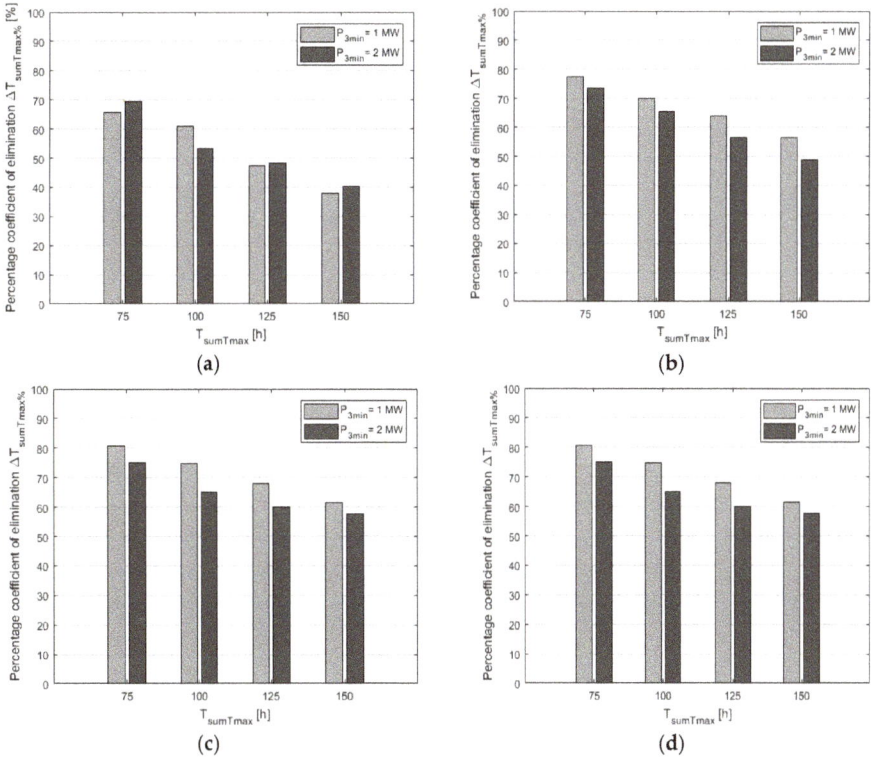

Figure 10. Changes in the percentage value of the coefficient of elimination of power generation periods with power values below P_{3min} (Equation 9) for a farm with rated power P_{WFn} = 10 MW, power P_{3min} = 1 MW and P_{3min} = 2 MW and: (**a**) $T_{max(2)}$ = 10 min; (**b**) $T_{max(3)}$ = 15 min; (**c**) $T_{max(4)}$ = 20 min; (**d**) $T_{max(5)}$ = 25 min.

Table 1. Results of optimising calculations for selected configurations of a wind farm—kinetic storage system for a farm with rated power $P_{WFn(2)}$ = 10 MW and different time values T_{max}.

No.	P_{3min} (MW)	T_{max} (min)	$T_{sumTmax}$ (h)	$T_{sumTmax}$ ($A_{FESSmin}$) (h)	$T_{sumTmax}(0)$ (h)	$\Delta T_{sumTmax}$ (%)	$A_{FESSmin}$ (kWh)	Turbine's Number/ Power(−)/(MW)	P_{WFnr} (MW)	ΔA (MWh)
				$T_{max(2)}$ = 10 min						
1	1	10	75	73.7	214.3	65.6	125	10/0.9	9	21.1
2	1	10	100	96.2	247.1	61.0	72	5/1.8	9	31.6
3	1	10	125	122.5	232.1	47.2	36	4/2.3	9.2	38.5
4	1	10	150	144.8	225.5	37.8	25	6/1.8	10.8	47.6
5	2	10	75	73.4	241.6	69.6	108	10 0.9	9	37.3
6	2	10	100	92.7	198.0	53.2	100	4/2.3	9.2	49.3
7	2	10	125	120.5	231.2	48.1	72	3/3	9	60.4
8	2	10	150	144.2	241.5	40.2	36	6/1.8	10.8	72.0
				$T_{max(3)}$ = 15 min						
9	1	15	75	74.9	329.8	77.3	630	12/0.9	10.8	25.7
10	1	15	100	93.3	312.2	70.1	240	10/0.9	9	30.0
11	1	15	125	118.1	327.6	63.9	125	4/2.3	9.2	36.9
12	1	15	150	142.7	327.6	56.4	72	4/2.3	9.2	42.8
13	2	15	75	74.9	283.9	73.6	1470	4/2.3	9.2	50.9
14	2	15	100	98.2	283.9	65.4	400	4/2.3	9.2	63.4
15	2	15	125	123.6	283.9	56.4	175	4/2.3	9.2	75.6
16	2	15	150	145.2	283.9	48.9	108	4/2.3	9.2	84.5
				$T_{max(4)}$ = 20 min						
17	1	20	75	74.7	384.9	80.6	1800	12/0.9	10.8	37.8
18	1	20	100	99.5	397.7	74.9	684	12/0.9	10.8	34.1
19	1	20	125	123.3	384.9	67.9	350	10/0.9	9	40.3
20	1	20	150	147.9	384.9	61.6	175	10/0.9	9	46.7
21	2	20	75	74.9	302.8	75.3	2670	11/0.9	9.9	52.0
22	2	20	100	98.8	285.6	65.4	840	10/0.9	9	62.9
23	2	20	125	124.9	314.2	60.2	474	12/0.9	10.8	76.5
24	2	20	150	149.5	353.0	57.7	288	4/2.3	9.2	96.7
				$T_{max(5)}$ = 25 min						
25	1	25	75	74.9	447.2	83.3	5400	10/0.9	9	29.3
26	1	25	100	99.3	482.8	79.4	1440	4/2.3	9.2	39.2
27	1	25	125	124.9	447.2	72.1	750	10/0.9	9	44.0
28	1	25	150	148.7	447.2	66.8	390	10/0.9	9	54.6
29	2	25	75	74.9	355.5	78.9	6200	11/0.9	9.9	54.4
30	2	25	100	99.8	355.5	71.9	2340	11/0.9	9.9	70.3
31	2	25	125	124.7	334.1	62.7	1584	10/0.9	9	87.3
32	2	25	150	148.9	334.1	55.5	500	10/0.9	9	95.8
				$T_{max(6)}$ = 30 min						
33	1	30	75	-	-	-	-	-	-	-
34	1	30	100	99.3	552.0	82.1	2100	6/1.8	10.8	54.0
35	1	30	125	124.3	552.0	77.5	1188	6/1.8	10.8	65.0
36	1	30	150	148.9	552.0	73.0	780	6/1.8	10.8	75.9
37	2	30	75	-	-	-	-	-	-	-
38	2	30	100	99.8	426.9	76.6	4050	12/0.9	10.8	70.9
39	2	30	125	124.9	398.8	68.7	2070	11/0.9	9.9	88.3
40	2	30	150	149.7	378.9	60.5	1075	11/0.9	9	103.0

5.4. Discussion

Following a thorough analysis of the results it was concluded that the applied optimisation methods, the parameters of the control algorithm of the energy flow between a wind farm, energy storage system and the power grid, and the system model were correct for the characteristics of the analysed task. Its solution had high reproducibility in minimising the capacity of the kinetic storage $A_{FESSmin}$ (Figure 6a) and fast computation times (Figure 6b). In the analysed examples, the tasks were solved between the 30th and 40th generations.

Based on the completed studies, it was established that the relationship between the minimum capacity of the energy storage system $A_{FESSmin}$ meeting the required criteria of a wind farm—kinetic energy storage system and time T_{max} was non-linear (Figures 7 and 8). It was demonstrated that the dependence points $A_{FESSmin} = f(T_{max})$ could be approximated by polynomial functions of order 3. The values of the determination coefficients identified for the functions amount to over 0.97, confirming the good fit of a polynomial model for the calculation points. A non-linear character was also observed for changes in the capacity $A_{FESSmin}$ as a function of the optimisation time constraint $T_{sumTmax} - A_{FESSmin} = f (T_{sumTmax})$ (Figure 9).

The application of higher constraints (lower values of $T_{sumTmax}$) amounting to 150, 125, 100, and 75 h/year contributed to an increase in the required capacity of energy storage $A_{FESSmin}$. The increase was diversified for different values of time T_{max} and power P_{3min}, but between subsequent values of $T_{sumTmax}$ it usually ranged from 1.5 to 3.0 times (Figure 8). In the case of optimising calculations made for time $T_{max} = 30$ minutes and the strongest constraint $T_{sumTmax} = 75$ h, the task was not solved (Figure 8) for neitherpower $P_{3min} = 10\%P_{WFn}$ nor for $P_{3min} = 20\%P_{WFn}$. Tests were repeated many times and an additional detailed analysis of the system operation with optimised parameters (type and number of turbines) was performed with a simultaneous increase in the power storage capacity. It was demonstrated that $T_{sumTmax}$ could not be limited at the level of 75 h/year by further increasing storage capacity. This was caused by the characteristics of the annual changes in wind speed for the reference geographical location (especially for the mean annual wind speed, which amounts to 4.1 m/s at an altitude of 1 m AGL, and to 6.2 m/s at 100 m AGL), and the nominal parameters of the analysed turbines and energy storages systems.

Depending on the geographical location of the farm and the parameters of the turbines and storage systems, it is possible that no solutions can occur for different values of time T_{max}. Therefore, it is important to identify the maximum value of time for the assumed wind farm power P_{WFn}, location and basic parameters of the system for which elimination of power generation periods with power values below P_{3min} is possible on the $T_{sumTmax}$ level. Still, the developed algorithm and application require significant modifications, alongside additional studies. The authors plan to address this in future works on the cooperation between wind farms and kinetic energy storages.

The level of the guaranteed output power P_{3min} (Figures 7 and 8) in periods lasting up to T_{max} greatly affects the minimum energy storage capacity $A_{FESSmin}$. Before establishing the parameter value, one should analyse the actual wind farm operating conditions, especially its mean working power. Increasing the power P_{3min} above the value results in a significant increment in the required energy storage capacity and may mean that the storage unit cannot be fully charged. Consequently, the system is not able to eliminate all assumed energy deficits lasting up to T_{max} and does not meet a certain group of constraints $T_{sumTmax}$ (e.g., Table 1—results for $T_{max} = 30$ min and constraints $T_{sumTmax} = 75$ h). A lack of solutions for the cases analysed in Table 1 applies to a farm with low mean annual power of ca. 1 MW. In cases where solutions were found, the optimum mean annual power values of the wind farms ranged from about 1.5 MW to 3 MW.

The optimisation of the kinetic energy storage capacity helps to achieve high percentage values of the coefficient of elimination of power generating periods when the power value is below P_{3min} ($\Delta T_{sumTmax\%}$ — Equation (9)) which amount to 40–80% (mean 53%) of the analysed cases depending on the value of time T_{max} and constraint value $T_{sumTmax}$ (Figure 10). The value of the indicator $\Delta T_{sumTmax\%}$ increased as the system operation constraints become more stringent (shorter times $T_{sumTmax}$), entailing a significant increase in the required storage capacity. When the value of the constraint time is reduced twice (from $T_{sumTmax} = 150$ h to $T_{sumTmax} = 75$ h), the optimised values of the energy storage capacity were 5 to 14 times higher, depending on time T_{max}.

Values of power not supplied to the power grid ΔA (Equation (10)) in the system operation periods with power below P_{3min} and time up to T_{max}, analysed in a one-year period (Table 1), were established for the optimised structures of a farm with a kinetic energy storage system. The power value increased as time T_{max} and constraint time $T_{sumTmax}$ increase. In the case of the analysed farm with power $P_{WFn} = 10$ MW, the values were low (Table 1) compared to the total energy generated in a one-year period and they did not exceed 0.5%.

It should be emphasised that it was possible to achieve the presented results of a wind farm—kinetic energy storage system optimisation due to a short (ca. 47 s) period of wind speed averaging. For a typical value of 10 min, it was not possible to precisely simulate the system cooperation with the power grid due to an approximated analysis of the energy storage operation. In such cases, the results for the minimum energy storage capacity $A_{FESSmin}$ suffered from a much higher error. Based on the authors' experience in the modelling of energy storage using RES, they suggest that the averaging period of wind speed measurements and other parameters related to power generation in RES—irradiation in particular—should be shorter than the period used currently and should amount to one or a few seconds.

6. Conclusions

The paper presents a new algorithm for identifying the minimum capacity of a kinetic energy storage system which helps to improve (on the assumed level) the reliability indicators of electrical energy supplies from an unstable source (i.e., wind farm) to the power grid. The concept of the method can be transferred onto different types of energy storage and unstable sources. In such a case, the relevant mathematical and numerical models of the system elements need to be replaced and preliminary studies conducted to establish the optimisation algorithm parameters.

The analyses helped to draw an important conclusion concerning changes in the values of reliability of power supplies from a wind farm to the power grid. It is possible to improve the value and stabilise the operation of the wind farm at relatively low capacities of energy storage (in cases of moderate parameters of the analysed system and optimisation constraints). It is still important to take such measures towards the majority of farms located close to one another, as it will help to predict the level of power generated from such sources more precisely. Any increase in the requirements concerning the system operating parameters results in a significant (up to several times) increase in the required capacity of the storage, and hence means higher investment and operating costs.

The authors of the paper focused on an essential matter related to the relative increase in the installed power of unstable sources in the generating systems operating in many countries. The situation causes a stochastic variance of electric power generation. The application of the presented methods and results can help to stabilise the output power of unstable sources at limited costs of necessary investments. Local compensation of temporary power deficiencies by storage system included in the structure of large wind farms helps to limit power transmission losses and to improve the total energy efficiency of the system.

Author Contributions: A.T. and L.K. conceived the base of the paper, discussed the optimization method of wind farm with flywheel energy storage system, elaborated summary. A.T. designed models of power system with wind turbine and energy storage, developed computer system, made calculations, prepared drawings. L.K. analyzed the results of simulation and optimization, provided some valuable suggestions, revised the paper.

Funding: This research was funded by Polish Government, grant number [04/42/DSPB/0431].

Conflicts of Interest: The authors declare no conflict of interest.

References

1. Trzmiel, G. Determination of a mathematical model of the thin-film photovoltaic panel (CiS) based on measurement data. *Eksploat. Niezawodn.* **2017**, *19*, 516–521. [CrossRef]
2. Koh, L.H.; Peng, W.; Tseng, K.J.; Gao, Z. Reliability evaluation of electric power systems with solar photovoltaic & energy storage. In Proceedings of the 2014 International Conference on Probabilistic Methods Applied to Power Systems, Durham, UK, 7–10 July 2014; pp. 1–5.
3. Renewables 2016 Global Status Report. Available online: http://www.ren21.net (accessed on 15 May 2018).
4. *Renewable Energy in Europe—2017 Update. Recent Growth and Knock-on Effects*; EEA Report, No 23/2017; European Environment Agency (EEA): Copenhagen, Denmark, 2017. [CrossRef]
5. Europe 2020 Indicators. Available online: http://ec.europa.eu/eurostat/portal/page/portal/europe_2020_indicators/ (accessed on 15 May 2018).
6. Kasprzyk, L. Modelling and analysis of dynamic states of the lead-acid batteries in electric vehicles. *Eksploat. Niezawodn.* **2017**, *19*, 229–236. [CrossRef]

7. Price, J.E.; Sheffrin, A. Adapting California's energy markets to growth in renewable resources. In Proceedings of the 2010 IEEE PES General Meeting, Providence, RI, USA, 25–29 July 2010; pp. 1–8.
8. Komarnicki, P.; Lombardi, P.; Styczynski, Z. Economics of Electric Energy Storage Systems. In *Electric Energy Storage Systems*; Springer: Berlin/Heidelberg, Germany, 2017; pp. 181–194.
9. Amiryar, M.E.; Pullen, K.R.; Nankoo, D. Development of a High-Fidelity Model for an Electrically Driven Energy Storage Flywheel Suitable for Small Scale Residential Applications. *Appl. Sci.* **2018**, *8*, 453. [CrossRef]
10. Baghaee, H.R.; Mirsalim, M.; Gharehpetian, G.B.; Talebi, H.A. Fuzzy unscented transform for uncertainty quantification of correlated wind/PV microgrids: Possibilistic–probabilistic power flow based on RBFNNs. *IET Renew. Power Gener.* **2017**, *11*, 867–877. [CrossRef]
11. Zheng, D.; Semero, Y.K.; Zhang, J.; Wei, D. Short-term wind power prediction in microgrids using a hybrid approach integrating genetic algorithm, particle swarm optimization, and adaptive neuro-fuzzy inference systems. *IEEJ Trans. Electr. Electron. Eng.* **2018**. [CrossRef]
12. Ding, Y.; Wang, P.; Chang, L.P. Reliability evaluation of electric power systems with high wind power penetration. In Proceedings of the 2009 8th International Conference on Reliability, Maintainability and Safety, Chengdu, China, 20–24 July 2009; pp. 24–26.
13. Sideratos, G.; Hatziargyriou, N. An advanced statistical method for wind power forecasting. *IEEE Trans. Power Syst.* **2007**, *22*, 258–265. [CrossRef]
14. Chang, W.Y. A literature review of wind forecasting methods. *J. Power Energy Eng.* **2014**, *2*, 161–168. [CrossRef]
15. Kassa, Y.; Zhang, J.H.; Zheng, D.H.; Wei, D. Short term wind power prediction using ANFIS. In Proceedings of the 2016 1st IEEE International Conference on Power and Renewable Energy, Shanghai, China, 21–23 October 2016; pp. 388–393.
16. Zhu, Y.; Zang, H.; Cheng, L.; Gao, S. Output Power Smoothing Control for a Wind Farm Based on the Allocation of Wind Turbines. *Appl. Sci.* **2018**, *8*, 980. [CrossRef]
17. Mazzeo, D.; Oliveti, G.; Baglivo, C.; Congedo, P.M. Energy reliability-constrained method for the multi-objective optimization of a photovoltaic-wind hybrid system with battery storage. *Energy* **2018**, *156*, 688–708. [CrossRef]
18. Alemany, J.; Kasprzyk, L.; Magnago, F. Effects of binary variables in mixed integer linear programming based unit commitment in large-scale electricity markets. *Electr. Power Syst. Res.* **2018**, *160*, 429–438. [CrossRef]
19. Lombardi, P.; Röhrig, C.; Rudion, K.; Marquardt, R.; Müller-Mienack, M.; Estermann, A.S.; Voropai, N.I. An A-CAES pilot installation in the distribution system: A technical study for RES integration. *Energy Sci. Eng.* **2014**, *2*, 116–127. [CrossRef]
20. Baghaee, H.R.; Mirsalim, M.; Gharehpetian, G.B. Multi-objective optimal power management and sizing of a reliable wind/PV microgrid with hydrogen energy storage using MOPSO. *J. Intell. Fuzzy Syst.* **2017**, *32*, 1753–1773. [CrossRef]
21. Kaviani, A.; Baghaee, H.R.; Riahy, G.H. Optimal sizing of a stand-alone wind/photovoltaic generation unit using particle swarm optimization. *Simul. Trans. Soc. Model. Simul.* **2009**, *85*, 89–99.
22. Acuña, L.G.; Padilla, R.V.; Mercado, A.S. Measuring reliability of hybrid photovoltaic-wind energy systems: A new indicator. *Renew. Energy* **2017**, *106*, 68–77. [CrossRef]
23. Jiang, X.; Zhang, Z.; Wang, J. Studies on the reliability and reserve capacity of electric power system with wind power integration. In Proceedings of the 2012 Power Engineering and Automation Conference, Wuhan, China, 18–20 September 2012; pp. 1–4.
24. Tomczewski, A. Operation of a Wind Turbine-Flywheel Energy Storage System under Conditions of Stochastic Change of Wind Energy. *Sci. World J.* **2014**, *2014*, 643769. [CrossRef] [PubMed]
25. Kaabeche, A.; Belhamel, M.; Ibtiouen, R. Techno-economic valuation and optimization of integrated photovoltaic/wind energy conversion system. *Sol. Energy* **2011**, *85*, 2407–2420. [CrossRef]
26. Díaz-Gonzáleza, F.; Sumpera, A.; Gomis-Bellmunta, O.; Villafáfila-Roblesb, R. A review of energy storage technologies for wind power applications. *Renew. Sustain. Energy Rev.* **2012**, *16*, 2154–2171. [CrossRef]
27. Fuchs, G.; Lunz, B.; Leuthold, M.; Sauer, D.U. *Technology Overview on Electricity Storage*; Overview on the Potential and on the Deployment Perspectives of Electricity Storage Technologies; Institute for Power Electronics and Electrical Drives: Aachen, Germany, 2012.
28. Amiryar, M.E.; Pullen, K.R. A Review of Flywheel Energy Storage System Technologies and Their Applications. *Appl. Sci.* **2017**, *7*, 286. [CrossRef]

29. Paska, J. Reliability Issues in Electric Power Systems with Distributed Generation. *Rynek Energii* **2008**, *5*, 18–28. (In Polish)
30. *Glossary of Terms Used in Reliability Standards*; NERC: Swindon, UK, 2008.
31. *Power System Reliability Analysis*; Application Guide; CIGRE WG 03 of SC 38 (Power System Analysis and Techniques); e-cigre: Paris, France, 1987.
32. *Power System Reliability Analysis*; Composite Power System Reliability Evaluation; CIGRE Task Force 38-03-10; e-cigre: Paris, France, 1992.
33. *Reliability Assessment Guidebook*; version 2.1; NERC: Swindon, UK, 2010.
34. Castro Mora, J.; Calero Baro, J.M.; Riquelme Santos, J.M.; Burgos Payan, M. An evolutive algorithm for wind farm optimal design. *Neurocomputing* **2007**, *70*, 2651–2658. [CrossRef]
35. Goldberg, D.E. *Genetic Algorithms in Search, Optimization and Machine Learning*; Addison-Wesley Longman Publishing: Boston, MA, USA, 1988.
36. Michalewicz, Z.; Fogel, D.B. *How to Solve It: Modern Heuristics*, 2nd ed.; Springer: Berlin/Heidelberg, Germany, 2004.
37. Kasprzyk, L.; Tomczewski, A.; Bednarek, K.; Bugała, A. Minimisation of the LCOE for the hybrid power supply system with the lead-acid battery. In Proceedings of the 2017 International Conference Energy, Enviroment and Material Systems, Polanica-Zdroj, Poland, 13–15 September 2017.
38. Ismail, M.S.; Moghavvemi, M.; Mahlia, T.M.I. Genetic algorithm based optimization on modeling and design of hybrid renewable energy systems. *Energy Convers. Manag.* **2014**, *85*, 120–130. [CrossRef]
39. Al-Shamma'a, A.A.; Addoweesh, K.E. Techno-economic optimization of hybrid power system using genetic algorithm. *Int. J. Energy Res.* **2014**, *38*, 1608–1623. [CrossRef]
40. Bednarek, K. Electrodynamic calculations and optimal designing of heavy-current lines. *Prz. Elektrotech.* **2008**, *84*, 138–141.
41. Bednarek, K.; Nawrowski, R.; Tomczewski, A. An application of genetic algorithm for three phases screened conductors optimization. In Proceedings of the 2000 International Conference on Parallel Computing in Electrical Engineering, Trois-Rivieres, QC, Canada, 27–30 August 2000; pp. 218–222.
42. Bednarek, K.; Jajczyk, J. Effectiveness of optimization methods in heavy-current equipment designing. *Prz. Elektrotech.* **2009**, *85*, 29–32.
43. Kasprzyk, L.; Tomczewski, A.; Bednarek, K. Efficiency and economic aspects in electromagnetic and optimization calculations of electrical systems. *Prz. Elektrotech.* **2010**, *86*, 57–60.
44. Kasprzyk, L.; Nawrowski, R.; Tomczewski, A. Optimization of Complex Lighting Systems in Interiors with use of Genetic Algorithm and Elements of Paralleling of the Computation Process. In *Intelligent Computer Techniques in Applied Electromagnetics, Studies in Computational Intelligence*; Wiak, S., Krawczyk, A., Dolezel, I., Eds.; Springer: Berlin/Heilderberg, Germany; New York, NY, USA, 2008; Volume 116, pp. 21–29.
45. Nawrowski, R.; Tomczewski, A. Optimization of overall costs in designing complex lighting systems. *Acta Techn. CSAV* **2008**, *53*, 65–79.
46. Conn, A.R.; Gould, N.I.M.; Toint, P.L. A Globally Convergent Augmented Lagrangian Barrier Algorithm for Optimization with General Inequality Constraints and Simple Bounds. *Math. Comput.* **1997**, *66*, 261–288. [CrossRef]

applied
sciences

MDPI

Article

Optimal Power Reserve of a Wind Turbine System Participating in Primary Frequency Control

Abdullah Bubshait [1],* and Marcelo G. Simões [2],*

[1]   Electrical Engineering Department, King Faisal University, Alahsa 31982, Saudi Arabia
[2]   Electrical Engineering Department, Colorado School of Mines, Golden, CO 80401, USA
*   Correspondence: asbubshait@kfu.edu.sa (A.B.); msimoes@mines.edu (M.G.S.); Tel.: +1-720-499-4934 (A.B.)

Received: 29 September 2018; Accepted: 18 October 2018; Published: 23 October 2018

Abstract: Participation of a wind turbine (WT) in primary frequency control (PFC) requires reserving some active power. The reserved power can be used to support the grid frequency. To maintain the required amount of reserve power, the WT is de-loaded to operate under its maximum power. The objective of this article is to design a control method for a WT system to maintain the reserved power of the WT, by controlling both pitch angle and rotor speed simultaneously in order to optimize the operation of the WT system. The pitch angle is obtained such that the stator current of the permanent magnet synchronous generator (PMSG) is reduced. Therefore, the resistive losses in the machine and the conduction losses of the converter are minimized. To avoid an excessive number of pitch motor operations, the wind forecast is implemented in order to predict consistent pitch angle valid for longer timeframe. Then, the selected pitch angle and the known curtailed power are used to find the optimal rotor speed by applying a nonlinear equation solver. To validate the proposed de-loading approach and control method, a detailed WT system is modeled in Matlab/Simulink (The Mathworks, Natick, MA, USA, 2017). Then, the proposed control scheme is validated using hardware-in-the-loop and real time simulation built in Opal-RT (10.4.14, Opal-RT Inc., Montreal, PQ, Canada).

Keywords: de-loading; droop curve; hardware-in-the-loop; reserve power; primary frequency control; optimal control; wind forecast

---

1. Introduction

In power systems with less conventional generators (mainly synchronous), several functions must be added to the renewable resources to compensate for the grid requirements, such as frequency response, negative sequence, harmonics mitigation, and resynchronization and black-start ability [1].

For adjusting the frequency of a grid during frequency deviation, there are different control strategies and approaches that have been implemented and presented in the literature. These control methods can be categorized based on duration and capability with certain features and limitations [2–4].

The participation of wind turbines (WTs) in primary frequency control (PFC) has been discussed widely in several publications. Controllers based on rotor speed regulation or pitch angle regulation to de-load WT for PFC have been discussed in [5–11].

In [12,13], a coordination control was proposed between inertial, rotor speed, and pitch angle for WT, based on a doubly-fed induction generator (DFIG). This approach depends on a new classification of wind speed to select among the proposed controllers. Storing kinetic energy in the rotor shaft to be used during the PFC stage has been proposed [14,15]. An adjustable droop controller was proposed to improve PFC in WT, based on a DFIG in [7].

A design for a discrete linear quadratic Gaussian controller for PFC participation was introduced in [16,17]. Controlling WT for PFC along with energy storage systems was discussed in [18,19].

An optimization algorithm was proposed to provide a stable power reference considering wind speed prediction in [20]. This approach gives a possible range for the power reserve for a specific timeframe.

In this paper, the aim is to de-load the turbines' power by controlling both pitch angle and rotor speed simultaneously, in order to optimize the operation of the wind turbine. The blade's pitch angle is obtained such that the rotor speed is increased to reduce the power losses (copper losses) in the stator of the machine and the conduction losses of the converter. In addition, to avoid blade fluctuation and an excessive number of pitch operations, a wind forecast will be implemented in order to predict a consistent pitch angle that is valid for longer period. This can reduce the high wear on the pitch system by reducing the number of pitch usages over time [21]. Then, the selected pitch angle and the known curtailed power are used to find the optimal rotor speed by applying a nonlinear solver. With the variation of wind speed, the control algorithm updates the pitch angle set point adaptively. Therefore, the reference of the rotor speed must follow the variation in the pitch angle and the wind speed.

The paper is organized as follows. Section 2 presents the model of the wind turbine system and discusses the understanding of power curve for PFC. Section 3 presents the proposed de-loading approach. In Section 4, simulation and real time studies are performed to validate the proposed method. The conclusion is presented in Section 5.

2. Wind Turbine System

The system considered in this paper is direct-drive, based on PMSG. The WT system is decoupled from the power system using a back-to-back power converter. The model of the WT and PMSG used for this study are presented in [22–25]. Detailed discussion of the model will be elaborated on in the following subsections.

2.1. Understanding the Wind Power Curve for Primary Frequency Control

To understand the relationship between pitch angle and rotor speed with the response of the torque and stator current of the machine, the model of the wind turbine and PMSG should be presented first.

The wind turbine model is a function of two main variables, the pitch angle (β) and the rotor speed (ω_r), as given by [25]:

$$P_W = \frac{1}{2} C_P(\lambda, \beta) A_W \rho_{air} V_W^3 \tag{1}$$

where A_W is the swept area of the WT and the power coefficient $C_P(\lambda, \beta)$ is expressed as

$$C_P(\lambda, \beta) = c_1 \left(\frac{c_2}{\gamma} - c_3 \beta - c_4 \right) e^{\frac{-c_5}{\gamma}} + C_6 \lambda \tag{2}$$

The coefficients c_1 to c_6 were taken from [25] ($c_1 = 05176$, $c_2 = 116$, $c_3 = 0.4$, $c_4 = 5$, $c_5 = 21$, $c_6 = 0.0068$), and the tip speed ratio is defined as

$$\lambda = \frac{R_W \omega_r}{V_W} \tag{3}$$

Here, R_w represents the radius of the wind turbine, and V_W represents the wind speed.

From the curve characteristics of the mechanical power Equation (1) and the power coefficient of Equation (2), the maximum power is achieved at the zero pitch angle and optimum tip speed ratio for underrated wind speed conditions. Figure 1 shows the characteristics of the wind turbine power for different pitch angles and different rotor speeds for a specific wind turbine. As the pitch angle increases, the power produced by the WT decreases. This helps identify the operating point of the WT system for de-loading purposes. In addition, the rotor speed can be used to reach a specific production of power. To implement de-loading to a WT system, the power is curtailed by a certain amount, (ΔP), using either the pitch angle or rotor speed controller. In Figure 1, the power is curtailed to ($P_{curt.}$), providing a reserve power of ΔP.

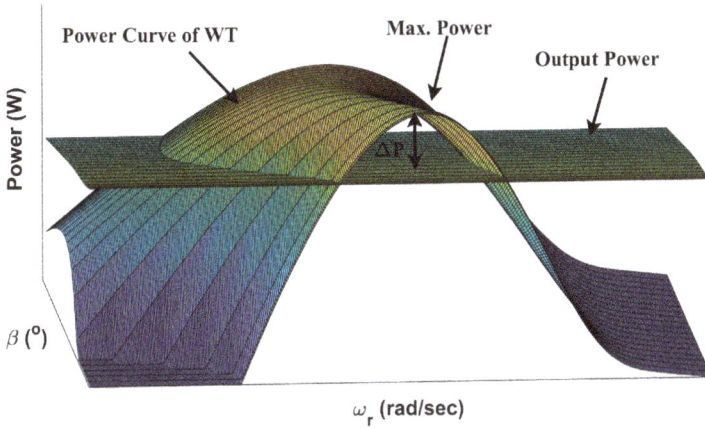

Figure 1. Power curve and output power versus rotor speed and pitch angle.

The power curve can be represented in terms of rotor speed for different pitch angles for several wind speeds. This will allow us to select an optimal operating point of rotor speed and pitch angle for the purpose of PFC. Figure 1 represents a three-dimensional plot of the power curve. During regular operation of WT, the maximum power is tracked. However, for PFC, the sub-optimal operating point is selected to reserve the required power in the WT system. The output power is plotted using a flat surface, where the intersections of the two surfaces demonstrate all possible sets of pitch angles and rotor speed values that can be used.

2.2. Machine Losses and Rotor Speed

The machine considered in this paper is based on PMSG. The detailed model of the machine is presented in [23]. The equation of the electromechanical torque is given as

$$T_e = \frac{3P}{4} \left\{ \psi_m + (L_d - L_q) I_d \right\} I_q \tag{4}$$

Assuming a round rotor type (i.e., $L_q = L_d$), the quadrature current can be given as

$$I_q = \frac{4 \, T_e}{3 \, P \, \psi_m} \tag{5}$$

where P is the number of poles of the machine and ψ_m is the flux linkage. L_q and L_d represent the quadrature and direct inductance, respectively, and I_q and I_d represent the quadrature and direct currents of the d–q reference frame of the PMSG.

At constant power production, a higher rotor speed means lower torque. From Equation (5), it is clear that I_q is proportional to the torque (i.e., lower torque results in lower current). This contributes to the resistive losses of the PMSG. The resistive (copper) losses in the stator equivalent circuit of the PMSG can be expressed as

$$P_{R,losses} = \frac{3}{2} \, R_s \left(I_q^2 + I_d^2 \right) \tag{6}$$

where R_S represents the stator resistance of the PMSG. The copper losses are directly related to the current of the quadrature axis and the direct axis. Thus, by controlling the quadrature current of the machine ($I_d = 0$), the resistive losses can be reduced by involving pitch angle control for PFC.

2.3. Conduction Losses of Insulated Gate Bipolar Transistor Switches

The power converter used is based on insulated gate bipolar transistor (IGBT) switches. Power losses of such switches are conduction losses and switching losses. The conduction losses occur when the switch is turned on and conducts current.

To evaluate the conduction losses of the IGBT, a voltage drop and a small resistor are added into the series with an ideal switch. The simple expression of the conduction losses is given as

$$P_{cond.} = V_{on}I_s + R_{on}I_s^2 \tag{7}$$

where I_s is the current crossing the IGBT switch during conducting, and V_{on} and R_{on} represent the voltage drop and the resistance of the switch, respectively

The conduction losses of an IGBT depends on the temperature, which causes variation of the voltage drop across the switch. Also, the loss depends on the current passing through the switch. Therefore, decreasing the current through the switch will reduce the conduction losses. The conduction losses of the semiconductor devices can be evaluated using information from the data sheet from the manufacturer, considering the thermal model of the switch.

2.4. Wind Data and Speed Prediction

Using wind speed sensor provides a reference for the power tracking controller. With the availability of wind data, a wind forecast can be implemented to predict the wind speed. This allows us to estimate the behavior of the turbine and select the optimal operating point, ensuring control stability. Using a Light Detection and Ranging (LIDAR) system improves the accuracy of the measurements of wind speed. The National Renewable Energy Lab (NREL) investigated the use of LIDAR in feedforward control in WT systems [26].

Knowledge of future wind speed can be utilized in the algorithm to enhance wind turbine operation. In the literature, there are several approaches to estimate the average of the wind speed [27]. Here, we implement the auto-regression model (AR) to generate and predict the mean of wind speed as

$$y_{t+1} = \sum_{i=0}^{n} \Phi_i y_{t-i} + \epsilon_{t+1} \tag{8}$$

where y_t and y_{t-i} represent the current and previous data (wind speed), respectively, Φ_i is the autoregressive parameter of the model, and ϵ_t is a Gaussian noise. The predicted wind speed is used to predict the future set points that required to maintain the reserved power for next timeframe ($t + 1$), as will be discussed in the next section.

Maximizing power using a speed controller is preferable during underrated wind speeds. However, in some cases the pitch controller can be involved to ensure smooth power production near rated wind speeds. In addition, combining a pitch angle controller with a speed controller can improve the efficiency of the machine, as will be discussed in the following section.

3. Proposed De-Loading Approach

The WT system considered in this paper is a direct drive configuration, where the power converter is directly connected to the machine's stator winding. In such a system, the rotor speed is fully controlled by the converter, while the blades are controlled by a pitch angle control system.

The reserve power for PFC is achieved by controlling both pitch angle and rotor speed simultaneously, in order to optimize the operation of the WT system. The goal is to minimize the power losses in the PMSG given in (6) and (7). This is accomplished by involving a pitch angle controller during underrated wind speeds.

Initially, with the prediction of wind speed, the pitch angle is obtained such that the rotor speed is increased to reduce the power losses in the stator of the PMSG. Predicted wind speed is averaged to avoid blade fluctuation and an excessive number of pitch motor operations. Therefore, averaged pitch angle is selected to ensure consistent operation of the WT, while minimizing resistive losses of the machine.

First, the measured wind speed is averaged and used to calculate the maximum power that can be achieved by the turbine. The wind forecast is needed to estimate the maximum possible power in advance $(t + k)$ to help reduce the number of pitch operations. For a given specific reserve power, the output power is obtained and used to determine the required pitch angle instantaneously.

The optimum operating point of pitch angle and rotor speed is determined using the steps as follows:

1. Calculate the current and future maximum available power as

$$P_{MPPT} = \frac{1}{2} C_P (\lambda_{opt.}, 0) A_W \rho_{air} V_W^3 \tag{9}$$

2. Calculate the output power as

$$P_{out} = P_{MPPT} - \Delta P + P_{PFC} \tag{10}$$

where P_{PFC} is the output power of the droop curve and ΔP is the reserved power.
3. Set $\beta = 0$ and solve the nonlinear power equation for ω_r as:

$$F(\omega_r) = 0 \tag{11}$$

$$F(\omega_r) = \frac{1}{2} A_W \rho_{air} V_W^3 \left\{ c_1 \left(\frac{c_2}{\gamma} - c_3 \beta - c_4 \right) e^{\frac{-c_5}{\gamma}} + C_6 \lambda \right\} - P_{out} \tag{12}$$

$$\frac{1}{\gamma} = \frac{1}{\lambda + 0.08 \times \beta} - \frac{0.035}{1 + \beta^3} \tag{13}$$

$$\lambda = \frac{R_W \omega_r}{V_W} \tag{14}$$

4. Increment β and repeat Step 3 until β_{max}.
5. Select an optimal solution such that:

$$\min_{\omega_r, \beta} (P_{R,losses} + P_{cond.})$$

All solutions found by the solver in Step 3 provide the same output power as demonstrated in Figure 2. However, the optimal solution, $S^* = \{\beta^*, \omega_r^*\}$, is selected such that pitch angle is as maximum as possible. As a result, the required rotor speed of the WT is increased and obtained by the nonlinear solver in Step 3. This will minimize the resistive power in the machine as in Equation (6) and the conduction losses in Equation (7). The algorithm must consider operation limits, such as

$$\begin{cases} \omega_{r,min} \leq \omega_r \leq \omega_{r,MPPT} \\ 0 \leq \beta \leq \beta_{max} \end{cases} \tag{15}$$

Figure 3 summarizes the algorithm used to generate the optimum operating point of pitch angle and rotor speed. Note that the output of the droop curve can be involved in the equation, in order to define the rotor speed for the amount of power necessary to be injected during a frequency drop. Otherwise, the PFC power is considered to be zero.

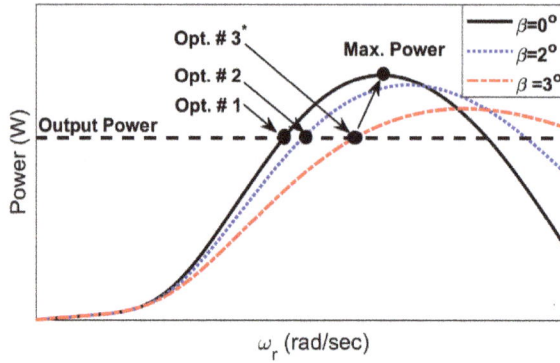

Figure 2. Demonstrating all possible operating points.

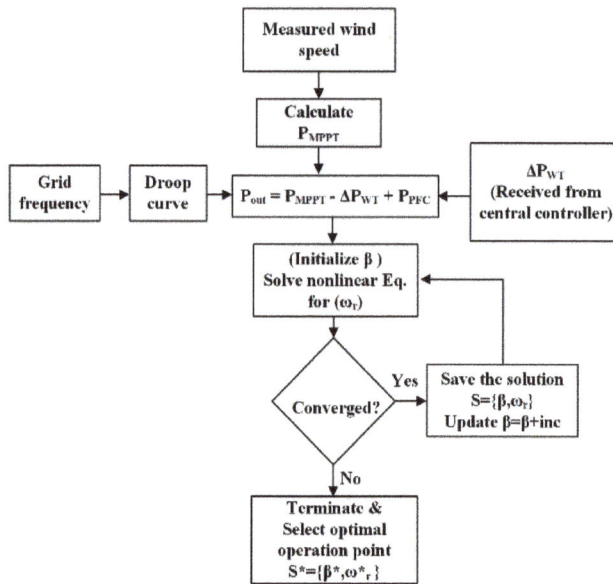

Figure 3. Proposed de-loading approach.

The proposed algorithm with the control loop is shown in Figure 4. The pitch controller is based on a conventional method, where the pitch angle is controlled considering the pitch servo transfer function [28]. The rotor speed is controlled using a designed proportional integral (PI) controller with an inner loop to control the current.

The controller must ensure a smooth and reliable response to frequency drop. Therefore, the controller is required to adjust the frequency variation by considering the output of a droop curve.

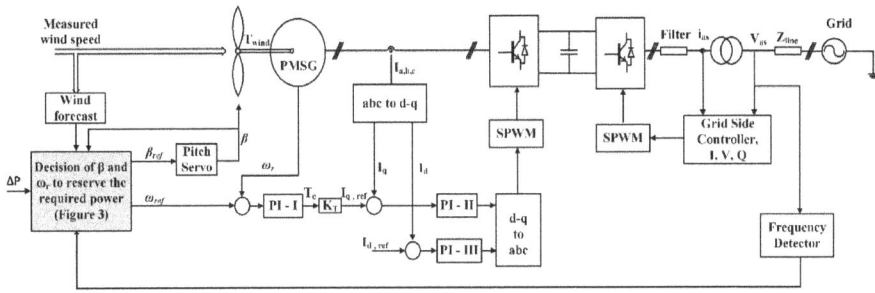

Figure 4. Wind turbine system with a control loop, including the proposed de-loading method.

4. Simulation and Hardware-in-the-Loop Results

The parameters of the system considered in this study including WT, PMSG are presented in Table 1. The wind turbine has a nominal rotor speed of 25.88 rpm at a 12 m/s rated wind speed. The rated power of the machine and WT is 2 MW.

Table 1. Wind turbine and permanent magnet synchronous generator (PMSG) Parameters.

Wind Turbine	Values
Nominal wind speed	12 m/s
Nominal output power	2.0 MW
Air density	1.225 kg/m^3
Wind turbine radius	35.5 m
Nominal rotor speed	25.88 rpm
Stator resistance of PMSG	0.821 mΩ
Armature inductance of PMSG	1.5731 mH
Flux linkage of PMSG	7.8264 V.s
Machine side switching frequency, $f_{M,s}$	1.5 kHz
DC link voltage reference, V_{DC}	1450 V

4.1. Simulation Studies

For simulation study, different scenarios were implemented to achieve de-loading. The first scenario was based on varying the rotor speed only, while keeping pitch angle constant. The second scenario was performed using the proposed method, but allowing pitch angle variation without wind speed forecasting. In the last scenario, the proposed de-loading method was implemented to obtain an optimal pitch angle that is valid for a longer time frame, and updating the rotor speed instantaneously.

The simulation experiment was repeated again to perform de-loading for PFC purposes. In this scenario, the pitch angle was kept constant (i.e., 0 degrees), while the de-loading was performed using a rotor speed controller. The maximum power point tracking (MPPT) was modified to track the output power by subtracting the maximum power by the required power to be reserved. Then, the reference rotor speed was calculated to provide the required reserve power.

The proposed algorithm was used to de-load the WT, in order to maintain an adequate amount of power for PFC use. In this scenario, the wind profile was assumed to be predicted by wind forecast. The averaged estimated wind speed was used to find the optimum pitch angle for a longer period. Then, the current and future wind speed is used to estimate the optimal possible pitch angle for that time frame. The rotor speed was extracted using the nonlinear relation, using Equation (10).

4.1.1. Long-Term (One Day) Simulation

First, a simulation of one day of wind profile was performed. In this study, simplified model of the WT and PMSG were used to expedite the simulation. The wind profile used in this study is shown in Figure 5. Figure 6 shows the maximum power available and the de-loaded power. The dotted line represents the WT power achieved by rotor speed only. The WT power obtained by coordinating between rotor speed and pitch angle is presented by the dashed line. It is clear that both approaches provide the same amount of reserve power (ΔP). In this study, the reserve is fixed at 0.3 MW for all wind profiles.

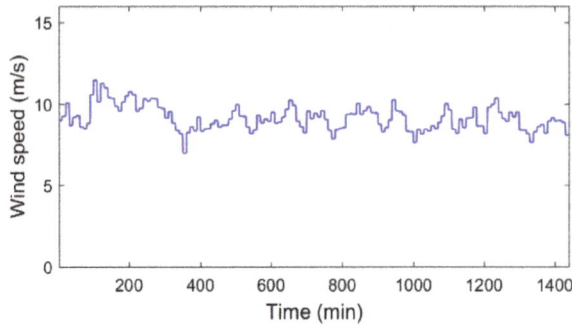

Figure 5. Wind profile (long-term simulation).

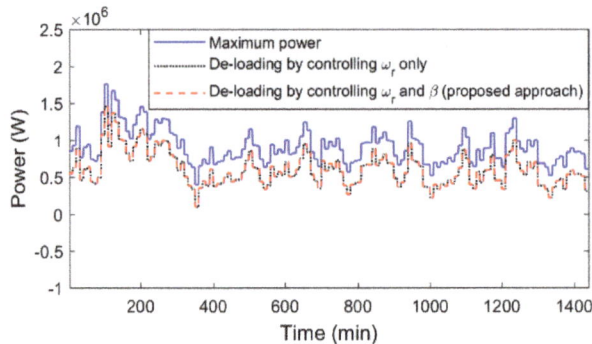

Figure 6. Mechanical power for the maximum power point tracking (MPPT) and de-loaded scenarios.

Although every method gives the same reserve, the rotor speed is different, as shown in Figure 7. The rotor speed is increased by involving the pitch angle. This is because of the higher pitch angle that results in lower torque of the WT. Therefore, to maintain the same reserve, the rotor speed is increased. As a result, the quadrature current of the PMSG (I_q) is reduced, as shown in Figure 8. In this simulation, we controlled the current of the direct axis to be zero. Therefore, the only current that appears in the losses is the current of the q-axis. From Figure 8, it is clear that with a 0° pitch angle, the current is higher for whole period considered in this simulation. Current produced by the proposed approach reduces the power losses in the stator of the machine and the conduction losses of the switch, as given in (6) and (7). In Figure 9, the pitch angle is represented using the dashed line.

The proposed strategy has the ability to utilize a wind speed forecast and predict the optimal solution. In this case, one can combine the optimal operating point of the current and future state in order to minimize the operation of the pitch angle. For the same wind profile, one model included the wind forecast and the other model did not consider the wind prediction in the decision. The pitch angle changed 82 times for the model without a forecast. In contrast, in the case where there was a wind forecast, the pitch angle changed 64 times, as shown in Figure 10. In both cases, the power produced by the WT was the same.

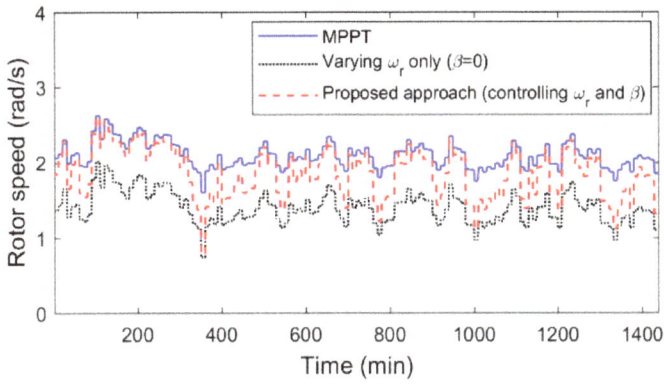

Figure 7. Rotor speed of a wind turbine (WT).

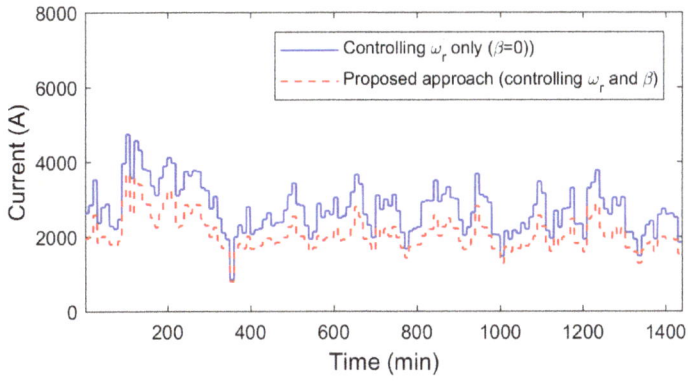

Figure 8. Stator current (I_q).

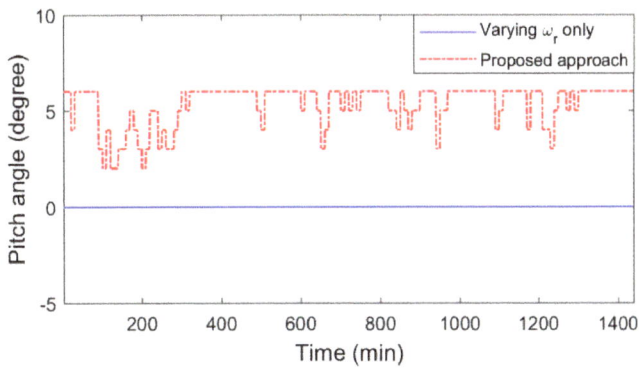

Figure 9. Pitch angle (β).

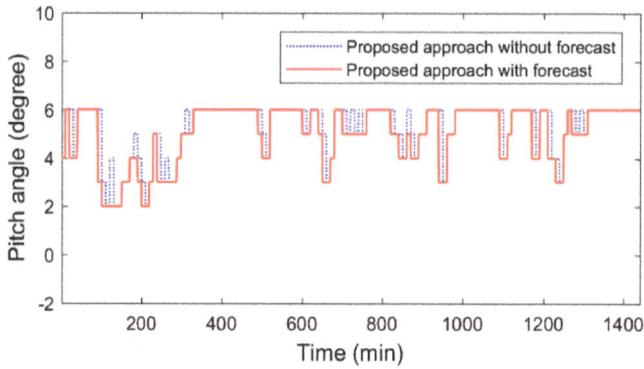

Figure 10. The pitch angle with a wind forecast and without a wind forecast.

4.1.2. Short-Term Simulation

To test the wind system with the designed controller, a study was performed to validate the proposed method, using a detailed model of the whole system. The proposed WT system and control scheme shown in Figure 4 was modeled in Matlab/Simulink.

The aim of this short-term simulation was to validate the performance of the proposed algorithm, and examine its accuracy and response to wind variation. Therefore, the approach should give the reference for the pitch angle and rotor speed as quickly as possible.

In this simulation study, the wind speed was scaled to be in seconds instead of minutes. The wind turbine system was tested with the wind profile shown in Figure 11. Like the one-day simulation, in order to demonstrate the performance of the proposed de-loading algorithm, two different approaches were performed, with different methods of de-loading. First, the rotor speed only was used as a tool to maintain the reserved power. Then, the proposed de-loading approach was implemented using the same wind profile for the purpose of comparison.

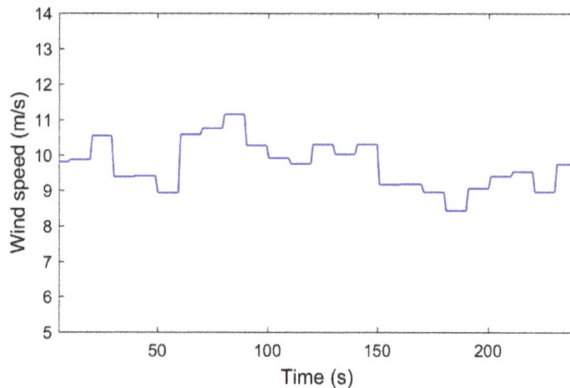

Figure 11. Wind profile (short-term simulation).

Initially, the MPPT was implemented to be used as a baseline. Then two scenarios discussed in the previous section were examined. Figure 12 shows the maximum and de-loaded power for the two methods. The rotor speed for the different methods are demonstrated in Figure 13. Both approaches provide the same amount of reserve power (ΔP). As a one-day simulation, the reserve was fixed at

0.3 MW for all wind speeds. For the rotor speed method, the rotor speed was reduced compared to the other method. This low speed results in higher stator current, as shown in Figure 14. The participation of the pitch angle controller is demonstrated in Figure 15.

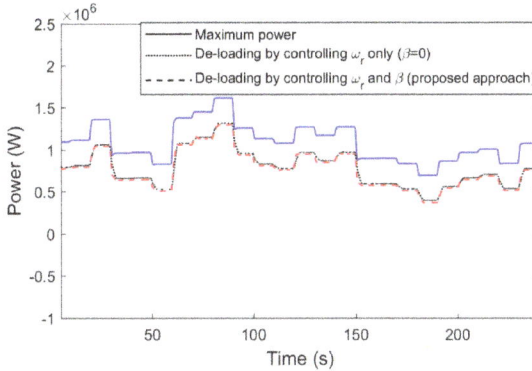

Figure 12. Mechanical power for the MPPT and de-loaded scenarios (short-term simulation).

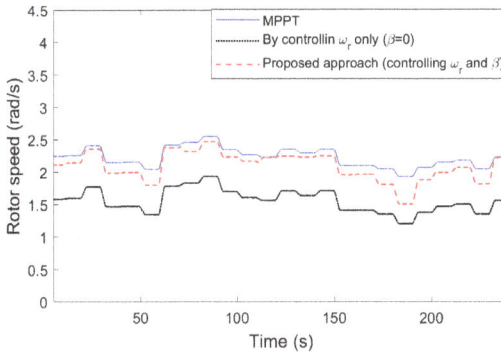

Figure 13. Rotor speed (short-term simulation).

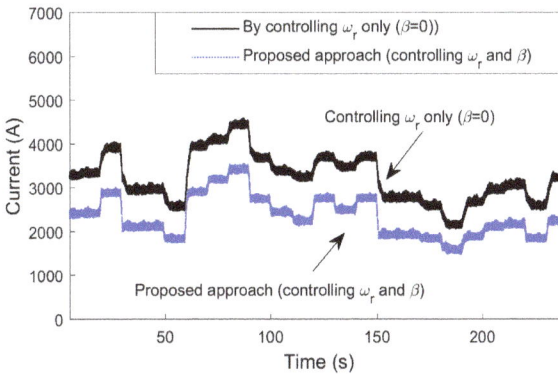

Figure 14. Stator current, I_q (short-term simulation).

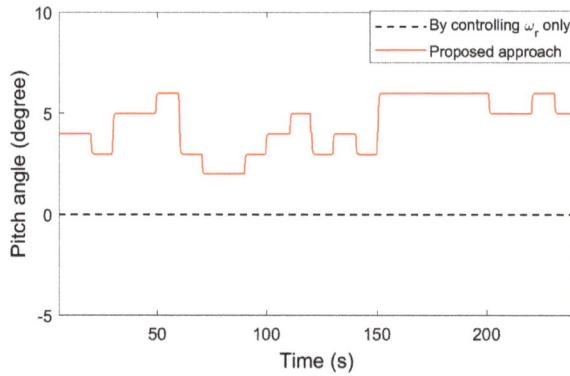

Figure 15. Pitch angle (short-term simulation).

The high current on the stator causes an increase in the resistive losses and conduction losses, as demonstrated in Table 2. The table presents the averaged resistive losses and conduction losses for the whole period (240 s). The resistive losses are calculated using Equation (6), where the d-axis current is controlled to be fixed at 0 A. The IGBT characteristics were taken from the ASEA Brown Boveri (ABB) data sheet that matches the rating of the WT system, (5SNA 3600E170300 HiPak IGBT module). The conduction losses were calculated using web-based software (SEMIS) developed by Plexim (Zürich, Switzerland) for ABB data-sheet characteristics, in order to consider the detailed thermal module of the converter.

Table 2. Power losses for different scenarios.

Losses	MPPT	Controlling ω_r Only	Proposed Approach
Resistive losses (kW)	12.577	13.238	7.063
Conduction losses (kW)	12.77	12.87	8.13

4.2. Real-Time and Hardware-in-Loop Results

To validate the response and the performance of the proposed approach with the designed controllers, hardware-in-the-loop (HIL) was implemented, as shown in Figure 16. The HIL study provides a real-time benchmark to test the proposed method and the designed controllers. The WT system is modeled and compiled inside an Opal-RT real time simulator. Then, C-code was generated for the controller of the machine side converter, and was implemented inside a Texas Instruments (TI) Digital signal processor (DSP), (TI TMS230F28335).

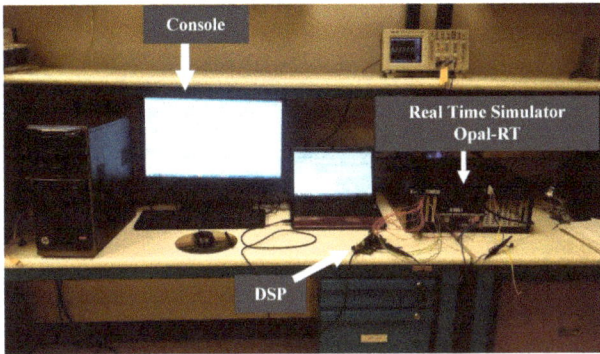

Figure 16. Hardware-in-the-loop implementation.

The controller (DSP) receives the rotor speed and the machine's current as an analog signal (fixed-sample) from Opal-RT. Then, the controller processes the signals and generates pulse width modulation (PWM) pulses to control the switches of the converter. The sampling frequency of the DSP is selected to be 9 kHz. The s-domain PI controllers were discretized using the backward Euler conversion method. The inner loop (current) controller is ensured to be faster than the outer loop (speed) controller.

In this study, the controller and proposed method was tested with the wind speed profile shown in Figure 17. Like the simulation studies, two different approaches for de-loading were implemented, using the same wind profile for comparison. A baseline of MPPT was tested first, and the maximum power was achieved. Then, the de-loading was implemented using the proposed method summarized in Figure 3, using the optimal pitch angle and rotor speed set. Then, de-loading was achieved using rotor speed only. A fixed reserve power of 0.3 MW was set for both approaches.

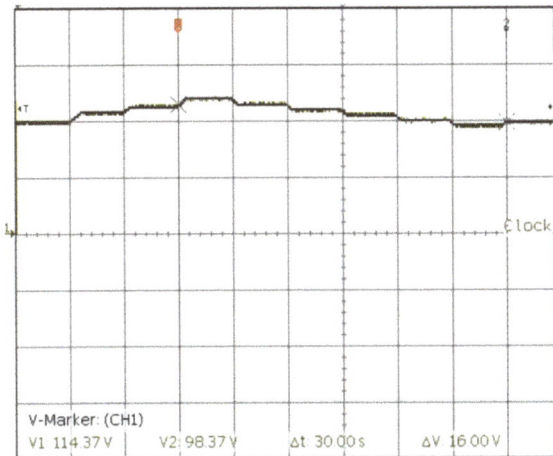

Figure 17. Wind profile (hardware-in-the-loop).

For consistency, the same wind speed and time frame were used for all scenarios. First, the controller received the reference to perform the MPPT. Then, the same controller was fed with a rotor reference to de-load the WT system. Finally, the reference generated by the proposed method was used to obtain the reference set points (rotor speed and pitch angle).

The results obtained from real time simulator and HIL are demonstrated in Figures 18–20. For the proposed experimental method, the three-phase current for machine is shown in Figure 18. The rotor speed (measured and reference) and *q*-axis current for all cases are shown in Figure 19. It is clear that the proposed method reduces the q-axis current. As a result, the power loss is decreased. The rotor speed in the plot is in (rad/s), and the gain is 1. For instant, the rotor speed at the first 5-s time slot was 2.3 (rad/s) for the MPPT case. The rotor speed that resulted from the proposed method is 2.2 (rad/s), whereas, the rotor speed is 1.6 (rad/s) when the pitch angle is kept at 0.

The gain for the q-axis current is 2000. Therefore, the current at the MPPT case is 3400 A for the first 5-s time slot. For the proposed and fixed pitch angle methods, the currents are 2600 A and 3600 A, respectively. Figure 20 shows the power generated from the WT system. Both approaches provide the required reserve power, as can be seen from the plot. The proposed method involves a pitch angle controller, as shown in Figure 20b.

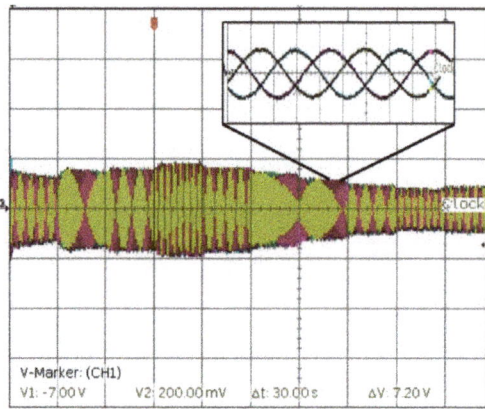

Figure 18. Three-phase machine current.

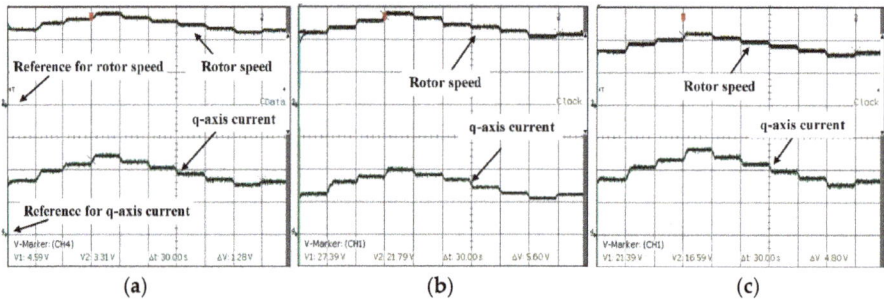

Figure 19. Rotor speed and q-axis current (I_q gain is 2000): (**a**) maximum power, (**b**) de-loading using proposed method, and (**c**) de-loading by controlling rotor speed only.

Appl. Sci. **2018**, *8*, 2022

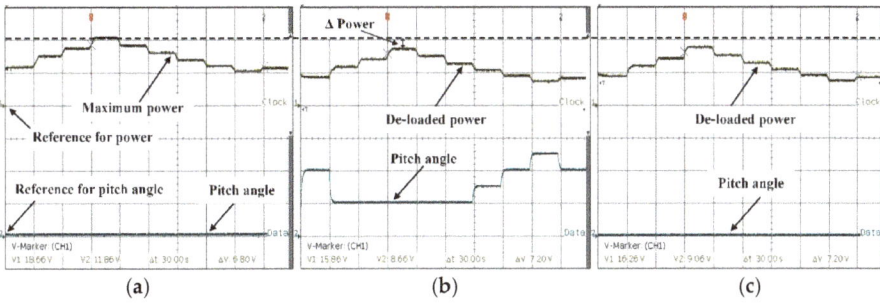

Figure 20. Active power (gain is 1×10^6) and pitch angle (gain is 2): (**a**) maximum power, (**b**) de-loading using proposed method, and (**c**) de-loading by controlling rotor speed only.

5. Conclusions

This paper is aimed to design an optimal control method for a WT system intended to provide PFC to support power system stability. The proposed method helps maintain reserve power to be used to support the stability of the power system, while reducing power losses in the WT system. The reserved power of the WT is achieved by controlling both pitch angle and rotor speed simultaneously. The main contribution of the proposed method is to reduce the current of the PMSG, and as a result, the copper losses and conduction losses are reduced. Comprehensive case studies were implemented in simulation, as well as HIL with a real-time simulator. The results showed good performance of the proposed algorithm. The method is able to maintain the required reserve power for different wind speeds. The copper losses were calculated and compared with conventional method to de-load the power of WT. In addition, the conduction losses were evaluated using web-based software developed for ABB data sheet characteristics. The copper and conduction losses were reduced significantly compared to the conventional rotor speed control method. De-loading of a WT system under its maximum capacity means that some power is not being used. Therefore, implementing this optimal strategy to a large wind farm can improve the overall power generation by reducing a significant amount of power loss.

Author Contributions: The authorship is equally shared, where M.G.S. served as the doctoral adviser for A.B. in his Ph.D. program at Colorado School of Mines.

Funding: This research received no external funding.

Conflicts of Interest: The authors declare no conflict of interest.

References

1. Muljadi, E.; Gevorgian, V.; Singh, M.; Santoso, S. Understanding inertial and frequency response of wind power plants. In Proceedings of the 2012 IEEE Power Electronics and Machines in Wind Applications, Denver, CO, USA, 16–18 July 2012; pp. 1–8.
2. Satpathy, G.; Mehta, A.K.; Kumar, R.; Baredar, P. An overview of various frequency regulation strategies of grid connected and stand-alone wind energy conversion system. In Proceedings of the International Conference on Recent Advances and Innovations in Engineering, ICRAIE, Jaipur, India, 9–11 May 2014.
3. Motamed, B.; Chen, P.; Persson, M. Comparison of primary frequency support methods for wind turbines. In Proceedings of the 2013 IEEE Grenoble Conference PowerTech, Grenoble, France, 16–20 June 2013.
4. Wang, Y.; Bayem, H.; Giralt-devant, M.; Silva, V.; Guillaud, X.; Francois, B. Methods for Assessing Available Wind Primary Power Reserve. *IEEE Trans. Sustain. Energy* **2015**, *6*, 272–280. [CrossRef]
5. Singh, M.; Gevorgian, V.; Muljadi, E.; Ela, E. Variable-speed wind power plant operating with reserve power capability. In Proceedings of the 2013 IEEE Energy Conversion Congress and Exposition, Denver, CO, USA, 15–19 September 2013; pp. 3305–3310.

6. Margaris, I.D.; Papathanassiou, S.A.; Hatziargyriou, N.D.; Hansen, A.D.; Sorensen, P. Frequency Control in Autonomous Power Systems with High Wind Power Penetration. *IEEE Trans. Sustain. Energy* **2012**, *3*, 189–199. [CrossRef]
7. Vidyanandan, K.V.; Senroy, N. Primary Frequency Regulation by Deloaded Wind Turbines Using Variable Droop. *IEEE Trans. Power Syst.* **2013**, *28*, 837–846. [CrossRef]
8. Pradhan, C.; Bhende, C. Enhancement in Primary Frequency Contribution using Dynamic Deloading of Wind Turbines. *IFAC-PapersOnLine* **2015**, *48*, 13–18. [CrossRef]
9. El Mokadem, M.; Courtecuisse, V.; Saudemont, C.; Robyns, B.; Deuse, J. Experimental study of variable speed wind generator contribution to primary frequency control. *Renew. Energy* **2009**, *34*, 833–844. [CrossRef]
10. Singarao, V.Y.; Rao, V.S. Frequency responsive services by wind generation resources in United States. *Renew. Sustain. Energy Rev.* **2016**, *55*, 1097–1108. [CrossRef]
11. Moutis, P.; Papathanassiou, S.A.; Hatziargyriou, N.D. Improved load-frequency control contribution of variable speed variable pitch wind generators. *Renew. Energy* **2012**, *48*, 514–523. [CrossRef]
12. Zhang, Z.-S.; Sun, Y.-Z.; Lin, J.; Li, G.-J. Coordinated frequency regulation by doubly fed induction generator-based wind power plants. *IET Renew. Power Gener.* **2012**, *6*, 38. [CrossRef]
13. Wu, Z.; Gao, W.; Wang, J.; Gu, S. A coordinated primary frequency regulation from Permanent Magnet Synchronous Wind Turbine Generation. In Proceedings of the 2012 IEEE Power Electronics and Machines in Wind Applications, Denver, CO, USA, 16–18 July 2012; pp. 1–6.
14. Žertek, A.; Member, S.; Verbi, G.; Member, S.; Pantoš, M. A Novel Strategy for Variable-Speed Wind Turbines ' Participation in Primary Frequency Control. *IEEE Trans. Sustain. Energy* **2012**, *3*, 791–799. [CrossRef]
15. Erlich, I.; Wilch, M. Primary frequency control by wind turbines. In Proceedings of the IEEE PES General Meeting, Providence, RI, USA, 25–29 July 2010; pp. 1–8.
16. Camblong, H.; Nourdine, S.; Vechiu, I.; Tapia, G. Control of wind turbines for fatigue loads reduction and contribution to the grid primary frequency regulation. *Energy* **2012**, *48*, 284–291. [CrossRef]
17. Camblong, H.; Vechiu, I.; Etxeberria, A.; Martinez, M.I. Wind turbine mechanical stresses reduction and contribution to frequency regulation. *Control Eng. Pract.* **2014**, *30*, 140–149. [CrossRef]
18. Diaz-Gonzalez, F.; Hau, M.; Sumper, A.; Gomis-Bellmunt, O. Coordinated operation of wind turbines and flywheel storage for primary frequency control support. *Int. J. Electr. Power Energy Syst.* **2015**, *68*, 313–326. [CrossRef]
19. Miao, L.; Wen, J.; Xie, H.; Yue, C.; Lee, W.-J. Coordinated Control Strategy of Wind Turbine Generator and Energy Storage Equipment for Frequency Support. *IEEE Trans. Ind. Appl.* **2015**, *51*, 2732–2742. [CrossRef]
20. Van de Vyver, J.; de Kooning, J.D.M.; Meersman, B.; Vandoorn, T.L.; Vandevelde, L. Optimization of constant power control of wind turbines to provide power reserves. In Proceedings of the 2013 48th International Universities' Power Engineering Conference (UPEC), Dublin, Ireland, 2–5 September 2013.
21. Fleming, P.A.; Aho, J.; Buckspan, A.; Ela, E.; Zhang, Y.; Gevorgian, V.; Scholbrock, A.; Pao, L.; Damiani, R. Effects of power reserve control on wind turbine structural loading. *Wind Energy* **2016**, *19*, 453–469. [CrossRef]
22. Lubosny, Z. *Wind Turbine Operation in Electric Power Systems*; Springer: New York, NY, USA, 2003.
23. Krishnan, R. *DC Motor Drives Permanent Magnet Synchronous and Brushless DC Motor Drives*; Taylor & Francis Group: Thames, UK, 2010.
24. Bubshait, A.S.; Simões, M.G.; Mortezaei, A.; Busarello, T.D.C. Power quality achievement using grid connected converter of wind turbine system. In Proceedings of the 2015 IEEE Industry Applications Society Annual Meeting, Addison, TX, USA, 18–22 October 2015; pp. 1–8.
25. Heier, S. *Grid Integration of Wind Energy Conversion Systems*; John Wiley & Sons, Ltd.: Hoboken, NJ, USA, 1998.
26. Dunne, F.; Simley, E.; Pao, L.Y. *LIDAR Wind Speed Measurement Analysis and Feed-Forward Blade Pitch Control for Load Mitigation in Wind Turbines*; National Renewable Energy Lab. (NREL): Golden, CO, USA, 2011.

27. Lei, M.; Shiyan, L.; Chuanwen, J.; Hongling, L.; Yan, Z. A review on the forecasting of wind speed and generated power. *Renew. Sustain. Energy Rev.* **2009**, *13*, 915–920. [CrossRef]
28. Senjyu, T.; Sakamoto, R.; Kinjo, T.; Urasaki, N.; Funabashi, T.; Fujita, H.; Sekine, H. Output power leveling of wind turbine generator by pitch angle control using adaptive control method. *IEEE Trans. Energy Convers.* **2006**, *21*, 467–475. [CrossRef]

applied sciences

MDPI

Article

Stability Augmentation of a Grid-Connected Wind Farm by Fuzzy-Logic-Controlled DFIG-Based Wind Turbines

Md. Rifat Hazari [1,*], Mohammad Abdul Mannan [2], S. M. Muyeen [3], Atsushi Umemura [1], Rion Takahashi [1] and Junji Tamura [1]

[1] Department of Electrical and Electronic Engineering, Kitami Institute of Technology (KIT), 165 Koen-cho, Kitami, Hokkaido 090-8507, Japan; umemura@mail.kitami-it.ac.jp (A.U.); rtaka@mail.kitami-it.ac.jp (R.T.); tamuraj@mail.kitami-it.ac.jp (J.T.)
[2] Department of Electrical and Electronic Engineering, American International University-Bangladesh (AIUB), Ka-66/1, Kuratoli Road, Kuril, Khilkhet, Dhaka 1229, Bangladesh; mdmannan@aiub.edu
[3] Department of Electrical and Computer Engineering, Curtin University, Perth, WA 6845, Australia; sm.muyeen@curtin.edu.au
* Correspondence: rifat.hazari@gmail.com; Tel.: +81-157-26-9266

Received: 30 November 2017; Accepted: 19 December 2017; Published: 24 December 2017

Abstract: Wind farm (WF) grid codes require wind generators to have low voltage ride through (LVRT) capability, which means that normal power production should be resumed quickly once the nominal grid voltage has been recovered. However, WFs with fixed-speed wind turbines with squirrel cage induction generators (FSWT-SCIGs) have failed to fulfill the LVRT requirement, which has a significant impact on power system stability. On the other hand, variable-speed wind turbines with doubly fed induction generators (VSWT-DFIGs) have sufficient LVRT augmentation capability and can control the active and reactive power delivered to the grid. However, the DFIG is more expensive than the SCIG due to its AC/DC/AC converter. Therefore, the combined use of SCIGs and DFIGs in a WF could be an effective solution. The design of the rotor-side converter (RSC) controller is crucial because the RSC controller contributes to the system stability. The cascaded control strategy based on four conventional PI controllers is widely used to control the RSC of the DFIG, which can inject only a small amount of reactive power during fault conditions. Therefore, the conventional strategy can stabilize the lower rating of the SCIG. In the present paper, a new control strategy based on fuzzy logic is proposed in the RSC controller of the DFIG in order to enhance the LVRT capability of the SCIG in a WF. The proposed fuzzy logic controller (FLC) is used to control the reactive power delivered to the grid during fault conditions. Moreover, reactive power injection can be increased in the proposed control strategy. Extensive simulations executed in the PSCAD/EMTDC environment for both the proposed and conventional PI controllers of the RSC of the DFIG reveal that the proposed control strategy can stabilize the higher rating of the SCIG.

Keywords: squirrel cage induction generator (SCIG); doubly fed induction generator (DFIG); fuzzy logic controller (FLC); PI controller; low voltage ride through (LVRT); power system

1. Introduction

Emerging environmental concerns and attempts to curtail the dependence on fossil fuel resources are bringing renewable energy resources into the mainstream of the electric power sector. Among the various renewable resources, wind power is the most promising from both technical and economic standpoints. The new global total for wind power at the end of 2015 was 432.9 GW, which represents a cumulative market growth of more than 17% [1]. By 2030, wind power could reach 2110 GW and supply up to 20% of the global electricity [2]. This large penetration of wind power into the existing grid has introduced some vulnerabilities to the power grid. In order to maintain the stability of the power system and ensure smooth operation, low voltage ride through (LVRT) requirements have been imposed around the world [3]. In the event of a fault, LVRT mandates that wind farms (WFs) stay connected to the grid in order to support the grid in the same manner as conventional synchronous generators (SGs).

Most wind turbines are constructed using fixed-speed wind turbines with squirrel cage induction generators (FSWT-SCIGs). SCIGs have some advantageous characteristics, such as simplicity, robust construction, low cost, and operational simplicity [4]. However, FSWT-SCIGs are connected directly to the grid and have no LVRT capabilities during voltage dips [4]. Moreover, the FSWT-SCIG requires a large reactive power in order to recover air gap flux when a short circuit fault occurs in the power system. If sufficient reactive power is not supplied, the electromagnetic torque of the SCIG decreases significantly. As a result, the rotor speed of the SCIG increases significantly and can make the power system unstable [4]. Reactive power compensation is a major issue, especially for FSWT-SCIGs. A capacitor bank is usually used to meet the reactive power compensation requirement of an SCIG. However, the SCIG requires more reactive power during fault conditions than in the steady state, and the capacitor bank is not able to supply more reactive power during transient conditions.

A static synchronous compensator (STATCOM) [5], superconducting magnetic energy storage (SMES) [6], and an energy capacitor system (ECS) [7], for example, are installed in WFs with FSWT-SCIGs in order to improve the LVRT capability during a fault condition. However, the overall system cost increases.

On the other hand, variable-speed wind turbines with doubly fed induction generators (VSWT-DFIGs) have some advantageous characteristics, such as light weight, higher output power and efficiency, lower cost, variable-speed operation, and smaller size. In addition to the lower power electronic converter rating required by the DFIG, compared to permanent magnet synchronous generators (PMSGs) [8], the recent price upsurge of permanent magnet materials has given the DFIG another advantage over the PMSG [9]. In addition, the DFIG has better system stability characteristics than the SCIG during fault conditions, because of its capability for independent control of active and reactive power delivered to the grid [10]. By taking advantage of DFIG reactive power control, it is possible to stabilize the SCIG in a WF. Thus, reactive power compensation can be implemented at lower cost. The partial converter is connected to the rotor terminal of the DFIG via slip rings. The converter consists of a rotor-side converter (RSC) and a grid-side converter (GSC). As reported in previous studies [10–12], various control strategies can be adopted for both the RSC and the GSC. However, the design procedure of the RSC is very crucial because it is controlling active and reactive power delivered to the grid.

Some auxiliary hardware circuits have been used to help the DFIG to improve the LVRT requirement. For example, the rotor crowbar circuit is used in the rotor terminals to isolate the RSC from the rotor circuit [13,14]. However, the rotor crowbar circuit converts the DFIG to a simple induction machine, which absorbs reactive power from the grid. A chopper circuit and parallel capacitors are used to smooth the DC-link voltage by dissipating the excessive power in the DC-link circuit [15,16]. Dynamic braking resistors connected to the stator [17] and a bridge type fault current limiter [18] are used to limit the stator and rotor overcurrents. A series-connected converter [19] and a dynamic voltage restorer [20] are used to keep the stator voltage constant under grid faults. In previous studies [21,22], static VAR compensators or STATCOMs were used to supply extra reactive power to the grid during grid faults. Although the LVRT capability is enhanced through various types of equipment [13–22], this equipment requires additional converters or equipment, which increases the complexity and cost of the wind turbine system and decreases its reliability.

The cascaded control system for the RSC described in [23] is also used to improve LVRT capability, where several PI controllers are used in the inner and outer loops. However, due to changes in the parameters of the grid during fault conditions, the conventional PI controller with a fixed gain is not sufficient to ensure the system stability of a large power system. The setting of the parameters of the PI controllers used in cascaded control is cumbersome, especially in power system applications that are difficult to express as a mathematical model or a transfer function. In [23], a Taguchi approach for optimum design of PI controllers in a cascaded control scheme was presented. However, this cascaded control strategy with the conventional PI controller in the inner loop cannot provide a large amount of reactive power. Thus, the strategy can stabilize only lower ratings of the SCIG. Therefore, using an fuzzy logic controller (FLC) in the inner loop of the rotor-side controller to more efficiently provide reactive power during fault periods is convenient. The FLC can handle nonlinear systems very effectively because it offers variable gain during transient conditions. Thus, the DFIG controlled by the FLC can stabilize a larger amount of SCIG. Moreover, the overall system cost can be decreased by incorporating a lower rating of the DFIG along with a higher rating of the SCIG. This is one of the novel features of the present paper.

Therefore, the main contribution of the present paper is the design of a new control strategy based on fuzzy logic in the inner loop of the rotor-side controller for the DFIG to improve the LVRT capability and increase the capacity of the SCIG-based WF. Detailed modeling and control strategies of the overall system are presented. In order to evaluate the effectiveness of the proposed controller, transient and dynamic analyses are performed. Real wind-speed data measured on Rishiri Island, Hokkaido, Japan are considered in the dynamic analysis.

The transient performance of the overall system composed of SGs, an FLC-controlled DFIG, and an SCIG is compared with that composed of a DFIG with the conventional PI-controlled RSC presented in [23]. Finally, the proposed control strategy is found to be very effective for ensuring the stability of a large power system. Moreover, the capacity of the installed SCIG can be increased.

The remainder of the present paper is organized as follows. Section 2 presents the wind turbine model. Section 3 presents the DFIG model, and the design procedure of the proposed FLC is introduced in Section 4. Section 5 deals with the power system model. Section 6 briefly describes the LVRT requirements for wind power. The simulation results and a discussion of the performance of the proposed and conventional methods are presented in Section 7. Finally, Section 8 summarizes the findings and concludes the paper.

2. Wind Turbine Model

In the wind turbine model, the aerodynamic power output is given as follows [4]:

$$P_w = 0.5\rho\pi R^2 V_w{}^3 C_p(\lambda, \beta), \tag{1}$$

where P_w is the captured wind power, ρ is the air density (KG/m^3), R is the radius of the rotor blade (m), V_w is the wind speed (m/s), and C_p is the power coefficient.

The value of C_p can be calculated as follows [10]:

$$C_p(\lambda, \beta) = c_1\left(\frac{c_2}{\lambda_i} - c_3\beta - c_4\right)e^{\frac{-c_5}{\lambda_i}} + c_6\lambda \tag{2}$$

$$\frac{1}{\lambda_i} = \frac{1}{\lambda - 0.08\beta} - \frac{0.035}{\beta^3 + 1} \tag{3}$$

$$\lambda = \frac{w_r R}{V_w} \tag{4}$$

$$T_w = \frac{P_w}{w_r}, \tag{5}$$

where T_w is the wind turbine torque, β is the pitch angle, and λ is the tip speed ratio. Moreover, c_1 through c_6 are the characteristic coefficients of the wind turbine ($c_1 = 0.5176$, $c_2 = 116$, $c_3 = 0.4$, $c_4 = 5$, $c_5 = 21$, and $c_6 = 0.0068$) [24], and w_r is the rotational speed of the wind turbine (rad/s).

The C_p vs λ characteristics shown in Figure 1 are obtained using Equation (2) with different values of the pitch angle (β). When β is equal to zero degrees, the optimum power coefficient (C_{popt}) is 0.48, and the optimum tip speed ratio (λ_{opt}) is 8.1.

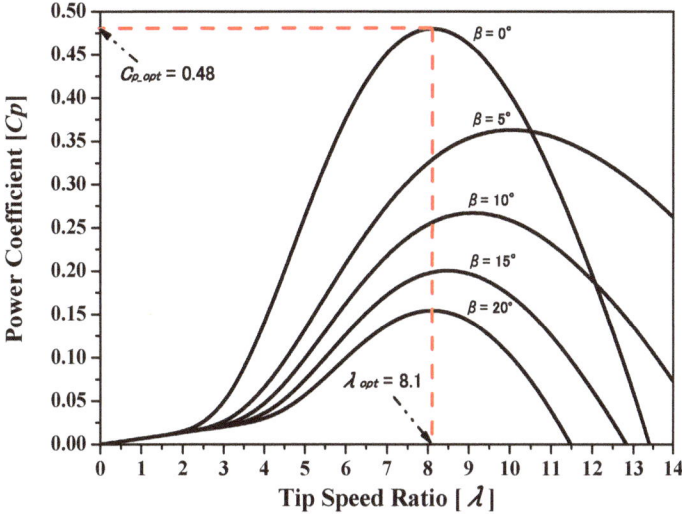

Figure 1. C_p vs λ characteristics of the wind turbine for various pitch angles.

Figures 2 and 3 show the models of the blade pitch control system for FSWT and VSWT [25], respectively. In FSWT, the pitch control system is used to control the power output of the SCIG so as not to exceed the rated power. In VSWT, the rotor speed of DFIG is regulate by the pitch controller so as not to exceed the rated speed. The control loop of the pitch actuator is represented by a first-order transfer function with a pitch rate limiter. A PI controller is used to manage the tracking error.

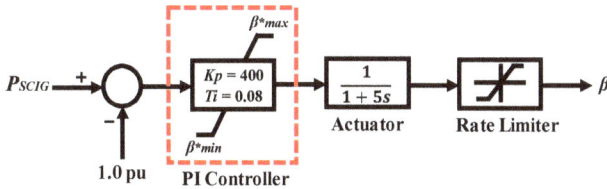

Figure 2. Pitch controller for fixed-speed wind turbine (FSWT).

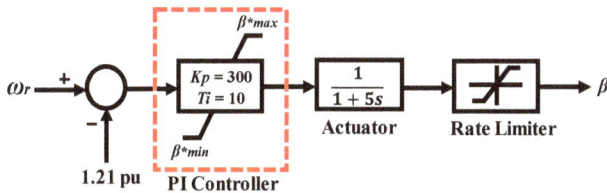

Figure 3. Pitch controller for variable-speed wind turbine (VSWT).

Figure 4 shows the maximum power point tracking (MPPT) curve for the VSWT-DFIG.

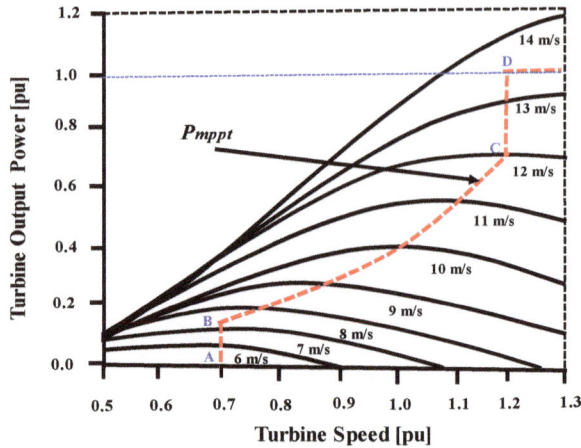

Figure 4. Wind turbine characteristics for the doubly fed induction generator (DFIG) with maximum power point tracking (MPPT).

3. DFIG Model

The configuration of the VSWT-DFIG system, along with its control system, is shown in Figure 5. The model consists of a wind turbine model with aerodynamic characteristics, a pitch controller, a wound rotor induction generator (WRIG), and an AC/DC/AC converter based on two levels of insulated gate bipolar transistors (IGBTs), which are controlled by the rotor-side controller and the grid-side controller, respectively.

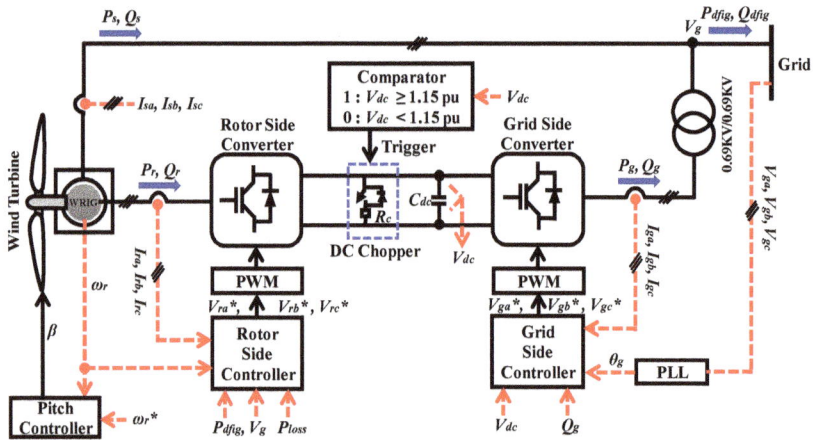

Figure 5. Configuration of the VSWT-DFIG system.

The wind turbine drives the WRIG to convert wind power into electrical power. The rotational speed (ω_r) is obtained from the rotor of the WRIG. A pitch controller is used to control the blade pitch angle of the wind turbine in order to reduce the output power when the rotational speed exceeds the rated speed. The WRIG model available in the PSCAD library is used in the present study [26]. The rotor position (θ_r) is derived from the rotor of the WRIG. As indicated by the configuration of the VSWT-DFIG system, the stator terminal is directly connected to the grid system. The AC/DC/AC converter is installed between the rotor of the WRIG and the grid system. The rating of the converter is 30% of the WRIG rating. The pulse width modulation (PWM) technique is used to generate the necessary gate pulses for driving the AC/DC/AC converter. The carrier frequency is taken as 3.0 kHz. The RSC is connected to the rotor winding of the WRIG, which provides variable frequency excitation depending on the wind-speed condition. The GSC is connected to the grid system through a transformer. A protection system with a DC chopper is installed in the DC-link circuit. The DC chopper is controlled by the comparator block, which triggers the DC chopper switch when the DC-link voltage becomes greater than or equal to the predefined limit ($V_{dc} \geq 1.15$ pu).

3.1. Conventional Rotor-Side Controller

The conventional cascaded controller for the RSC is presented in [23]. This controller consists of four conventional PI controllers to compensate different error signals. The reference reactive power ($Q_{dfig}{}^{*}$) is set to zero for unity power factor operation. The active power and reactive power delivered to the grid are controlled using q-axis and d-axis rotor currents, respectively.

3.2. Proposed Rotor-Side Controller

The proposed controller for the RSC is depicted in Figure 6. This controller consists of three PI controllers and one FLC. The main motivation behind using one FLC in the inner loop of the cascaded controller is maximization of the reactive power injection. The FLC offers variable gain depending on the system parameters. Due to the variable gain, the FLC can inject reactive power (Q_{dfig}) more effectively in the fault condition. Thus, the grid voltage can quickly be retraced back to the nominal value. Moreover, the FLC can stabilize a higher rating of the SCIG as compared to the conventional PI-based controller of the RSC in the inner loop.

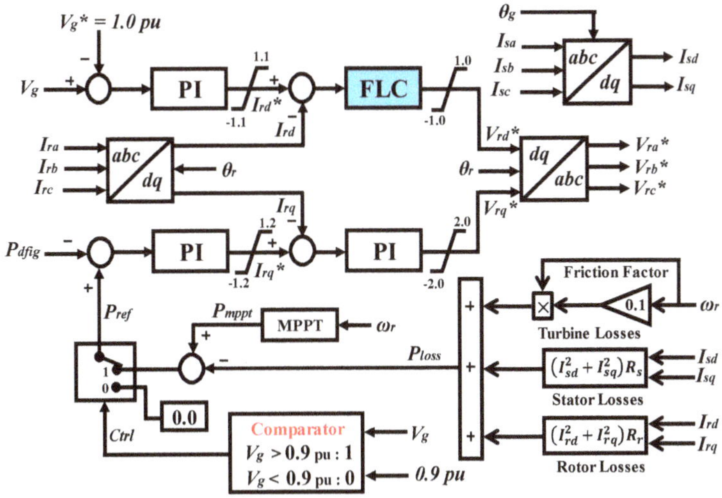

Figure 6. Proposed rotor-side controller.

The active power (P_{dfig}) and reactive power (Q_{dfig}) outputs of the DFIG are controlled by regulating the rotor winding current. The reference active power (P_{ref}) is calculated by subtracting the losses (P_{loss}) from the MPPT output (P_{mppt}). In the upper loop portion, the grid voltage (V_g) is taken as feedback to regulate the terminal voltage constant at 1.0 pu. The q-axis current (I_{rq}) controls the active power delivered to the grid, and the d-axis current (I_{rd}) controls the reactive power delivered to the grid.

In the normal operating condition ($V_g > 0.9$ pu), the RSC regulates the active power delivered to the grid. During a fault condition ($V_g < 0.9$ pu), a comparator sends a signal so that active power transfer to the grid becomes zero. By controlling the power in this manner, the reactive power injected to the grid can be maximized.

The detailed design procedure of the FLC will be discussed in Section 4.

3.3. Grid-Side Controller

The controller for the GSC is depicted in Figure 7. This controller consists of four PI controllers to compensate different error signals. The GSC reactive power (Q_g) and DC-link voltage (V_{dc}) are controlled through d-axis (I_{gd}) and q-axis (I_{gq}) current components, respectively. The reactive power reference is set to zero, and the DC-link voltage reference is set to 1.0 pu (1.2 kV).

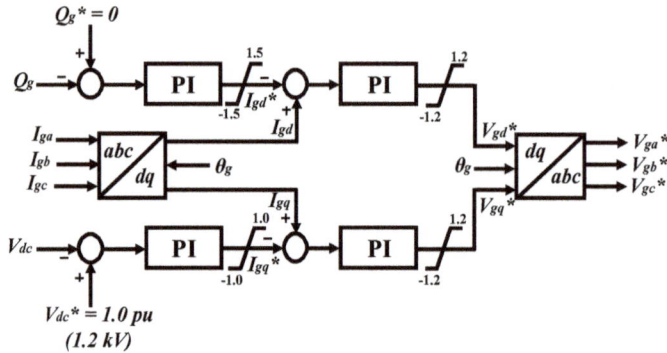

Figure 7. Grid-side controller.

4. Fuzzy Logic Controller Design

Figure 8 shows a block diagram of the proposed FLC. The FLC is composed of fuzzification, a membership function, a rule base, a fuzzy inference, and defuzzification, as shown in Figure 9.

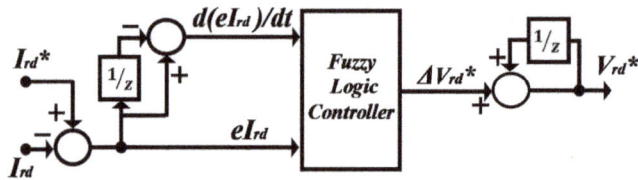

Figure 8. Proposed fuzzy logic controller (FLC).

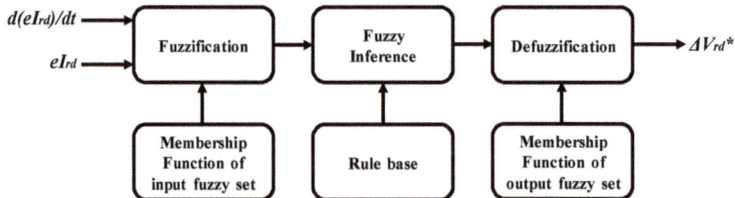

Figure 9. Internal structure of the FLC.

In order to design the proposed FLC, the error of the rotor d-axis current (eI_{rd}) and rate of change of the eI_{rd} ($d[eI_{rd}]/dt$) are considered as the controller inputs. The reference rotor d-axis voltage ($V_{rd}{}^*$) is chosen as the controller output. In Figure 8, $1/z$ is one sampling time delay.

The triangular membership functions with overlap used for the input and output fuzzy sets are shown in Figure 10, where linguistic variables are indicated as NB (Negative Big), NM (Negative Medium), NS (Negative Small), ZO (Zero), PS (Positive Small), PM (Positive Medium), and PB (Positive Big).

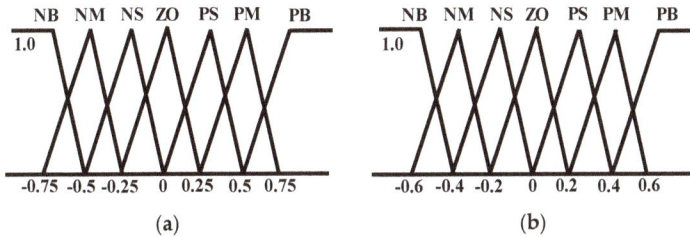

Figure 10. Membership functions for the FLC: (a) Inputs (eI_{rd}, $d[eI_{rd}]/dt$); (b) output ($V_{rd}{}^*$).

The rules of fuzzy mapping of the input variables to the output are given in the following form:

IF <eI_{rd} is PB> and <$d(eIrd)/dt$ is NS> THEN <$V_{rd}{}^*$ is PS>
IF <eI_{rd} is NM> and <$d(eIrd)/dt$ is NS> THEN <$V_{rd}{}^*$ is NB>

The entire rule base is listed in Table 1, which includes a total of 49 rules.

Table 1. Fuzzy rules.

	$V_{rd}{}^*$	$d(eI_{rd})/dt$						
		PB	PM	PS	ZO	NS	NM	NB
	Positive Big (PB)	PB	PB	PM	PM	PS	ZO	ZO
	Positive Medium (PM)	PB	PM	PM	PS	ZO	NS	PS
	Positive Small (PS)	PM	PM	PS	ZO	NS	NM	PS
eI_{rd}	Zero (ZO)	PM	PS	ZO	NS	NM	NM	PM
	Negative Small (NS)	PS	ZO	NS	NM	NM	NB	PM
	Negative Medium (NM)	ZO	NS	NM	NM	NB	NB	PB
	Negative Big (NB)	NS	NM	NM	NB	NB	NB	PB

In the present study, Mamdani's max-min method is used as the inference mechanism [27]. The center of gravity method is used for defuzzification in order to obtain $V_{rd}{}^*$ [28].

5. Power System Model

The power system model used for transient stability analysis is shown in Figure 11. The model is composed of a nine-bus main system [29] and a WF. The main system is composed of three conventional power plants: two thermal power plants (SG1 and SG2) and one hydropower plant (SG3). Both SG1 and SG3 are operated under automatic generation control (AGC), and SG2 is operated under governor-free (GF) control. The parameters of the SGs are listed in Table 2. The IEEE type AC4A excitation system model shown in Figure 12 is considered for all SGs [30]. Table 3 lists the parameters taken from [30]. Figure 13 shows a block diagram of the reheat steam turbine governor system used in the thermal power plants (SG1 and SG2) [30]. The hydro turbine governor model system used for the hydropower plant (SG3) is shown in Figure 14 [30]. The parameters of both turbine systems are presented in Table 4 [30]. For AGC operation, an integral controller is installed on the governor system for both SG1 and SG3.

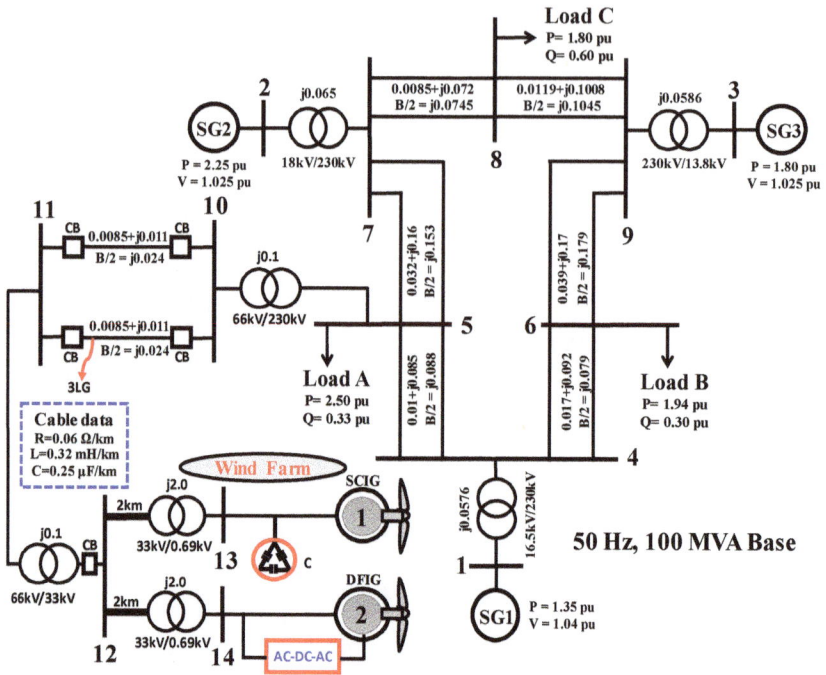

Figure 11. Power system model.

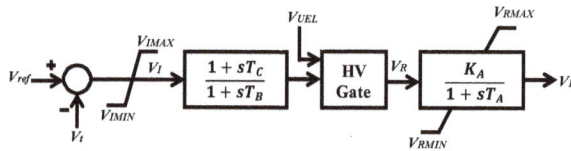

Figure 12. IEEE type AC4A excitation system model.

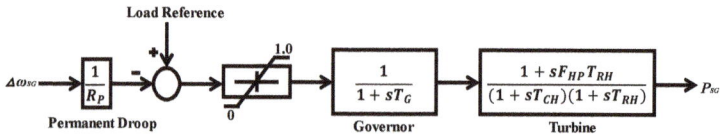

Figure 13. Steam turbine governor model.

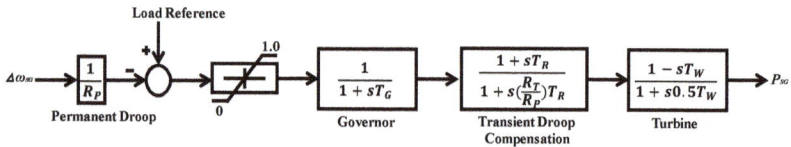

Figure 14. Hydro turbine governor model.

Table 2. Parameters of synchronous generators (SGs).

Parameter	SG1 (Thermal)	SG2 (Thermal)	SG3 (Hydro)
Rated Power	150 MVA	250 MVA	200 MVA
Voltage	16.5 kV	18 kV	13.8 kV
R_a	0.003 pu	0.003 pu	0.003 pu
X_l	0.1 pu	0.1 pu	0.1 pu
X_d	2.11 pu	2.11 pu	1.20 pu
X_q	2.05 pu	2.05 pu	0.700 pu
X'_d	0.25 pu	0.25 pu	0.24 pu
X''_d	0.21 pu	0.21 pu	0.20 pu
X''_q	0.21 pu	0.21 pu	0.20 pu
T'_{do}	6.8 s	7.4 s	7.2 s
T''_{do}	0.033 s	0.033 s	0.031 s
T''_{qo}	0.030 s	0.030 s	0.030 s
H	4.0 s	4.0 s	4.0 s

Table 3. Typical values of IEEE type AC4A excitation system.

Parameter	Value
K_A	200
T_A	0.04
T_B	12
T_C	1.0

Table 4. Typical values of turbine parameters.

Steam Turbine		Hydraulic Turbine	
Parameter	Value	Parameter	Value
R_p	0.05	R_p	0.05
T_G	0.2 s	T_G	0.2 s
T_{CH}	0.3 s	R_T	0.38 s
T_{RH}	7.0 s	T_R	5.0 s
F_{HP}	0.3	T_W	1.0 s

The integral controller on selected units for AGC is shown in Figure 15 [30]. The output of the AGC supplies the power load reference of the governor system depending on the speed deviation of the SG ($\Delta\omega_{sg}$). The integral gain K_i is set to 6.

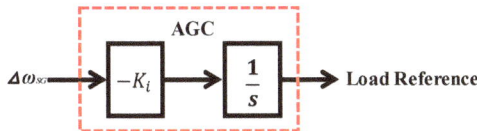

Figure 15. Controller for automatic generation control (AGC).

A WF is connected to the main system at bus 5, as shown in Figure 11, and consists of one VSWT-DFIG and one FSWT-SCIG. In order to reduce computational time, each wind generator is represented as an aggregated equivalent single machine [31,32]. The total capacity of the WF is 100 MW. A capacitor bank (C) is used for reactive power compensation of the SCIG. The value of C is chosen such that the power factor of the SCIG-based wind generator becomes unity at the rated operating condition. The base power of the system is 100 MVA, and the rated frequency is 50 Hz. The parameters of the DFIG and the SCIG are presented in Table 5.

Table 5. Parameters of wind generators.

Doubly Fed Induction Generator (DFIG)		Squirrel Cage Induction Generator (SCIG)	
MVA	27, 28, 58 and 59	MVA	41, 42, 72, 73
R_s	0.007 pu	R_1	0.01 pu
R_r	0.005 pu	X_1	0.1 pu
L_{is}	0.171 pu	X_m	3.5 pu
L_{rl}	0.156 pu	R_{21}	0.035 pu
L_m	2.9 pu	R_{22}	0.014 pu
-	-	X_{21}	0.03 pu
-	-	X_{22}	0.089 pu
-	-	H	1.5 s

6. LVRT Requirement for Wind Power

The requirement of LVRT for wind power is depicted in Figure 16 [33]. The WF must remain connected to the grid if the voltage drop is within the defined r.m.s. value and its duration is also within the defined period, as shown in the figure. If the voltage of the connection point recovers to 90% of the rated voltage within 1.5 s following the voltage drop, all wind turbines within the WF shall stay online without tripping.

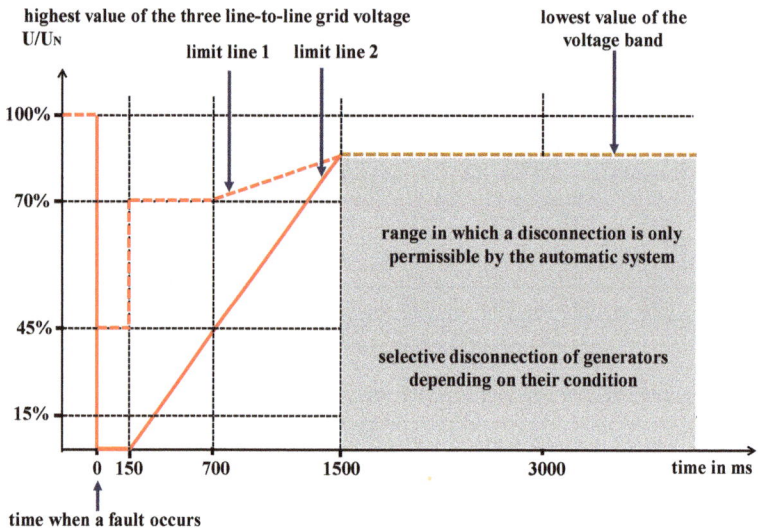

Figure 16. Low voltage ride through (LVRT) requirement for wind farm (WF).

7. Simulation Results and Discussions

7.1. Transient Stability Analysis

Simulation analysis is performed on the model system shown in Figure 11 using PSCAD/EMTDC software. The FORTRAN language is incorporated into PSCAD/EMTDC in order to implement FLC as new component. The simulation time is chosen as 10 s. The triple-line-to-ground (3LG) fault near bus 11 is considered to be a network disturbance, as shown in Figure 11. The fault occurs at 0.1 s. The duration of the fault is 0.1 s. The circuit breakers (CBs) on the faulted line are opened at 0.2 s in order to isolate the faulty line from the power system. The CBs are reclosed at 1.0 s based on the consideration that the fault has been cleared. The wind speed data applied to each wind turbine is maintained constant at the rated speed based on the assumption that the wind speed does not change dramatically within this small period of time. Simulation analyses are carried out for both the proposed and conventional rotor-side controllers reported in [23] in order to demonstrate the effectiveness of the proposed control system. The simulation results are presented and discussed in the following subsections.

7.1.1. Analysis Using the Conventional Rotor-Side Controller

Two cases are considered using the conventional rotor-side controller. The parameters for conventional PI controllers are chosen based on the method presented in the literature [23]. The power rating of each wind generator in Case 01 is DFIG = 59 MW and SCIG = 41 MW (total: 100 MW), and, in Case 02, DFIG = 58 MW and SCIG = 42 MW (total: 100 MW). Different power ratings of the wind generators are chosen, because the objective is to stabilize the maximum possible rating of SCIG by using lowest possible rating of DFIG, while the total capacity of WF is kept constant at 100 MW. In this present study, it is calculated by running the simulation for multiple times with different combinations of power ratings of the wind generators.

Figure 17a,b show the responses of reactive powers, which indicates that the DFIG can provide the necessary reactive power during the severe symmetrical 3LG fault in Case 01. As a result, the connection point voltage recovers to the rated value quickly in Case 01, as shown in Figure 18a. However, in Case 02, the DFIG does not provide the necessary reactive power during the fault condition. Thus, the connection point voltage cannot recover to the rated value. Since the connection point voltage does not satisfy the standard grid code of Figure 16 in Case 02, the WF is disconnected from the power system by opening CBs near bus 12 at 2 s. The rotor speed responses of both wind generators are stable in Case 01, but unstable in Case 02, as shown in Figure 19.

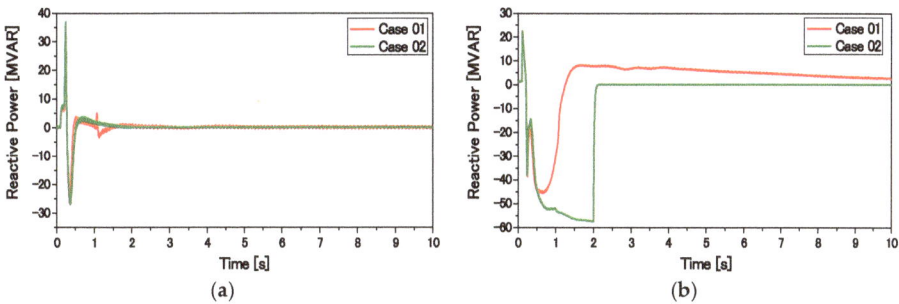

Figure 17. Reactive power output of wind generators: (**a**) DFIG; (**b**) SCIG.

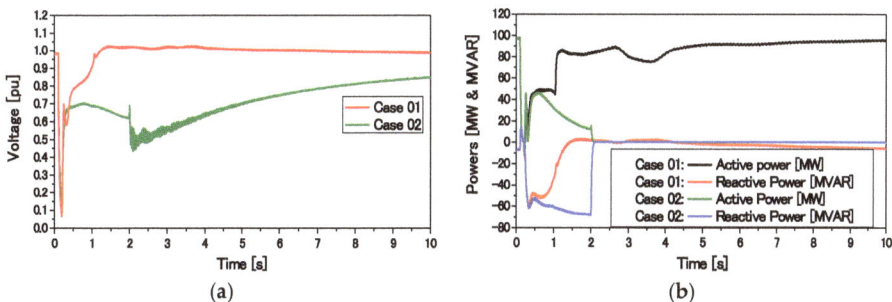

Figure 18. Individual response of the WF at bus 12: (**a**) Voltage at connection point; (**b**) total active and reactive power at connection point.

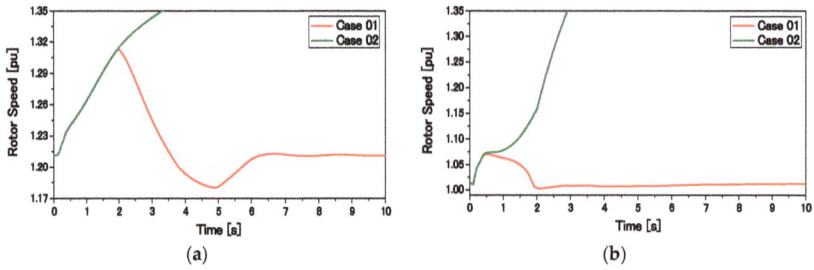

Figure 19. Rotor speed response of wind generators: (**a**) DFIG; (**b**) SCIG.

Figure 20 shows the active power output of DFIG and SCIG, respectively. The active power can recover to the nominal value in Case 01 for both wind generators, but failed to recover to the nominal value in Case 02. Moreover, the DC-link voltage of the DFIG becomes more stable in Case 01, as compared to Case 02, as shown in Figure 21a.

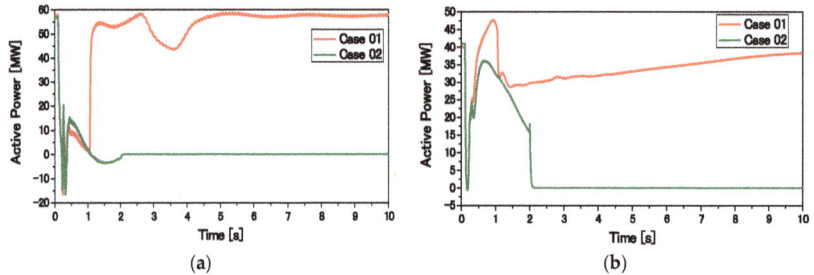

Figure 20. Active power output of wind generators: (**a**) DFIG; (**b**) SCIG.

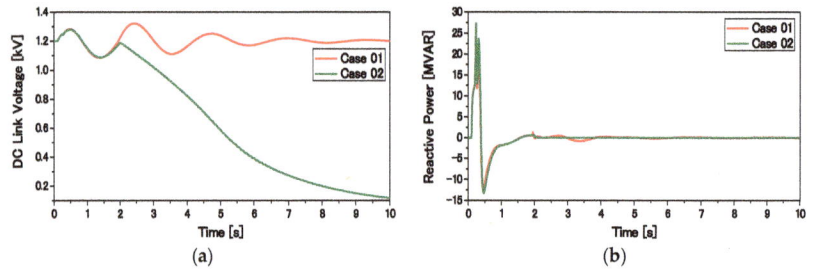

Figure 21. Individual responses of the DFIG: (**a**) DC-link voltage; (**b**) reactive power output of GSC.

Figure 22a,b show the active power output and rotor speed responses, respectively, of the conventional power plants (SGs). The active power and rotational speed of the SGs can return to the initial condition in Case 01. However, the active power of the SGs in Case 02 increases significantly after the WF has been disconnected, resulting in a rotor speed drop of the SGs. It is clear that the system becomes unstable in Case 02, which can also be seen from Figure 23, where the system frequency collapses in Case 02 after the WF has been disconnected.

Therefore, the lowest power rating of the DFIG with the conventional rotor-side controller is 59 MW in order to stabilize the 41 MW SCIG. The DFIG can also stabilize the SGs.

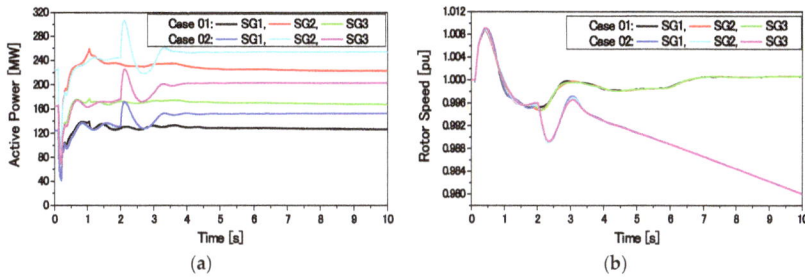

Figure 22. Individual responses of conventional SGs: (**a**) active power; (**b**) rotor speed.

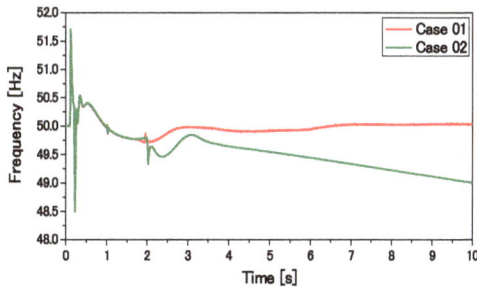

Figure 23. Frequency responses of the power system.

7.1.2. Analysis Using the Proposed Rotor-Side Controller

Two cases are considered using the proposed rotor-side controller shown in Figure 6. The power rating of each wind generator in Case 01 is DFIG = 28 MW and SCIG = 72 MW (total: 100 MW), and, in Case 02, DFIG = 27 MW and SCIG = 73 MW (total: 100 MW).

Figure 24a,b show the responses of reactive powers, which indicate that the DFIG can provide the necessary reactive power during the severe symmetrical 3LG fault in Case 01. As a result, the connection point voltage quickly recovers to the rated value in Case 01, as shown in Figure 25a. However, in Case 02, the DFIG does not provide the necessary reactive power during the fault condition, and thus, the connection point voltage cannot be back to the rated value. Since the connection point voltage does not satisfy the standard grid code of Figure 16 in Case 02, the WF is disconnected from the power system by opening CBs near bus 12 at 2 s. The rotor speed responses of both wind generators are stable in Case 01, but unstable in Case 02, as shown in Figure 26.

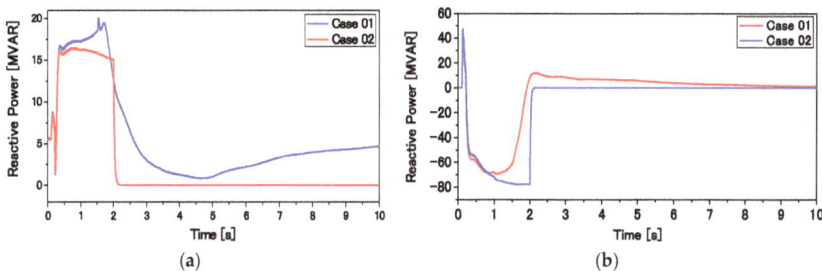

Figure 24. Reactive power output of wind generators: (**a**) DFIG; (**b**) SCIG.

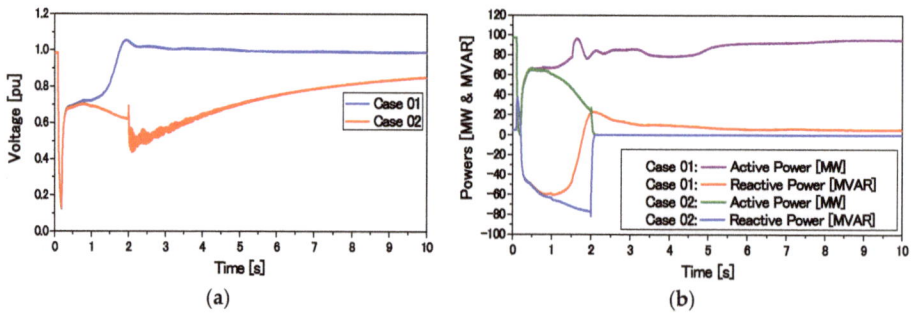

Figure 25. Individual response of the WF at bus 12: (**a**) Voltage at connection point; (**b**) total active and reactive power at connection point.

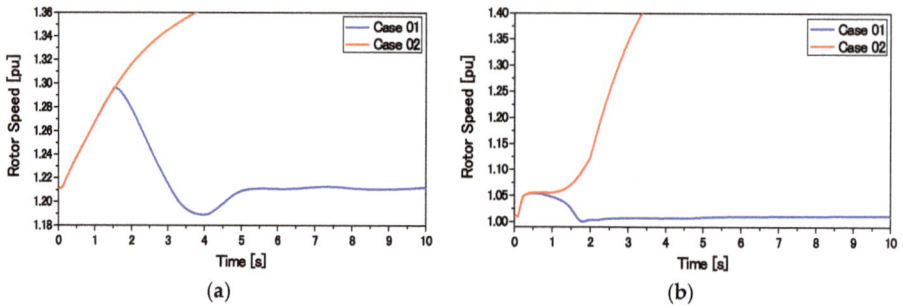

Figure 26. Rotor speed response of wind generators: (**a**) DFIG; (**b**) SCIG.

Figure 27 shows the active power output of DFIG and SCIG, respectively. The active power can recover to the nominal value in Case 01 for both wind generators, but failed to recover to the nominal value in Case 02. Moreover, the DC-link voltage of the DFIG becomes more stable in Case 01, as compared to Case 02, as shown in Figure 28a. Figure 29a,b show the active power output and rotor speed responses, respectively, of the conventional power plants (SGs). The active power and rotational speed of the SGs can return to the initial condition in Case 01. However, the active power of the SGs in Case 02 increases significantly after the WF has been disconnected, resulting in a rotor speed drop of the SGs. It is clear that the system becomes unstable in Case 02, which can also be seen from Figure 30, where the system frequency collapses in Case 02 after the WF has been disconnected.

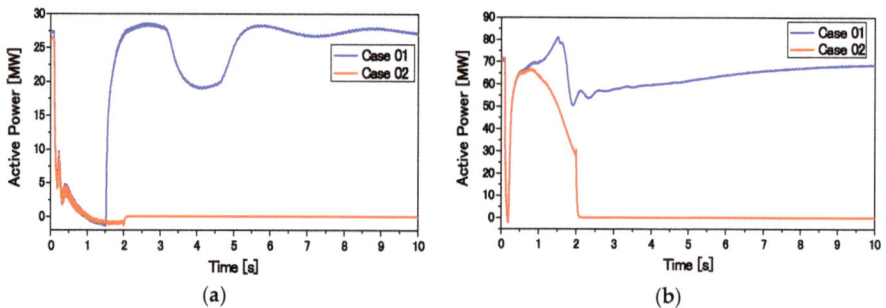

Figure 27. Active power output of wind generators: (**a**) DFIG; (**b**) SCIG.

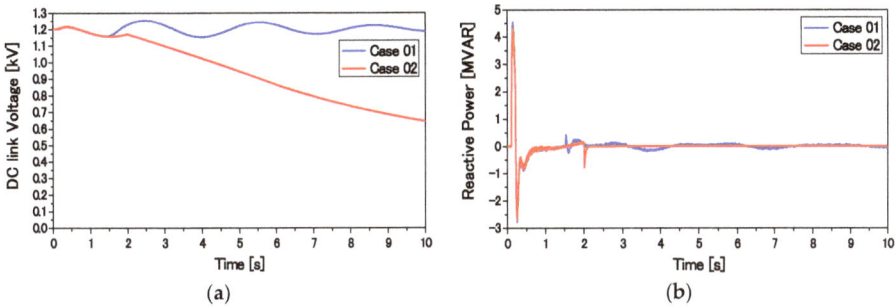

Figure 28. Individual responses of the DFIG: (**a**) DC-link voltage; (**b**) reactive power output of GSC.

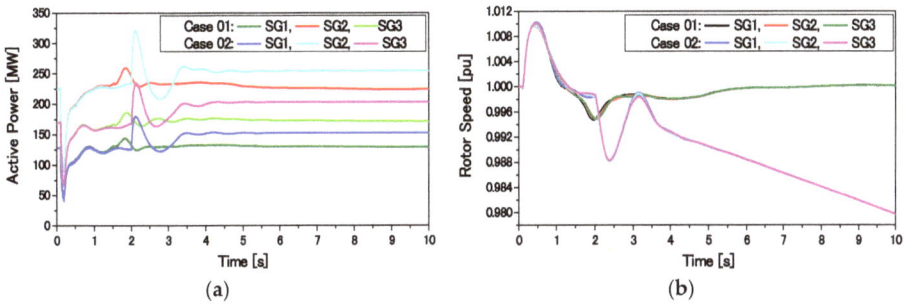

Figure 29. Individual responses of conventional SGs: (**a**) active power; (**b**) rotor speed.

Therefore, the lowest power rating of the DFIG with the proposed rotor-side controller is 28 MW in order to stabilize the 72 MW SCIG. The DFIG can also stabilize the SGs. Finally, the reactive power output of DFIG (capacity is 28 MW) for both proposed and conventional rotor-side controllers is depicted in Figure 31. The reactive power output for the proposed FLC-controlled DFIG is larger and more efficient than the conventional PI-controlled DFIG.

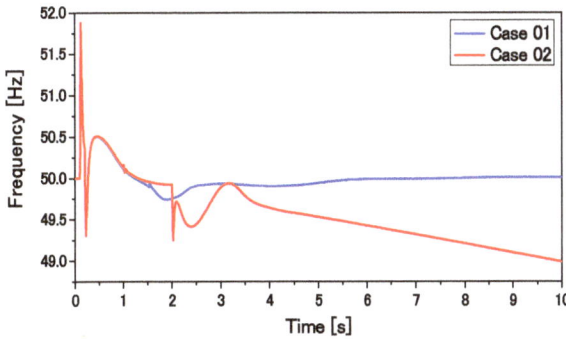

Figure 30. Frequency responses of the power system.

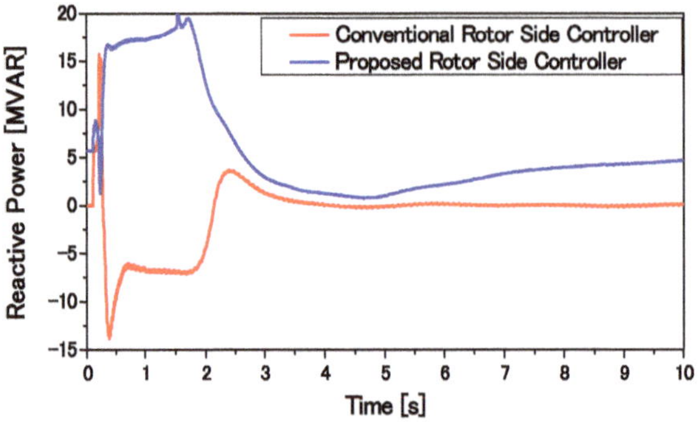

Figure 31. Reactive power output of DFIG.

7.2. Dynamic Performance Analysis Using the Proposed Rotor-Side Controller

In order to evaluate the dynamic performance of the proposed system, the real wind speed data measured at Rishiri Island, Hokkaido, Japan, shown in Figure 32, is used in the simulation. The power system model shown in Figure 11 is considered in this dynamic analysis. The capacities of the DFIG and the SCIG are 28 MW and 72 MW (The total capacity of the WF is 100 MW), respectively. Because this power ratings of the wind generators are stable case for the proposed system as presented in Section 7.1.2.

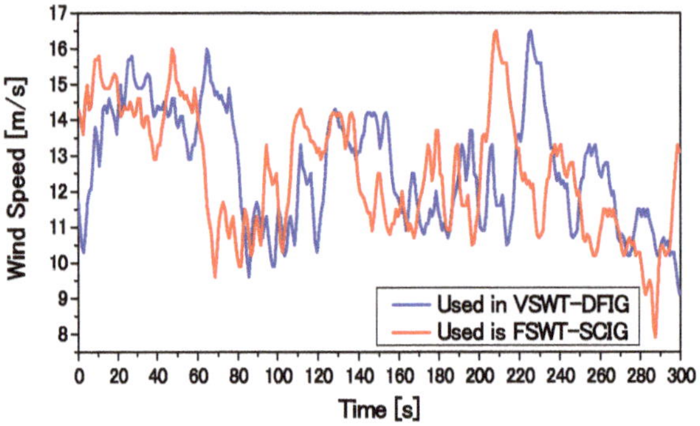

Figure 32. Wind speed data.

Figure 33 shows the reactive power output of wind generators. The DFIG provides the necessary reactive power to the SCIG for voltage regulation. Thus, the connection point voltage at bus 12 is approximately constant, as shown in Figure 34. Figure 35 shows the active power outputs of the VSWT-DFIG and the FSWT-SCIG. The DC-link voltage of the DFIG is maintained constant, as shown in Figure 36. The variation of the DC-link voltage is very small, even though there are wide fluctuations in the wind speed. Figure 37 shows the responses of the blade pitch angle. The increase in the blade pitch angle will help to reduce the mechanical power extraction from the wind turbines. The total active and reactive power output of the wind generators at bus 12 is shown in Figure 38.

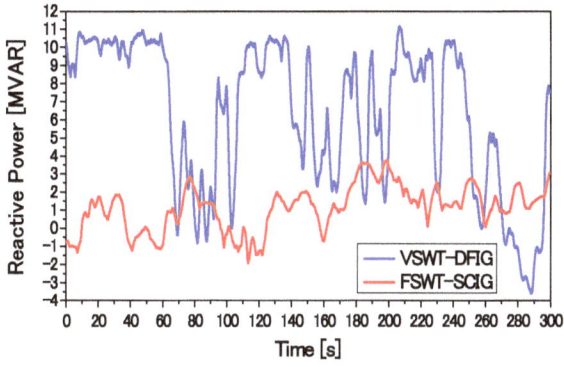

Figure 33. Reactive power output of wind generators.

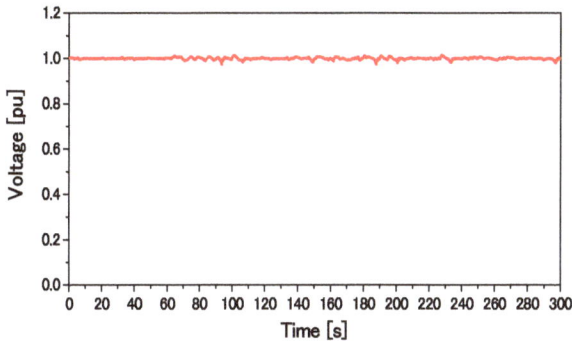

Figure 34. Voltage response at the connection point of wind generators.

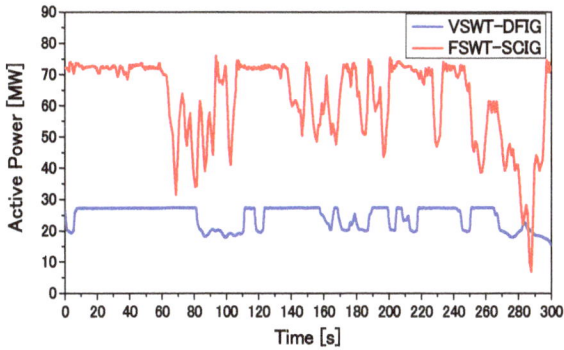

Figure 35. Active power output of wind generators.

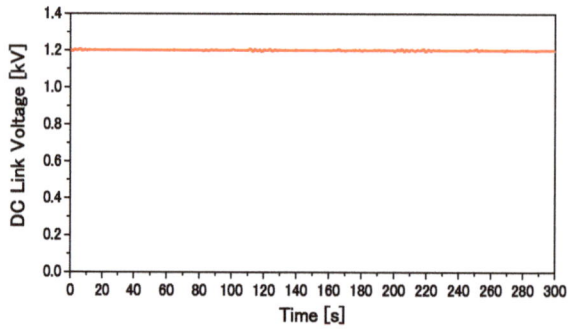

Figure 36. DC-link voltage response of DFIG.

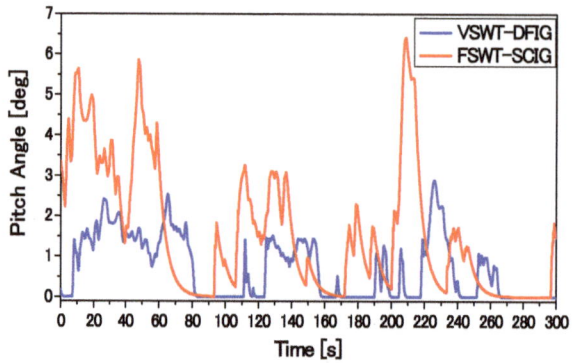

Figure 37. Pitch angle of wind generators.

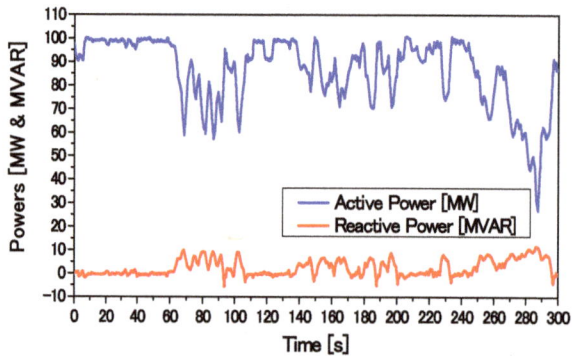

Figure 38. Total active and reactive power output at bus 12.

The active power output of conventional power plants (SGs) and the power system frequency response are shown in Figures 39 and 40. The conventional SGs adjust their active power output according to the fluctuating power injected from the WF. Moreover, the frequency variation lies within the permissible limit in Japan (±0.2 Hz).

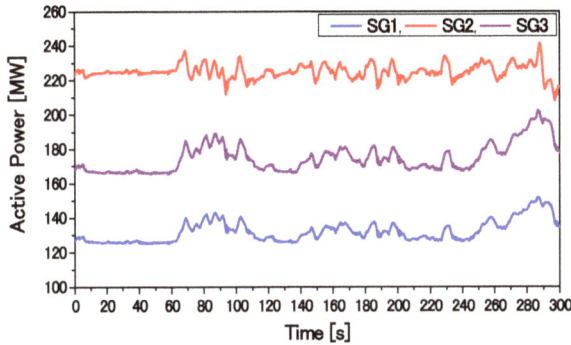

Figure 39. Active power output of SGs.

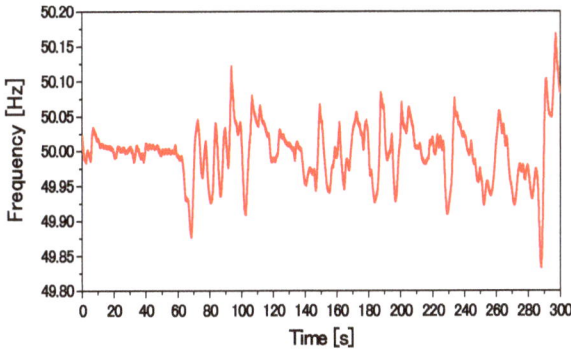

Figure 40. Power system frequency response.

7.3. Discussion

The transient simulation analyses in Sections 7.1.1 and 7.1.2 reveal that the necessary power rating of the DFIG to stabilize the SCIG in the WF, as well as to prevent conventional SGs from becoming out of step during a 3LG fault, is much lower in the case of the proposed rotor-side controller than in the case of the conventional rotor-side controller, where the total capacity of the DFIG and the SCIG is 100 MW. Table 6 summarizes the results, which reveal that, for stable operation of the WF and SGs, the lowest power rating of the DFIG is 28 MW for the proposed method and 59 MW for the conventional method.

Table 6. Performances of the proposed and conventional rotor-side controllers of DFIG.

Controller for RSC	DFIG (MW)	SCIG (MW)	Total Capacity of WF (MW)	WF Condition (After the Fault)	SGs Condition (After the Fault)
Proposed FLC	28	72	100	stable	stable
Conventional PI controller	59	41	100	stable	stable
Proposed FLC	27	73	100	unstable	out of step
Conventional PI controller	58	42	100	unstable	out of step

The dynamic simulation analysis confirmed that the proposed FLC-controlled DFIG can effectively inject reactive power and thus maintain the terminal voltage constant under a randomly varying wind speed.

8. Conclusions

In order to enhance the LVRT performance of the SCIG-based WF, partial installation of the DFIG with the new rotor-side controller based on the FLC is proposed in the present study. Moreover, a comparative study of the proposed and conventional rotor-side controllers is carried out. Based on the simulation results and performance analyses, the following points are of notable significance regarding the proposed method:

1. The proposed FLC-controlled DFIG of a lower power rating can stabilize the larger power rating of SCIG as well as conventional SGs during fault conditions.
2. The installation cost can be decreased by incorporating a small number of VSWT-DFIGs with the proposed controller and a large number of FSWT-SCIGs into a WF.
3. The proposed FLC controlled DFIG system can maintain its terminal voltage at constant under normal operating conditions by effectively injecting reactive power into the grid.

Therefore, if the proposed DFIG with a relatively small power rating is installed at a WF composed mainly of SCIGs, its LVRT capability, as well as the stability of a connected power system, can be enhanced.

Acknowledgments: The present study was supported by a Grant-in-Aid for Scientific Research (B) from the Ministry of Education, Science, Sports, and Culture of Japan.

Author Contributions: Md. Rifat Hazari and Junji Tamura developed the theoretical concepts, designed the power system model and proposed RSC controller, and performed the simulation analysis. Md. Rifat Hazari and S. M. Muyeen developed the fuzzy logic controller. Mohammad Abdul Mannan, S. M. Muyeen, Atsushi Umemura, Rion Takahashi, and Junji Tamura revised the manuscript. Md. Rifat Hazari wrote the manuscript.

Conflicts of Interest: The authors declare no conflict of interest.

Abbreviations

FSWT	fixed-speed wind turbine
SCIG	squirrel cage induction generators
WF	wind farm
LVRT	low-voltage ride through
VSWT	variable-speed wind turbine
DFIG	doubly fed induction generators
RSC	rotor-side converter
GSC	grid-side converter
FLC	fuzzy logic controller
SG	synchronous generator
STATCOM	static synchronous compensator
SMES	superconducting magnetic energy storage
ECS	energy capacitor system
PMSG	permanent magnet synchronous generator
MPPT	maximum power point tracking
WRIG	wound rotor induction generator
IGBT	insulated gate bipolar transistor
PWM	pulse width modulation
NB	negative big
NM	negative medium
NS	negative small
ZO	zero
PS	positive small
PM	positive medium

PB	positive big
AGC	automatic generation control
GF	governor free
3LG	triple-line-to-ground fault
CB	circuit breaker
P_w	captured wind power
ρ	air density (KG/m^3)
R	radius of the rotor blade (m)
V_w	wind speed (m/s)
C_p	power coefficient
T_w	wind turbine torque
β	pitch angle
λ	tip speed ratio
Cp_{opt}	optimum power coefficient
λ_{opt}	optimum tip speed ratio
ω_r	rotational speed
θ_r	rotor position
V_{dc}	DC-link voltage
$V_{dc}{}^*$	reference DC-link voltage
C_{dc}	DC-link capacitor
P_{dfig}	active power output of DFIG
Q_{dfig}	reactive power output of DFIG
P_{ref}	reference active power
$Q_{dfig}{}^*$	reference reactive power
P_{loss}	power losses
P_{mppt}	MPPT output
V_g	grid voltage
$V_g{}^*$	grid voltage reference
I_{sa}, I_{sb}, I_{sc}	stator currents for phases A, B, and C
I_{sd}, I_{sq}	stator d-axis and q-axis currents
I_{ra}, I_{rb}, I_{rc}	rotor currents for phases A, B, and C
I_{rd}, I_{rq}	rotor d and q axis currents
I_{ga}, I_{gb}, I_{gc}	grid currents for phases A, B, and C
I_{gd}, I_{gq}	grid d-axis and q-axis currents
eI_{rd}	error of rotor d-axis current
$d(eI_{rd})/dt$	change in the error of the rotor d-axis current
$1/z$	one sampling time delay

References

1. Global Wind Energy Council (GWEC). Annual Market Update 2015, Global Wind Report. 2015. Available online: http://www.gwec.net/ (accessed on 15 October 2017).
2. Global Wind Energy Council (GWEC). Global Wind Energy Outlook 2016: Wind Power to Dominate Power Sector Growth. 2016. Available online: http://www.gwec.net/ (accessed on 15 October 2017).
3. Tsili, M.; Papathanassiou, S. A Review of Grid Code Technical Requirements for Wind Farms. *IET Renew. Power Gener.* **2009**, *3*, 308–332. [CrossRef]
4. Muyeen, S.M.; Tamura, J.; Murata, T. *Stability Augmentation of a Grid Connected Wind Farm*; Springer: London, UK, 2009.
5. Suul, J.A.; Molinas, M.; Undeland, T. STATCOM-based Indirect Torque Control of Induction Machines during Voltage Recovery after Grid Faults. *IEEE Trans. Power Electron.* **2010**, *25*, 1240–1250. [CrossRef]
6. Yu, J.; Duan, X.; Tang, Y.; Yuan, P. Control Scheme Studies of Voltage Source Type Superconducting Magnetic Energy Storage (SMES) Under Asymmetrical Voltage. *IEEE Trans. Appl. Supercond.* **2002**, *12*, 750–753.
7. Muyeen, S.M.; Takahashi, R.; Ali, M.H.; Murata, T.; Tamura, J. Transient Stability Augmentation of Power Systems including Wind Farms using ECS. *IEEE Trans. Power Syst.* **2008**, *23*, 1179–1187. [CrossRef]

8. Tripathi, S.; Tiwari, A.; Singh, D. Grid-integrated Permanent Magnet Synchronous Generator based Wind Energy Conversion Systems: A Technology Review. *Renew. Sustain. Energy Rev.* **2015**, *51*, 1288–1305. [CrossRef]

9. Boldea, I.; Tutelea, L.; Blaabjerg, F. High Power Wind Generator Designs with Less or No PMs: An Overview. In Proceedings of the International Conference on Electrical Machines and Systems (ICEMS), Hangzhou, China, 22–25 October 2014; pp. 1–14.

10. Rosyadi, M.; Umemura, A.; Takahashi, R.; Tamura, J.; Uchiyama, N.; Ide, K. Simplified Model of Variable Speed Wind Turbine Generator for Dynamic Simulation Analysis. *IEEJ Trans. Power Energy* **2015**, *135*, 538–549. [CrossRef]

11. Okedu, K.; Muyeen, S.M.; Takahashi, R.; Tamura, J. Wind Farm Stabilization by using DFIG with Current Controlled Voltage Source Converter taking Grid Codes into Consideration. *IEEJ Trans. Power Energy* **2012**, *132*, 251–259. [CrossRef]

12. Bourdoulis, M.K.; Alexandridis, A.T. Direct Power Control of DFIG Wind Systems Based on Nonlinear Modeling and Analysis. *IEEE J. Emerg. Sel. Top. Power Electron.* **2014**, *2*, 764–775. [CrossRef]

13. Pannell, G.; Atkinson, D.J.; Zahawi, B. Minimum-threshold Crowbar for a Fault-ride-through Grid-code-Compliant DFIG Wind Turbine. *IEEE Trans. Energy Convers.* **2010**, *25*, 750–759. [CrossRef]

14. Vidal, J.; Abad, G.; Arza, J.; Aurtenechea, S. Single Phase DC Crowbar Topologies for Low Voltage Ride Through Fulfillment of High-power Doubly Fed Induction Generator-based Wind Turbines. *IEEE Trans. Energy Convers.* **2013**, *28*, 768–781. [CrossRef]

15. Pannell, G.; Zahawi, B.; Atkinson, D.J.; Missailidis, P. Evaluation of the Performance of a DC-link Brake Chopper as a DFIG Low-Voltage Fault-Ride-Through Device. *IEEE Trans. Energy Convers.* **2013**, *28*, 535–542. [CrossRef]

16. Huchel, L.; Moursi, M.S.E.; Zeineldin, H.H. A Parallel Capacitor Control Strategy for Enhanced FRT Capability of DFIG. *IEEE Trans. Sustain. Energy* **2015**, *6*, 303–312. [CrossRef]

17. Okedu, K.E.; Muyeen, S.M.; Takahashi, R.; Junji, T. Wind Farms Fault Ride Through using DFIG with New Protection Scheme. *IEEE Trans. Sustain. Energy* **2012**, *3*, 242–254. [CrossRef]

18. Guo, W.; Xiao, L.; Dai, S.; Xu, X.; Li, Y.; Wang, Y. Evaluation of The Performance of BTFCLs for Enhancing LVRT Capability of DFIG. *IEEE Trans. Power Electron.* **2015**, *30*, 3623–3637. [CrossRef]

19. Zhang, S.; Tseng, K.J.; Choi, S.S.; Nguyen, T.D.; Yao, D.L. Advanced Control of Series Voltage Compensation to Enhance Wind Turbine Ride Through. *IEEE Trans. Power Electron.* **2012**, *27*, 763–772. [CrossRef]

20. Ibrahim, A.O.; Nguyen, T.H.; Lee, D.C.; Kim, S.C. A Fault Ride through Technique of DFIG Wind Turbine Systems using Dynamic Voltage Restorers. *IEEE Trans. Energy Convers.* **2011**, *26*, 871–882. [CrossRef]

21. Amaris, H.; Alfonso, M. Coordinated Reactive Power Management in Power Networks with Wind Turbines and FACTS Devices. *Energy Convers. Manag.* **2011**, *52*, 2575–2586. [CrossRef]

22. Qiao, W.; Harley, R.G.; Venayagamoorthy, G.K. Coordinated Reactive Power Control of a Large Wind Farm and a STATCOM using Heuristic Dynamic Programming. *IEEE Trans. Energy Convers.* **2009**, *24*, 493–503. [CrossRef]

23. Hasanien, H.M.; Muyeen, S.M. A Taguchi Approach for Optimum Design of Proportional-Integral Controllers in Cascaded Control Scheme. *IEEE Trans. Power Syst.* **2013**, *28*, 1636–1644. [CrossRef]

24. Matlab Documentation Center. Available online: http://www.mathworks.co.jp/jp/help/ (accessed on 20 December 2016).

25. Liu, J.; Rosyadi, M.; Umemura, A.; Takahashi, R.; Tamura, J. A Control Method of Permanent Magnet Wind Generators in Grid Connected Wind Farm to Damp Load Frequency Oscillation. *IEEJ Trans. Power Energy* **2014**, *134*, 393–398. [CrossRef]

26. Manitoba HVDC Research Center. *PSCAD/EMTDC User's Manual*; Manitoba HVDC Research Center: Winnipeg, MB, Canada, 1994. Available online: https://hvdc.ca/uploads/ck/files/reference_material/EMTDC_User_Guide_v4_2_1.pdf (accessed on 21 October 2017).

27. Mamdani, E.M. Applications of Fuzzy Algorithms for Simple Dynamic Plants. *Proc. IEEE* **1974**, *21*, 1585–1588. [CrossRef]

28. Driankov, D.; Hellendoorn, H.; Reinfrank, M. *An Introduction to Fuzzy Control*; Springer: Berlin, Germany, 1993.

29. Anderson, P.M.; Fouad, A.A. *Power System Control & Stability*; John Wiley & Sons: Oxford, UK, 2008.

30. Kundur, P. *Power System Stability & Control*; McGraw-Hill Inc.: New York, NY, USA, 1994.

31. Ackerman, T. *Wind Power in Power System*; John Wiley & Sons: Oxford, UK, 2005.

32. WECC Renewable Energy Modeling Task Force. *WECC Wind Power Plant Dynamic Modeling Guide*; WECC Renewable Energy Modeling Task Force, 2014. Available online: https://www.wecc.biz/Reliability/WECC%20Wind%20Plant%20Dynamic%20Modeling%20Guidelines.pdf (accessed on 21 October 2017).

33. E. ON NETZ GmbH. *Grid Connection Regulation for High and Extra High Voltage*; E. ON NETZ GmbH: Essen, Germany, 2006.

![applied sciences logo] *applied sciences*

MDPI

Article

LPV Model Based Sensor Fault Diagnosis and Isolation for Permanent Magnet Synchronous Generator in Wind Energy Conversion Systems

Zhimin Yang [1,2], Yi Chai [1,2,*], Hongpeng Yin [1,2] and Songbing Tao [1,2]

[1]   Key Laboratory of Complex System Safety and Control, Ministry of Education, Chongqing University, Chongqing 400044, China; yangzhimin@cqu.edu.cn (Z.Y.); yinhongpeng@gmail.com (H.Y.); taosongbing@gmail.com (S.T.)

[2]   School of Automation, Chongqing University, Chongqing 400044, China

*   Correspondence: chaiyi@cqu.edu.cn; Tel.: +86-23-6510-2482

Received: 10 September 2018; Accepted: 1 October 2018; Published: 3 October 2018

Abstract: This paper deals with the current sensor fault diagnosis and isolation (FDI) problem for a permanent magnet synchronous generator (PMSG) based wind system. An observer based scheme is presented to detect and isolate both additive and multiplicative faults in current sensors, under varying torque and speed. This scheme includes a robust residual generator and a fault estimation based isolator. First, the PMSG system model is reformulated as a linear parameter varying (LPV) model by incorporating the electromechanical dynamics into the current dynamics. Then, polytopic decomposition is introduced for $\mathcal{H}_\infty$ design of an LPV residual generator and fault estimator in the form of linear matrix inequalities (LMIs). The proposed gain-scheduled FDI is capable of online monitoring three-phase currents and isolating multiple sensor faults by comparing the diagnosis variables with the predefined thresholds. Finally, a MATLAB/SIMULINK model of wind conversion system is established to illustrate FDI performance of the proposed method. The results show that multiple sensor faults are isolated simultaneously with varying input torque and mechanical power.

Keywords: fault diagnosis and isolation; multiple sensor faults; LPV observer; permanent magnet synchronous generator

---

1. Introduction

Due to the high power density and efficiency, permanent magnet synchronous generator based wind turbines are promising in wind conversion systems (WECSs) with variable speed operation and full-scale power delivery [1,2]. To fulfill control demands for maximum power point tracking (MPPT) and grid codes, closed-loop feedback control is designed, relying on the mechanical, current and voltage measurements. Any inaccurate measurements caused by sensor faults will cause the controller malfunction and performance degradation. According to industrial and field statistics [3–5], current sensor faults are a type of major faults resulting from the electromagnetic interference and high power density, which causes system shutdown and fragile components.

Fault diagnosis and isolation (FDI) schemes enable the control system to locate fault sensors and to compensate the fault further. For power converter systems, various diagnostic techniques are presented to handle current sensor FDI problems, including observer based, signal processing based and data-driven based methods. Model based diagnostic techniques are discussed most for power converter systems. A parallel observers based method is presented in [6] to diagnose stator and rotor current sensor faults in doubly fed induction generator (DFIG) system, but it requires open-loop operation while detecting the sensor fault. Similarly, a sensor FDI in [7] that a bank of observers are designed to generate residuals sensitive to sensor fault for DFIG based WECSs also requires open-loop operation until the fault is isolated. In [8], a geometric approach is presented to detect and isolate multiple sensor faults in induction motor (IM) drives. By utilizing the redundant properties of three-phase currents, two stationary frame based state space models are established to generate distinguished residuals sensitive to phase a and phase b sensor faults. In [9], the nonlinear model of DFIG is transformed into a Takagi–Sugeno (T-S) fuzzy model and a bank of observers based on the model are presented to generate residuals for sensor fault detection and isolation. To deal with both additive and multiplicative sensor faults, a generalized observer scheme is presented in [10] by combining $\mathcal{H}_-/\mathcal{H}_\infty$ filter with Kalman-like observer for DFIG systems.

Aforementioned schemes are presented for sensor FDI in IM drives while only a few model based sensor FDIs are proposed for permanent magnet synchronous motor (PMSM) and permanent magnet synchronous generator (PMSG) systems. In [11], a two-stage extended Kalman filter (EKF) and adaptive observer is presented to generate mechanical estimations for speed and rotor position sensor fault diagnosis. An adaptive EKF for position sensor fault diagnosis and tolerant scheme is presented in [12] for PMSM drive in electric vehicle (EV). In [13], a high-order sliding model based observer is proposed to detect and estimate rotor speed sensor fault in PMSM based EV. The authors later present a bank of observers based scheme for multiple sensor FDI [14]. However, it requires additional voltage sensors to establish the fault observers. In [15], an EKF based FDI is presented for the diagnosis of sensor fault in PMSM drives, but it can only isolate single sensor faults and does not additionally discuss about the influence of unknown disturbances on FDI performance. Furthermore, in [2], to diagnose additive and multiplicative faults for PMSG based WECSs, a two-stage model based method is proposed, in which time-varying Kalman filter (TVKF) and maximum-shift method are designed to generate robust residuals and evaluate these residuals.

According to the state-of-the-art analysis, model based FDI methods are rarely reported for simultaneous multiple current sensor FDI for PMSM and PMSG based applications. Nevertheless, current sensor FDI is necessary for control system in power converters to provide further information for fault tolerant control [2,8]. In this paper, an observer based scheme is presented to detect and isolate both additive and multiplicative faults in current sensors under varying torque and speed. The proposed method includes a robust residual generator and a fault estimation based isolator. The system model is established in the stationary reference frame and the nonlinear term with rotor position is transformed into a polytopic linear parameter varying (LPV) model. Based on the stability and convergency analysis, a gain-scheduled fault detector and isolator is designed in the form linear matrix inequalities (LMIs). The proposed gain-scheduled FDI scheme is capable of online monitoring three-phase currents and isolating multiple sensor faults with only one fault estimator. Comparing with the existing methods for current sensor isolation, this method does not require complex observer combination or a bank of observers and can isolate both additive and multiplicative faults. The contributions of this paper are concluded as follows:

(1) A scheme is proposed for detection and isolation of multiple sensor faults. Compared with the existing methods, the proposed method is capable of isolating three-phase current sensor faults while most existing schemes are presented to isolate faults in stationary frame or synchronous reference frame.

(2) The proposed isolator is based on a fault estimation scheme. Fault estimates contain all the fault information, which makes it possible to deal with both additive and multiplicative faults.

(3) All of the measurements are available in the control loop. No additional hardware or measurements are required. Furthermore, the proposed method is implemented in closed-loop operation.

The rest of this paper is organized as follows. Section 2 establishes PMSG and sensor fault model, and polytopic decomposition of the model. The gain-scheduled observer design for fault detection residuals generation is presented in Section 3. In Section 4, a fault estimation design scheme is presented for isolation of each phase current sensor fault. The simulation results are presented in Section 5 to illustrate the performance of proposed method. Finally, the conclusion is presented in Section 6.

2. Problem Statement

The system configuration of PMSG based WECS is shown in Figure 1. Typically, a full-scale back-to-back converter is designed as the interface between generator and the electrical grid. The field-oriented control (FOC) is employed to transfer maximum power generated by wind turbine while tracking the rotor speed reference that requires measured values of rotor position θ, rotor speed ω_r and three-phase currents i_a, i_b, i_c.

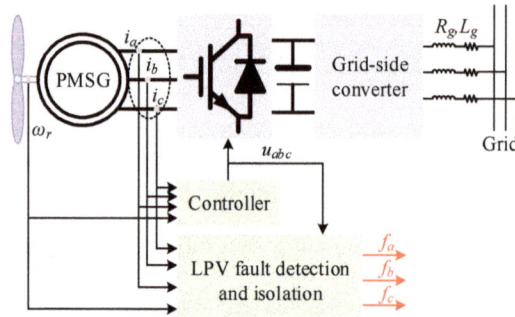

Figure 1. Control and fault diagnosis scheme for the PMSG system (reproduced from [2,16]).

Machine-side control is designed to implement an MPPT scheme for variable-speed WECSs. Each wind turbine operates in a certain wind speed region according to its ideal power curve. This leads a varying rotor speed or torque to feed the generator. As shown in Figure 2, generator stops in Regions I and IV while it continues to generate electrical power in Regions II and III.

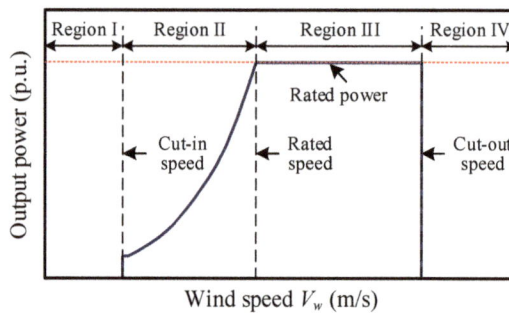

Figure 2. Operation regions of a PMSG based wind turbine (reproduced from [17]).

The mechanical signals and current measurements are crucial to ensure a stable and optimal operating condition of the WECSs. Any sensor malfunction will be fed back into the control system, which could cause performance reduction or even system downtime. Sensor faults are investigated in [11] and exhibited as: (1) sensor gain drop **Type a**; (2) bias in sensor measurement **Type b** and (3) complete sensor outage **Type c** . **Type b** and **Type a** faults can be modeled as an addictive fault in sensor measurements

$$y_m\left(k\right) = y_r\left(k\right) + f\left(k\right) \tag{1}$$

in which $y_m(k)$, $y_r(k)$ and $f(k)$ denote the faulty measurements, nominal values and fault signals respectively. **Type a** fault is the sensor gain degradation and modeled as a multiplicative fault in [2]

$$y_m(k) = \beta(k)\, y_r(k). \tag{2}$$

By defining $f(k) = (\beta(k) - 1)\, y_r(k)$, **Type a** fault in Equation (2) is rewritten as an additive fault with Formula (1). These three types of faults are uniformly modeled as additive faults.

Remark 1. *In* (1)*, $f(k)$ is unknown when $kT_s > t_{fault}$ ($\forall\ kT_s < t_{fault}$, $f(k) = 0$). A practical assumption of the sensor faults $f(k)$ is introduced in this paper: $f(k)$ is $\mathcal{L}_2-$ bounded $\|f(k)\| \leq \|\alpha(k)\|$ and $\alpha(k)$ is a known function. The upper bound of fault is essential for fault estimator design in Theorem* 2.

2.1. LPV Model of PMSG

The mathematical model of a surface-mounted PMSG can be expressed in the stationary reference frame as [14,18,19]

$$
\begin{aligned}
\frac{di_\alpha}{dt} &= -\frac{R_s}{L_s} i_\alpha + \frac{n_p \psi}{L_s} \sin(\theta)\, \omega_r + \frac{1}{L_s} u_\alpha \\
\frac{di_\beta}{dt} &= -\frac{R_s}{L_s} i_\beta - \frac{n_p \psi}{L_s} \cos(\theta)\, \omega_r + \frac{1}{L_s} u_\beta \\
\frac{d\omega_r}{dt} &= -\frac{3n_p \psi}{2J} \sin(\theta)\, i_\alpha + \frac{3n_p \psi}{2J} \cos(\theta)\, i_\beta - \frac{F}{J}\omega_r - \frac{1}{J} T_L \\
\frac{d\theta}{dt} &= n_p \omega_r,
\end{aligned}
\tag{3}
$$

where i_α, i_β and u_α, u_β are the currents [A] and voltages [V] of phases α and β in the stationary frame, respectively. In addition, θ is rotor electrical angle [rad]; ω_r denotes rotor velocity [rad/s]; F is the viscous friction coefficient [N·m·s/rad]; T_L is the load torque [N·m]; J is the inertia of the motor [Kg·m^2]; ψ is the magnetic flux of the motor [Wb]; R_s is the resistances of the phase winding [Ω]; L is the inductance of the phase winding [H]; n_p is the number of pairs of rotor poles.

By defining state variables $x = \left[i_\alpha, i_\beta, \omega_r, \theta\right]^T$ and measurements $y = \left[i_a, i_b, i_c, \omega_r, \theta\right]^T$, system (3) is expressed as a linear parameter varying model

$$
\begin{aligned}
\dot{x} &= A(\theta)\, x + B_u u + B_d d, \\
y &= Cx,
\end{aligned}
\tag{4}
$$

in which i_a, i_b and i_c denote three-phase currents of PMSG which are acquired by current sensors. Provided that the currents and voltages remain nearly constant at each sample time interval T_s. The Forward Euler Approximation method is introduced for discretization of a PMSG model

$$x(kT_s + T_s) = x(kT_s) + T_s \left\{ \frac{dx(t)}{dt} \right\}_{t=kT_s}, \tag{5}$$

where T_s is the sampling time. Accordingly, system (4) leads to the following discrete system model:

$$
\begin{aligned}
x(k+1) &= A(\theta_k)\, x(k) + B_u u(k) + B_d d(k), \\
y(k) &= Cx(k),
\end{aligned}
\tag{6}
$$

where $A(\theta_k)$, B_u, B_d and C are listed as Equation (A1) in Appendix A.

2.2. Polytopic Decomposition of the System Model

An LPV model of PMSG with sensor faults is presented as follows with bounded varying parameters:

$$x(k+1) = A(\theta_k) x(k) + B_u u(k) + B_d d(k),$$
$$y(k) = Cx(k) + F_f f(k), \tag{7}$$

where F_f is the fault distribution matrix. $A(\theta_k)$ contains two time-varying terms $\sin(\theta_k)$ and $\cos(\theta_k)$, an auxiliary variable is defined $\mu(\theta_k) = \begin{bmatrix} \mu_1(\theta_k) & \mu_2(\theta_k) \end{bmatrix}^T$,

$$\mu_1(\theta_k) = \sin(\theta_k),$$
$$\mu_2(\theta_k) = \cos(\theta_k). \tag{8}$$

It is obvious that $A(\theta_k)$ depends affinely on the parameter $\mu(\theta_k)$

$$A(\theta_k) = A_0 + \mu_1(\theta_k) A_1 + \mu_2(\theta_k) A_2, \tag{9}$$

where A_0, A_1 and A_2 are constant matrices. The time-varying parameter vector $\mu(\theta_k)$ is determined by the rotor electrical angle θ_k. Moreover, $\mu_1(\theta_k)$ and $\mu_2(\theta_k)$ are trigonometric function and bounded by the lower and upper bounds

$$\mu_1 \in \begin{bmatrix} \underline{\mu}_1 & \bar{\mu}_1 \end{bmatrix}$$
$$\mu_2 \in \begin{bmatrix} \underline{\mu}_2 & \bar{\mu}_2 \end{bmatrix} \tag{10}$$
$$\mu(\theta_k)^T \mu(\theta_k) = 1$$

in which $\underline{\mu}_1 = \underline{\mu}_2 = -1$ and $\bar{\mu}_1 = \bar{\mu}_2 = 1$. A convex polytope Θ with four vertices $\mu_{v,1}, \mu_{v,2}, \mu_{v,3}, \mu_{v,4}$ is defined to ensure that the trajectory of parameter $\mu(\theta_k)$ is enclosed:

$$\mu_{v,1} = \begin{bmatrix} \underline{\mu}_1 \\ \underline{\mu}_2 \end{bmatrix} \quad \mu_{v,2} = \begin{bmatrix} \underline{\mu}_1 \\ \bar{\mu}_2 \end{bmatrix}$$
$$\mu_{v,3} = \begin{bmatrix} \bar{\mu}_1 \\ \underline{\mu}_2 \end{bmatrix} \quad \mu_{v,1} = \begin{bmatrix} \bar{\mu}_1 \\ \bar{\mu}_2 \end{bmatrix}. \tag{11}$$

Consequently, parameter $\mu(\theta_k)$ can be expressed as a convex combination of the vertices with coordinates $\eta_k = \begin{bmatrix} \eta_{k,1} & \eta_{k,2} & \eta_{k,3} & \eta_{k,4} \end{bmatrix}^T$,

$$\mu(\theta_k) = \begin{bmatrix} \mu_{v,1} & \mu_{v,2} & \mu_{v,3} & \mu_{v,4} \end{bmatrix} \eta_k,$$
$$\eta_{k,1} + \eta_{k,2} + \eta_{k,3} + \eta_{k,4} = 1, \tag{12}$$

in which $\forall i = 1, \cdots, 4$, $\eta_{k,i} \geq 0$ and the parameter-dependent matrix $A(\mu)$ is rewritten by a combination of coordinate vector η_k

$$A(\mu) = \sum_{i=1}^{4} \eta_{k,i} A_{v,i}, \tag{13}$$

in which $A_{v,i} = A(\mu_{v,i})$ with $i = 1, \cdots, 4$. The system Equation (7) can be transformed into a polytopic form

$$x(k+1) = \sum_{i=1}^{4} \eta_{k,i} A_{v,i} x(k) + B_u u(k) + B_d d(k),$$
$$y(k) = Cx(k) + F_f f(k). \tag{14}$$

Since Equation (12) is an underdetermined equation, further constraints are required to solve this equation. In [20–22], a vertex expansion technique is presented to get a unique solution of η_k. Furthermore, this work decomposes $A(\theta_k)$ with this method.

2.3. Extended Bounded Real Lemma

This section extends the bounded real lemma to polytopic-LPV system, consider the following systems

$$G_{yw}(z, \rho) : \begin{cases} x(k+1) = A(\rho)x(k) + B(\rho)w(k), \\ y(k) = C(\rho)x(k) + D(\rho)w(k), \end{cases} \tag{15}$$

where $x(k)$, $w(k)$, $y(k)$ denote the state variables, disturbances and measurements, respectively. $\rho \in \mathcal{P}_\rho$ is a time-varying parameter vector with $\rho_i \in [\underline{\rho}_i, \bar{\rho}_i]$. Assuming that parameter space $\mathcal{P}_\rho$ is a convex hull, the system (15) can be presented as a polytopic form

$$\begin{bmatrix} A(\rho) & B(\rho) \\ C(\rho) & D(\rho) \end{bmatrix} \overset{\Delta}{=} \sum_{i=1}^{N} \eta_i \begin{bmatrix} A_i & B_i \\ C_i & D_i \end{bmatrix}, \tag{16}$$

where N is the number of vertices, $\sum_{i=1}^{N} \eta_i = 1$ and $\eta_i \geq 0$. The $\mathcal{H}_\infty$ performance is defined as Equation (17) to guarantee the asymptotically stability of system (15)

$$\|G_{yw}(z, \rho)\|_\infty = \sup_{\|w(k)\|_2 \neq 0} \frac{\|y(k)\|_2}{\|w(k)\|_2}. \tag{17}$$

An extended Bounded real lemma can be derived from the results in [23,24].

Lemma 1. *Given the system (15) and for all $\rho \in \mathcal{P}_\rho$, $G_{yw}(z, \rho)$ is asymptotically stable with $\|G_{yw}(z, \rho)\|_\infty < \gamma$, if there exists a symmetric positive definite matrix $\mathcal{P}$ satisfying that*

$$\begin{bmatrix} -\mathcal{P} & 0 & \mathcal{P}A(\rho) & \mathcal{P}B(\rho) \\ * & -\gamma I & C(\rho) & D(\rho) \\ * & * & -\mathcal{P} & 0 \\ * & * & * & -\gamma I \end{bmatrix} < 0. \tag{18}$$

Lemma 1 can be proved by definition of a Lyapunov function

$$V(x(k), \rho) = x^T(k)\mathcal{P}x(k)$$

such that

$$V(x(k+1), \rho) - V(x(k), \rho) + \gamma^{-1}\|y(k)\|_2^2 - \gamma\|w(k)\|_2^2 < 0 \tag{19}$$

for all $k = 0, 1, \cdots, k+1$. In this paper, a parameter-independent Lyapunov function is defined. In order to achieve a less conservative solution, the parameter-dependent matrix $\mathcal{P}(\rho)$ is designed in [25,26].

3. Current Sensor Fault Detection

In this section, an LPV observer based residual generator is presented to detect current sensor faults. The fault detection threshold is based on the $\mathcal{L}_2$ re-constructible condition proposed in [27].

3.1. Parameter-Dependent Observer Design

For system (14), a parameter-dependent observer based residual generator is designed to detect sensor fault

$$\begin{aligned} \hat{x}(k+1) &= A(\theta_k)\hat{x}(k) + B_u u(k) + L(\theta_k)r(k), \\ \hat{y}(k) &= C\hat{x}(k), \\ r(k) &= y(k) - \hat{y}(k), \end{aligned} \tag{20}$$

in which $\hat{x}(k)$ is state estimation, $r(k)$ is the desired residual to current sensor fault. $L(\theta_k)$ denotes the observer gain. By defining the state estimation error $e(k) = x(k) - \hat{x}(k)$, the error dynamics is obtained by substituting Equation (20) into system Equation (7)

$$e(k+1) = (A(\theta_k) - L(\theta_k)C)e(k) + B_d d(k) - L(\theta_k)F_f f(k),$$
$$r(k) = Ce(k) + F_f f(k). \tag{21}$$

For the fault-free case $f(k) = 0$, the error dynamics (21) become

$$e(k+1) = (A(\theta_k) - L(\theta_k)C)e(k) + B_d d(k),$$
$$r(k) = Ce(k). \tag{22}$$

The following theorem provides a method to determine the gain matrix $L(\theta_k)$ and to guarantee the stability and convergency of the proposed residual generator.

Theorem 1. *For the system (14) and residual generator (22), suppose that there exists a scalar $\gamma > 0$, positive definite matrix $P = P^T$ and real matrices U_i, for $i = 1, \cdots, 4$ such that*

$$\begin{bmatrix} -P & 0_{n_x \times n_x} & PA_{v,i} - U_iC & PB_d \\ * & -\gamma I_{n_x \times n_x} & C & 0_{n_x \times n_d} \\ * & * & -P & 0_{n_x \times n_d} \\ * & * & * & -\gamma I_{n_d \times n_d} \end{bmatrix} < 0. \tag{23}$$

Then, the residual generator is s asymptotically stable and the following holds

$$\sum_{k=0}^{\infty} r^T(k)r(k) = \gamma^2 \sum_{k=0}^{\infty} d^T(k)d(k) + \gamma V(0) \tag{24}$$

and parameter-dependent observer gain is given for $i = 1, \cdots, 4$

$$L(\theta_k) = \sum_{i=1}^{4} \eta_{k,i} L_i,$$
$$L_i = P^{-1} U_i. \tag{25}$$

Remark 2. *In this paper, only $A(\theta_k)$ in system (6) is parameter-dependent while B_u, B_d, C and F_f remain constant. Otherwise, varying matrices C and F_f may lead to a bilinear matrix inequality (BMI) of Equation (23). Further procedures are required to deal with such BMIs.*

Proof of Theorem 1. This proof contains two parts: one is to prove the stability of the residual generator and the other is to calculate the upper bound of residuals in Equation (24).

First, assume that Equation (23) holds. By substituting Equation (25) into Equation (23),

$$\begin{bmatrix} -P & 0 & P(A_{v,i} - L_iC) & PB_d \\ * & -\gamma I & C & 0 \\ * & * & -P & 0 \\ * & * & * & -\gamma I \end{bmatrix} < 0 \tag{26}$$

for all $i = 1, \cdots, N$, by multiplying Equation (26) with $\eta_{k,i}$ and sum to obtain

$$\begin{bmatrix} -P & 0 & P(A(\theta_k) - L(\theta_k)C) & PB_d \\ * & -\gamma I & C & 0 \\ * & * & -P & 0 \\ * & * & * & -\gamma I \end{bmatrix} < 0. \tag{27}$$

Letting $A(\rho) = A(\theta_k) - L(\theta_k)C$, $B(\rho) = B_d$, $C(\rho) = C$ and $D(\rho) = 0$, Equation (27) implies Equation (18). According to Lemma 1, residual generator (22) is asymptotically stable.

Second, consider the following Lyapunov function

$$V(k) = e^T(k)Pe(k),\tag{28}$$

where P is a positive definite matrix and $\Delta V(k) = V(k+1) - V(k)$. Noting that

$$\sum_{i=0}^{k}\Delta V(i) = V(k) - V(0)\tag{29}$$

for all $i = 0, 1, \cdots, k$, Equation (19) is summed as follows:

$$V(k+1) - V(0) + \gamma^{-1}\sum_{i=0}^{k}\|y(i)\|_2^2 - \gamma\sum_{i=0}^{k}\|d(i)\|_2^2 < 0.\tag{30}$$

Since $V(k+1) > 0$, Equation (24) is obtained by multiplying γ to inequality (30). The proof is completed. □

3.2. Current Sensor Fault Detection

Theorem 1 provides a scheme to design a robust observer for residual generation sensitive to current sensor faults. For the purpose of fault detection, a residual evaluation function is defined by a moving window $[1, N]$

$$J_d = \frac{1}{N}\sum_{i=l+1}^{l+N}r^T(i)r(i),\tag{31}$$

where N is the sampling length related to the current frequency. This paper follows $\mathcal{L}_2$ re-constructible condition [27] to set the evaluation threshold. Recalling Theorem 1, the error system (22) satisfies the $\mathcal{L}_2$ re-constructible condition with Equation (24). The detection threshold of evaluation function (31) is

$$J_{th} = \frac{1}{N}\left(\gamma^2 d_{\max}^2 + \sup_{x(0),\hat{x}(0)}\gamma V(x(0),\hat{x}(0))\right).\tag{32}$$

The detection logic is defined as follows:

- $J > J_{th}$, sensor fault alarm,
- $J \leq J_{th}$, no fault alarm.

Although the proposed residual generator is designed for fault detection of three-phase current sensor faults, it can be utilized to detect the component or actuator faults in the system. Nevertheless, this is not in the scope of this work.

4. Sensor Fault Isolation Scheme

This section deals with the fault isolation problem of a three-phase current sensor. Since it is difficult to isolate the sensor fault by direct residual analysis, a robust fault estimation based method is presented to generate distinguished residual sensitive to each phase current sensor fault. First, a parameter-dependent fault estimation observer is constructed for system (7)

$$\begin{aligned}\hat{x}(k+1) &= A(\theta_k)\hat{x}(k) + B_u u(k) + L(\theta_k)(y(k) - \hat{y}(k)),\\ \hat{y}(k) &= C\hat{x}(k) + F_f\hat{f}(k),\\ \hat{f}(k+1) &= \hat{f}(k) + \Gamma(\theta_k)(y(k) - \hat{y}(k)),\end{aligned}\tag{33}$$

where $\hat{x}(k) \in \mathcal{R}^{n_x}$, $\hat{y}(k) \in \mathcal{R}^{n_y}$ and $\hat{f}(k) \in \mathcal{R}^{n_f}$ are observer state, observer output and estimate of sensor faults. $L(\theta_k)$ and $\Gamma(\theta_k)$ are the gain matrices. Suppose that

$$\begin{bmatrix} L(\theta_k) \\ \Gamma(\theta_k) \end{bmatrix} = \sum_{i=1}^{4} \eta_{k,i} \begin{bmatrix} L_i \\ \Gamma_i \end{bmatrix}. \tag{34}$$

Let $e_x(k) = x(k) - \hat{x}(k)$ and $e_f(k) = f(k) - \hat{f}(k)$, the estimation error dynamics is expressed as

$$\begin{aligned} e_x(k+1) &= (A(\theta_k) - L(\theta_k)C)\,e_x(k) - L(\theta_k)F_f e_f(k) + B_d d(k), \\ e_f(k+1) &= -\Gamma(\theta_k)C e_x(k) + \left(I - \Gamma(\theta_k)F_f\right)e_f(k) + f(k+1) - f(k). \end{aligned} \tag{35}$$

Choose $\xi(k) = \begin{bmatrix} e_x^T(k), e_f^T(k) \end{bmatrix}^T$ and $\Delta f(k) = f(k+1) - f(k)$,

$$\begin{aligned} \xi(k+1) &= (\bar{A}(\theta_k) - \bar{L}(\theta_k)\bar{C})\,\xi(k) + \bar{B}_d \bar{d}(k), \\ e_f(k) &= \bar{C}_e \xi(k), \end{aligned} \tag{36}$$

with $\bar{d}(k) = \begin{bmatrix} d(k) \\ \Delta f(k) \end{bmatrix}$, $\bar{A}(\theta_k) = \begin{bmatrix} A(\theta_k) & 0_{n_x \times n_f} \\ 0_{n_f \times n_x} & I_{n_f} \end{bmatrix}$, $\bar{L}(\theta_k) = \begin{bmatrix} L(\theta_k) \\ \Gamma(\theta_k) \end{bmatrix}$, $\bar{B}_d = \begin{bmatrix} B_d & 0_{n_x \times n_f} \\ 0_{n_f \times n_d} & I_{n_f \times n_f} \end{bmatrix}$, $\bar{C} = \begin{bmatrix} C & F_f \end{bmatrix}$, $\bar{C}_e = \begin{bmatrix} 0_{n_f \times n_x} & I_{n_f \times n_f} \end{bmatrix}$.

Theorem 2. *If there exists a symmetric positive definite matrix P_1, real matrices $\bar{\mathcal{Y}}_{v,i}$ with appropriate dimensions, $i = 1, \cdots, 4$, positive scalar γ_1, such that the following linear matrix inequality holds*

$$\begin{bmatrix} -P_1 & 0_{(n_x+n_f)\times n_f} & P_1\bar{A}_{v,i} - \bar{\mathcal{Y}}_{v,i}\bar{C} & P\bar{B}_d \\ * & -\gamma_1 I_{n_f \times n_f} & \bar{C}_e & 0_{n_f \times (n_d+n_f)} \\ * & * & -P_1 & 0_{(n_x+n_f)\times(n_d+n_f)} \\ * & * & * & -\gamma_1 I_{(n_d+n_f)\times(n_d+n_f)} \end{bmatrix} < 0, \tag{37}$$

then fault estimaiton error dynamics (36) *satisfies $\mathcal{H}_\infty$ performance $\|e_f(k)\|_2^2 \le \gamma_1^2\|\bar{d}(k)\|_2^2$. The fault estimation observer gain matrix is $\bar{L}(\theta_k) = \sum_{i=1}^{4} \eta_{k,i}(\theta_k)\bar{L}_i$ with $\bar{L}_i = P_1^{-1}\bar{\mathcal{Y}}_{v,i}$.*

Proof of Theorem 2. According to the polytopic decompostion, parameter-dependent matrix $\bar{A}(\theta_k)$ and $\bar{L}(\theta_k)$ are as follows:

$$\bar{A}(\theta_k) = \sum_{i=0}^{4} \eta_{k,i}\bar{A}_{v,i}, \tag{38}$$
$$\bar{L}(\theta_k) = \sum_{i=0}^{4} \eta_{k,i}\bar{L}_{v,i}.$$

Consider the following Lyapunov function

$$V(k) = \xi^T(k)P_1\xi(k). \tag{39}$$

The cost function J_∞ is defined as

$$J_\infty = \sum_{k=0}^{\infty} \left[\Delta V(k) + \frac{1}{\gamma_1}e_f^T(k)e_f(k) - \gamma_1 \bar{d}^T(k)\bar{d}(k) \right] < 0. \tag{40}$$

By substituting $\Delta V(k)$ and $e_f^T(k)\,e_f(k)$ into Equation (40), we have

$$
J_\infty = \begin{bmatrix} \xi(k) \\ \bar{d}(k) \end{bmatrix}^T \begin{bmatrix} \bar{A}^T(\theta_k)\mathcal{P}_1 - \bar{C}^T L^T(\theta_k)\mathcal{P}_1 \\ \bar{B}_d^T \mathcal{P}_1 \end{bmatrix} \mathcal{P}_1^{-1} \begin{bmatrix} \mathcal{P}_1 \bar{A}(\theta_k) - \mathcal{P}_1 L(\theta_k)\bar{C} & \mathcal{P}_1 \bar{B}_d \end{bmatrix} \begin{bmatrix} \xi(k) \\ \bar{d}(k) \end{bmatrix}
$$
$$
+ \begin{bmatrix} \xi(k) \\ \bar{d}(k) \end{bmatrix}^T \begin{bmatrix} \frac{1}{\gamma_1}\bar{C}_e^T \bar{C}_e - \mathcal{P}_1 & 0 \\ 0 & -\gamma_1 I \end{bmatrix} \begin{bmatrix} \xi(k) \\ \bar{d}(k) \end{bmatrix}.
$$
(41)

Furthermore, Equation (41) can be rewritten as

$$
\begin{bmatrix} \bar{A}^T(\theta_k)\mathcal{P}_1 - \bar{C}^T L^T(\theta_k)\mathcal{P}_1 \\ \bar{B}_d^T \mathcal{P}_1 \end{bmatrix} \mathcal{P}_1^{-1} \begin{bmatrix} \mathcal{P}_1 \bar{A}(\theta_k) - \mathcal{P}_1 L(\theta_k)\bar{C} & \mathcal{P}_1 \bar{B}_d \end{bmatrix} + \begin{bmatrix} \frac{1}{\gamma_1}\bar{C}_e^T \bar{C}_e - \mathcal{P}_1 & 0 \\ 0 & -\gamma_1 I \end{bmatrix} < 0.
$$
(42)

Note that Equation (37) is a sufficient condition of (42). Thus, if Equation (37) holds, the estimation error system satisfies $\mathcal{H}_\infty$ performance $\|e_f(k)\|_2^2 \le \gamma_1^2 \|\bar{d}(k)\|_2^2$. The proof is completed. $\square$

This method presents a biased estimation of sensor fault $\hat{f}(k)$ with an upper bound $\gamma_1^2 \|\bar{d}(k)\|_2^2$ due to the disturbances $\bar{d}(k)$. However, it has the ability to locate the fault sensor phase by the tuned isolation threshold for each estimation $\hat{f}_a(k)$, $\hat{f}_b(k)$ and $\hat{f}_c(k)$.

Remark 3. *Before designing the proposed residual generator and fault estimator, it is necessary to check observability of the pair $(A(\theta_k),C)$ and $(\bar{A}(\theta_k),\bar{C})$. This paper checks this property by analyzing the observability of each pair on the vertices.*

5. Simulation Results and Discussion

To illustrate the proposed model based fault diagnosis for current sensor in machine side converter, a MATLAB/SIMULINK (Version R2018a, MathWorks Inc., Natick, MA, USA) model is developed referring to the real laboratory prototype. The parameter is listed in Table 1. A field-oriented control combined with a space vector modulation is applied to control the rotor-side converter. Both the wind conversion system and fault diagnosis algorithm are implemented in the SIMULINK environment. Observer gains can be obtained by solving Equations (23) and (37) with MATLAB LMI tool or Yalmip tool box.

Three types of sensor faults are designed to verify the performance:

- **Type a**: gain error in phase a sensor, only 80% of the measured value fed to the controller,
- **Type b**: bias fault in phase b sensor, 4 A is added to the measured value,
- **Type c**: disconnection of phase c sensor, the measurement output becomes zero.

Type a fault is modeled as multiplicative fault, **Type b** and **Type c** are additive faults. These faults are commonly presented in literature and practice applications.

Table 1. Parameters of the surface-mounted permanent magnet synchronous generator.

Quantity	Value	Quantity	Value
Magnet steel	NdFeB permanent magnet	Insulation class	Class F
Protection	IP54	Stator winding connection	Star connection
Rated voltage	110 V	Rated frequency	32.67 Hz
Stator resistance	0.3667 Ω	Rated power	2.5 kW
Stator inductance	3.29 mH	Rated speed	335 r/min
Flux linkage	0.283 Wb	DC-link voltage	300 V
Generator inertia	0.1133 Kg· m^2	Grid inductance	2 mH
Viscous damping	0.008 N·m·s	Grid resistance	0.19 Ω
Pole pairs	7	Grid voltage	110 V

The proposed method generates three fault isolation variables J_a, J_b, J_c and a detection flag f_d. f_a, f_b and f_c denote the isolation flags related to J_a, J_b and J_c. When the detection observer detects fault in the system, f_d changes from '0' to '1'. Only when the isolation variables J_a, J_b and J_c exceed the defined thresholds will the corresponding isolation flags change from '0' to '1'.

5.1. Performance for Single Sensor FDI with External Disturbance

The mechanical power fed to generator varies slowly to simulate real wind power in all simulation. Type a and type c fault occur at $t = 0.4$ s and $t = 2.0$ s as shown in Figure 3 and Figure 4. Fault detection variable f_d changes instantly and corresponding isolation flag f_a and f_c change subsequently at $t = 0.42$ s and $t = 2.01$ s. In addition, the mechanical power variation starts at $t = 0.5$ s, which causes sudden variation of the rotor speed. During this period, f_d and f_a remain higher than the threshold while f_b and f_c are lower than detection threshold, which indicates that the isolation algorithm is robust to the disturbances.

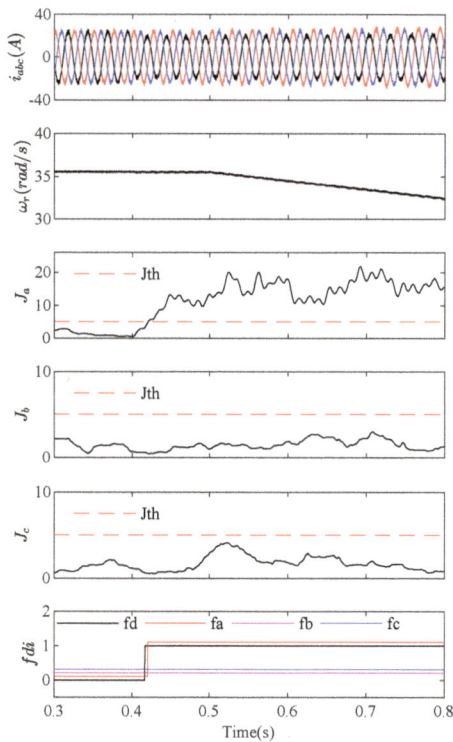

Figure 3. Simulation results on the single phase fault isolation, under varying mechanical disturbance (from top to bottom are phase a current i_{abc}, rotor speed ω_r, isolation variables f_a, f_b, f_c and FDI flags).

The current waveforms of the healthy phase will be distorted by the faulty sensor because of feedback control. As shown in Figure 4, **Type c** fault of phase c current sensor distorts phase a and phase b currents. When phase c fault is triggered at $t = 2$ s, fault detection flag f_d and isolation flag f_c change from '0' to '1' while f_a and f_b remain '0'. This reports that the proposed isolation variables are only sensitive to the corresponding fault phase and makes it possible to isolate all phase current sensor faults. The next section will show the multiple sensor fault diagnosis performance.

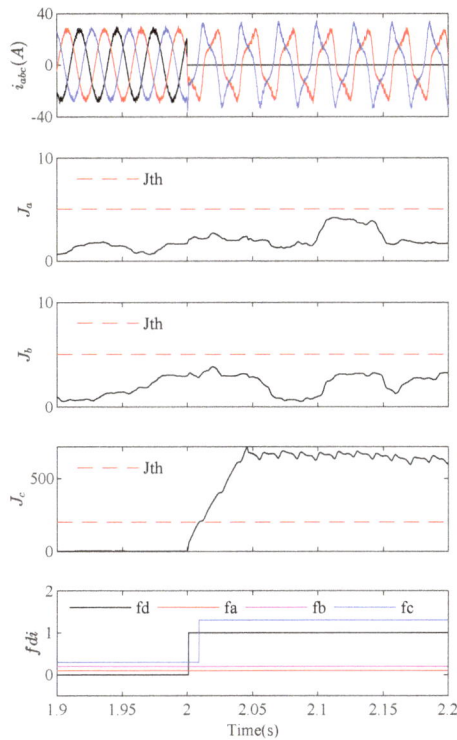

Figure 4. Simulation results for fault isolation of **Type c** fault in phase c (from top to bottom are three-phase current i_{abc}, isolation variables f_a, f_b, f_c and FDI flags f_d, f_a, f_b, f_c).

5.2. Multiple Fault Detection and Isolation

Multiple sensor FDI scenarios are presented as follows:

- **Type a** and **Type b** fault at $t = 0.4$ s and $t = 0.8$ s,
- **Type b** and **Type c** fault at $t = 0.4$ s and $t = 0.7$ s,
- **Type a** and **Type c** fault at $t = 0.4$ s and $t = 0.6$ s,
- Three type faults occur simultaneously at $t = 0.4$ s, $t = 0.7$ s and $t = 1.2$ s.

The FDI variable behaviors during multiple current sensor faults are shown in Figure 5–8. The fault detection flag f_d and isolation flags f_a, f_b, f_c change from '0' to '1' after the faults occurred. For two sensor fault scenarios, the isolation flag for healthy sensor remains '0', indicating that it operates normally.

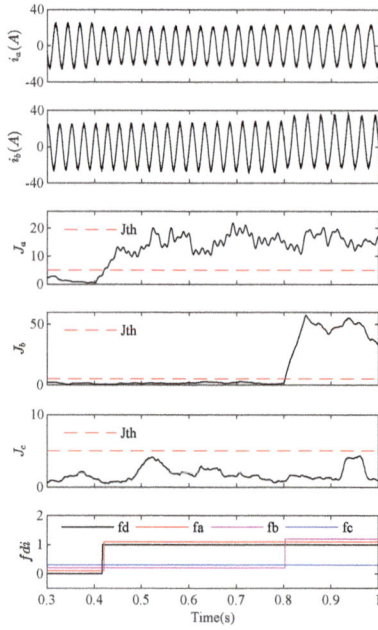

Figure 5. Simulation results on simultaneous isolation of phase a and phase b fault.

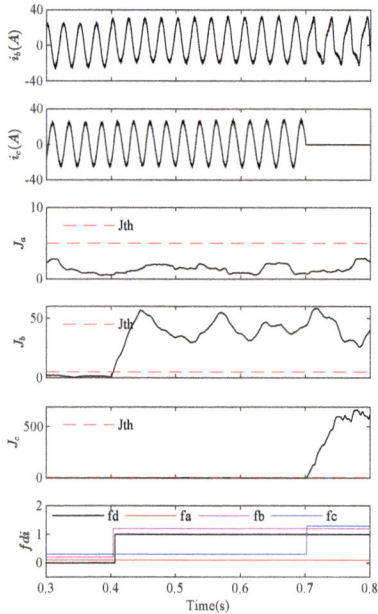

Figure 6. Simulation results on simultaneous isolation of phase b and phase c fault.

5.3. Comparison with the Existing Sensor FDIs

Table 2 presents a brief comparison of model-based current sensor FDIs. These schemes are proposed for an IM system [8,28–30], DFIG system [9,10] and PMSG system [2,31]. The schemes in [28–30] are presented to isolate single sensor fault. In these approaches, a bank of observers is established to monitor each sensor fault on the basis of the rest sensors are health. T-S fuzzy observer [9] and sliding mode observer [31] are proposed to isolate single type of faults in $\alpha - \beta$ and $d - q$ frame. The measurements of three-phase currents are utilized to generate estimation errors that can only isolate faults in a stationary frame or synchronous reference frame. The TVKF scheme is presented for the PMSG system in [2]. It utilizes a generalized likelihood ratio maximum-shift strategy to evaluate the faulty residuals generated by the frequency domain model, which results in higher computation complexity. This work is presented for multiple sensor fault isolation of PMSG or PMSM based system with a simple observer based algorithm framework. In general, the proposed method can detect and isolate multiple sensor faults simultaneously in a short diagnosis time. It employs mechanical signals and current signals in the control loop without any additional measurements and hardware circuits, and can be integrated into the control loop. In addition, it shows excellent performance in multiple types of faults including gain fault, biased fault, and disconnection fault. The fault detection threshold is related to disturbances with respect to Theorem 1 and Theorem 2. Furthermore, the maximum and minimum power for WECSs are specifically defined by the operational region shown in Figure 2.

Table 2. Comparison of a model-based current sensor FDI scheme.

FD Scheme	Measurements	Fault Types	Isolability	System Model	Detection Variables
Bank of observers [28]	1 voltage, 3 currents, 1 speed	Type c	Single	IM model in $\alpha - \beta$	Estimation errors of rotor flux and speed
EKF [29]	1 voltage, 2 currents, 1 speed	Type c	Single	IM model in $\alpha - \beta$	Estimation errors of phase currents
Adaptive observer [30]	1 voltage, 2 currents, 1 speed	Type c	Single	IM model in $d - q$	Fault inference based on current errors
Bank of observers [8]	1 voltage, 2 currents, 1 speed	Type a, Type b, Type c	2 faults	IM model in $\alpha - \beta$	Geometric residuals
TS fuzzy observers [9]	2 currents, 1 speed	Type b	2 faults	DFIG model in $\alpha - \beta$	Estimation errors of the states
Integrated filters [10] $\mathcal{H}\_/\mathcal{H}_\infty$	2 currents, 1 speed, 1 position	Type a, Type b, Type c	2 faults	DFIG model in $d - q$ and $\alpha - \beta$	Generalized likelihood ratio of residuals
TVKF [2]	2 currents, 1 speed	Type a, Type b, Type c	3 faults	PMSG model in harmonic domain	Generalized likelihood ratio of residuals
Sliding mode observer [31]	3 currents, 1 speed, 1 position	Type c	2 faults	PMSG model in $d - q$	Evaluation of estimation errors
This method	3 currents, 1 speed, 1 position	Type a, Type b, Type c	3 faults	PMSG model in $\alpha - \beta$	Evaluation of the fault estimates

5.4. Discussions

In this paper, only simulation results are presented to validate the FDI performance. The proposed scheme is designed for online operation and is independent from control strategies. Some essential issues are discussed with respect to the lack of experimental results.

a   The component parameters of simulation model come from the real laboratory prototype with rated power 2.5 kW. Its controller parameters are designed on the simulation file and can guarantee the control performance. The real waveforms and power characteristics are the same as those of simulation results. The observer design is a dual problem of controller design. Thus, the parameters designed in SIMULINK environment can be applied to the real experiments.

b   The threshold selection is the most challenging problem in implementing the proposed algorithm. In real application, the mechanical torque and measurement noise are different from the simulation configuration. This will be further introduced into the observer and error dynamics. These effects can be modeled as generalized unknown disturbances. The upper bound of the disturbances in real application is slightly different from simulation scenarios. However, this does not affect the performance since the upper bounds of disturbances and faults hold for real applications.

c   The harmonics is another issue for current sensor fault diagnosis. The influences of harmonics on system behavior need to be further discussed with respect to system parameter and dynamics variation. However, few results have been presented for dealing with this problem, even for the controller designs in [20–22]. Recalling the FDI schemes in Table 2, only the method in [2] utilizes the harmonic model of PMSG to diagnose additive and multiplicative faults in current sensors. The state space model and output equation are linear combinations of each order harmonic in frequency domain, which indicates that the residuals can be modeled as the combination of finite harmonics. The proposed FDI takes the time domain behaviors of residuals into consideration. The average value of each fault estimate is calculated with a sliding window. Current sensor faults are evaluated via the threshold function defined in Equation (32). From this perspective, the harmonics will not affect the residual evaluation in time domain analysis.

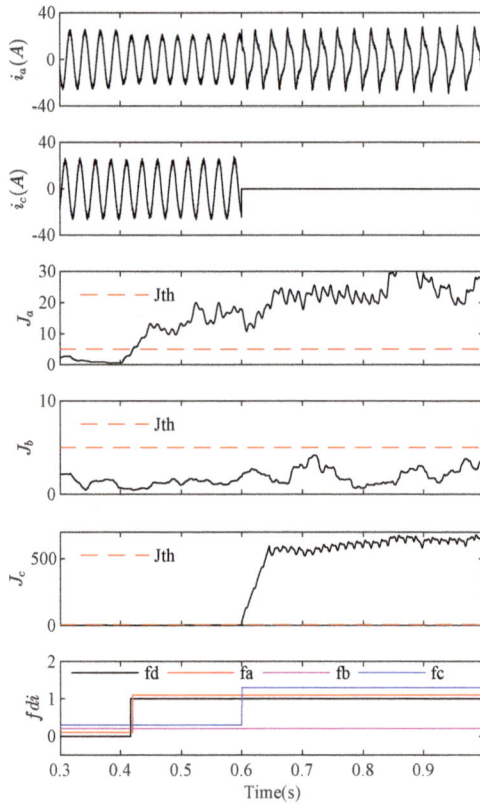

Figure 7. Simulation results on simultaneous isolation of phase a and phase c fault.

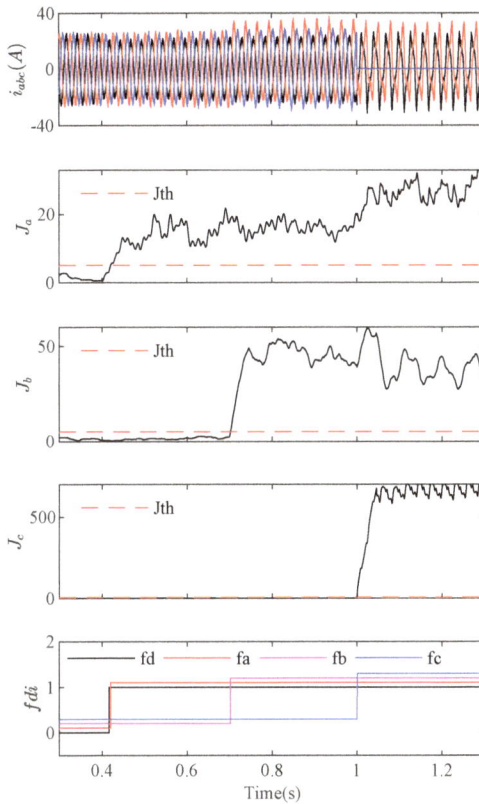

Figure 8. Simulation results on simultaneous isolation of three-phase sensor faults.

6. Conclusions

In this paper, a robust observer based sensor FDI scheme, targeting both additive and multiplicative faults, is presented for PMSG in WECSs. This method isolates multiple sensor faults via two procedures: robust residual generation based fault detection and fault estimation based isolation. The system is reformulated as a LPV model in the $\alpha - \beta$ frame by introducing electromechanical dynamics of PMSG. The polytopic decomposition technique is applied to obtain the parameter-dependent form of the system model by defining a convex polytope with four vertices. Furthermore, the gain-scheduled residual generator and fault estimator are designed by $\mathcal{H}_\infty$ synthesis in the form of LMIs. The proposed gain-scheduled FDI scheme is capable of online monitoring of three-phase currents and isolating multiple sensor faults under varying disturbances.

The proposed scheme is implemented in a MATLAB/SIMULINK environment and multiple sensor faults are isolated correctly. Due to the lack of experimental validation, the corresponding issues are discussed in Section 5.4, of which the challenging issue is the influence of disturbances and harmonics on threshold selection and system dynamics. In a real power conversion system, the diagnosis thresholds in (32) need to be investigated further by defining the augmented disturbances including measurement noise and parameter uncertainties. However, it does not affect the theoretical results of Theorem 1 and Theorem 2 since the observability of the pair $(A(\theta_k), C)$ and $(\bar{A}(\theta_k), \bar{C})$ is independent of disturbances. Since further results is lacked about the influence of harmonics on system dynamics, it is difficult to quantify these effects on the system. The results in [20–22] indicate that the gain-scheduled controller design based on the this LPV model suffers less from the harmonics. Observer design as the dual problem of controller design is less dependent on the harmonic problem.

In addition, fault estimates are sensitive enough to distinguish each phase current sensor fault but cannot be applied directly for the purpose of fault compensation because of the unknown disturbances. The future work will be focused on the unbiased fault estimation and fault tolerant control for PMSG based WECS system on the basis of this LPV modeling technology.

Author Contributions: Methodology and Writing—Original Draft, Z.Y.; Writing—Review and Editing, Y.C. and H.Y.; Data Curation, S.T.

Funding: This research was funded by the National Natural Science Foundation of China Grant No. 61633005 and 61773080, and Chongqing Nature Science Foundation of Fundamental Science and Frontier Technologies Grant No. cstc2015jcyjB0569, and the Scientific Reserve Talent Programs of Chongqing University Grant No. cqu2017CDHB1B04.

Acknowledgments: The authors would like to thank the editors and all of the anonymous reviewers for their valuable insights and comments.

Conflicts of Interest: The authors declare no conflict of interest.

Appendix A

The system matrices of Equation (6) are as follows:

$$
A(\theta_k) = \begin{bmatrix}
1 - \frac{R_s T_s}{L_s} & 0 & \frac{n_p \psi T_s}{L_s} \sin(\theta_k) & 0 \\
0 & 1 - \frac{R_s T_s}{L_s} & -\frac{n_p \psi T_s}{L_s} \cos(\theta_k) & 0 \\
-\frac{3 n_p \psi T_s}{2J} \sin(\theta_k) & \frac{3 n_p \psi T_s}{2J} \cos(\theta_k) & 1 - \frac{F T_s}{J} & 0 \\
0 & 0 & n_p T_s & 1
\end{bmatrix},
$$

$$
B_u = \begin{bmatrix} \frac{T_s}{L_s} & 0 \\ 0 & \frac{T_s}{L_s} \\ 0 & 0 \\ 0 & 0 \end{bmatrix}; \quad
B_d = \begin{bmatrix} 0 \\ 0 \\ -\frac{T_s}{J} \\ 0 \end{bmatrix}; \quad
C = \begin{bmatrix}
1 & 0 & 0 & 0 \\
-\frac{1}{2} & \frac{\sqrt{3}}{2} & 0 & 0 \\
-\frac{1}{2} & -\frac{\sqrt{3}}{2} & 0 & 0 \\
0 & 0 & 1 & 0 \\
0 & 0 & 0 & 1
\end{bmatrix}.
\tag{A1}
$$

References

1. Wu, C.; Guo, C.; Xie, Z.; Ni, F.; Liu, H. A Signal-Based Fault Detection and Tolerance Control Method of Current Sensor for PMSM Drive. *IEEE Trans. Ind. Electron.* **2018**, *65*, 9646–9657. [CrossRef]
2. Beddek, K.; Merabet, A.; Kesraoui, M.; Tanvir, A.A.; Beguenane, R. Signal-Based Sensor Fault Detection and Isolation for PMSG in Wind Energy Conversion Systems. *IEEE Trans. Instrum. Meas.* **2017**, *66*, 2403–2412. [CrossRef]
3. Song, Y.; Wang, B. Survey on Reliability of Power Electronic Systems. *IEEE Trans. Power Electron.* **2013**, *28*, 591–604. [CrossRef]
4. Yang, Z.; Chai, Y. A survey of fault diagnosis for onshore grid-connected converter in wind energy conversion systems. *Renew. Sustain. Energy Rev.* **2016**, *66*, 345–359. [CrossRef]
5. Yang, S.; Xiang, D.; Bryant, A.; Mawby, P.; Ran, L.; Tavner, P. Condition Monitoring for Device Reliability in Power Electronic Converters: A Review. *IEEE Trans. Power Electron.* **2010**, *25*, 2734–2752. [CrossRef]
6. Rothenhagen, K.; Fuchs, F.W. Current Sensor Fault Detection, Isolation, and Reconfiguration for Doubly Fed Induction Generators. *IEEE Trans. Ind. Electron.* **2009**, *56*, 4239–4245. [CrossRef]
7. Rothenhagen, K.; Fuchs, F.W. Doubly Fed Induction Generator Model-Based Sensor Fault Detection and Control Loop Reconfiguration. *IEEE Trans. Ind. Electron.* **2009**, *56*, 4229–4238. [CrossRef]
8. Aguilera, F.; de la Barrera, P.; Angelo, C.D.; Trejo, D.E. Current-sensor fault detection and isolation for induction-motor drives using a geometric approach. *Control Eng. Pract.* **2016**, *53*, 35–46. [CrossRef]
9. Abdelmalek, S.; Barazane, L.; Larabi, A.; Bettayeb, M. A novel scheme for current sensor faults diagnosis in the stator of a DFIG described by a T-S fuzzy model. *Measurement* **2016**, *91*, 680–691. [CrossRef]

10. Boulkroune, B.; Gálvez-Carrillo, M.; Kinnaert, M. Combined Signal and Model-Based Sensor Fault Diagnosis for a Doubly Fed Induction Generator. *IEEE Trans. Control Syst. Technol.* **2013**, *21*, 1771–1783. [CrossRef]
11. Akrad, A.; Hilairet, M.; Diallo, D. Design of a Fault-Tolerant Controller Based on Observers for a PMSM Drive. *IEEE Trans. Ind. Electron.* **2011**, *58*, 1416–1427. [CrossRef]
12. Mwasilu, F.; Jung, J. Enhanced Fault-Tolerant Control of Interior PMSMs Based on an Adaptive EKF for EV Traction Applications. *IEEE Trans. Power Electron.* **2016**, *31*, 5746–5758. [CrossRef]
13. Kommuri, S.K.; Defoort, M.; Karimi, H.R.; Veluvolu, K.C. A Robust Observer-Based Sensor Fault-Tolerant Control for PMSM in Electric Vehicles. *IEEE Trans. Ind. Electron.* **2016**, *63*, 7671–7681. [CrossRef]
14. Kommuri, S.K.; Lee, S.B.; Veluvolu, K.C. Robust Sensors-Fault-Tolerance With Sliding Mode Estimation and Control for PMSM Drives. *IEEE/ASME Trans. Mech.* **2018**, *23*, 17–28. [CrossRef]
15. Foo, G.H.B.; Zhang, X.; Vilathgamuwa, D.M. A Sensor Fault Detection and Isolation Method in Interior Permanent-Magnet Synchronous Motor Drives Based on an Extended Kalman Filter. *IEEE Trans. Ind. Electron.* **2013**, *60*, 3485–3495. [CrossRef]
16. Wang, Y.; Meng, J.; Zhang, X.; Xu, L. Control of PMSG-Based Wind Turbines for System Inertial Response and Power Oscillation Damping. *IEEE Trans. Sustain. Energy* **2015**, *6*, 565–574. [CrossRef]
17. Badihi, H.; Zhang, Y.; Hong, H. Wind Turbine Fault Diagnosis and Fault-Tolerant Torque Load Control Against Actuator Faults. *IEEE Trans. Control Syst. Technol.* **2015**, *23*, 1351–1372. [CrossRef]
18. Bifaretti, S.; Iacovone, V.; Rocchi, A.; Tomei, P.; Verrelli, C. Nonlinear speed tracking control for sensorless PMSMs with unknown load torque: From theory to practice. *Control Eng. Pract.* **2012**, *20*, 714–724. [CrossRef]
19. Tomei, P.; Verrelli, C.M. Observer-Based Speed Tracking Control for Sensorless Permanent Magnet Synchronous Motors With Unknown Load Torque. *IEEE Trans. Autom. Control* **2011**, *56*, 1484–1488. [CrossRef]
20. Kang, C.M.; Lee, S.; Chung, C.C. Discrete-Time LPV $\mathcal{H}_2$ Observer With Nonlinear Bounded Varying Parameter and Its Application to the Vehicle State Observer. *IEEE Trans. Ind. Electron.* **2018**, *65*, 8768–8777. [CrossRef]
21. Lee, Y.; Lee, S.; Chung, C.C. LPV $\mathcal{H}_\infty$ Control with Disturbance Estimation for Permanent Magnet Synchronous Motors. *IEEE Trans. Ind. Electron.* **2018**, *65*, 488–497. [CrossRef]
22. Lee, Y.; Shin, D.; Kim, W.; Chung, C.C. Nonlinear $\mathcal{H}_2$ Control for a Nonlinear System With Bounded Varying Parameters: Application to PM Stepper Motors. *IEEE/ASME Trans. Mech.* **2017**, *22*, 1349–1359. [CrossRef]
23. de Souza, C.E.; Barbosa, K.A.; Neto, A.T. Robust $\mathcal{H}_\infty$ filtering for discrete-time linear systems with uncertain time-varying parameters. *IEEE Trans. Signal Process.* **2006**, *54*, 2110–2118. [CrossRef]
24. Pandey, A.P.; de Oliveira, M.C. Discrete-time $\mathcal{H}_\infty$ control of linear parameter-varying systems. *Int. J. Control* **2018**. [CrossRef]
25. Pandey, A.P.; de Oliveira, M.C. A new discrete-time stabilizability condition for Linear Parameter-Varying systems. *Automatica* **2017**, *79*, 214–217. [CrossRef]
26. Pandey, A.P.; de Oliveira, M.C. On the Necessity of LMI-Based Design Conditions for Discrete Time LPV Filters. *IEEE Trans. Autom. Control* **2018**, *63*, 3187–3188. [CrossRef]
27. Li, L.; Ding, S.X.; Qiu, J.; Yang, Y.; Zhang, Y. Weighted Fuzzy Observer-Based Fault Detection Approach for Discrete-Time Nonlinear Systems via Piecewise-Fuzzy Lyapunov Functions. *IEEE Trans. Fuzzy Syst.* **2016**, *24*, 1320–1333. [CrossRef]
28. Yu, Y.; Wang, Z.; Xu, D.; Zhou, T.; Xu, R. Speed and Current Sensor Fault Detection and Isolation Based on Adaptive Observers for IM Drives. *J. Power Electron.* **2014**, *14*, 967–979. [CrossRef]
29. Zhang, X.; Foo, G.; Vilathgamuwa, M.D.; Tseng, K.J.; Bhangu, B.S.; Gajanayake, C. Sensor fault detection, isolation and system reconfiguration based on extended Kalman filter for induction motor drives. *IET Electr. Power Appl.* **2013**, *7*, 607–617. [CrossRef]

30. Najafabadi, T.A.; Salmasi, F.R.; Jabehdar-Maralani, P. Detection and Isolation of Speed-, DC-Link Voltage-, and Current-Sensor Faults Based on an Adaptive Observer in Induction-Motor Drives. *IEEE Trans. Ind. Electron.* **2011**, *58*, 1662–1672. [CrossRef]

31. Saha, S.; Haque, M.E.; Mahmud, M.A. Diagnosis and Mitigation of Sensor Malfunctioning in a Permanent Magnet Synchronous Generator Based Wind Energy Conversion System. *IEEE Trans. Energy Convers.* **2018**, *33*, 938–948. [CrossRef]

![applied sciences logo] *applied sciences*

MDPI

Article

Fault Studies and Distance Protection of Transmission Lines Connected to DFIG-Based Wind Farms

Bin Li [1], Junyu Liu [1,*], Xin Wang [2] and Lili Zhao [3]

[1] Smart Grid Key Laboratory of Ministry of Education, Tianjin University, Tianjin 300072, China; binli@tju.edu.cn

[2] Maintenance Company of Tianjin Electric Power Company, Tianjin 300250, China; wangxin_tju@163.com

[3] Electric Power Dispatching Control Center of Guizhou Power Grid Company, Guiyang 550002, China; lilizhao09@126.com

* Correspondence: liujunyu0811@126.com; Tel.: +86-022-2740-5477

Received: 12 February 2018; Accepted: 31 March 2018; Published: 5 April 2018

Featured Application: The work can improve technology level of relay protection equipment and distance protection reliability of transmission lines connected to wind farms.

Abstract: Doubly fed induction generator (DFIG) based wind farms are being increasingly integrated into power grids with transmission lines, and distance protection is usually used as either the main or the backup protection for the transmission line. This paper analyzes the composition of a DFIG short circuit current and indicates the existence of a rotor speed frequency component. By analyzing several real fault cases of the DFIG-based wind farms connected to transmission lines, the weak power supply system and current frequency deviation of the wind farm side are illustrated. When a fault occurs on the transmission line, the short circuit current on the wind farm side is small and its frequency may no longer be nominal due to the existence of rotor speed frequency component, whereas the voltage frequency remains nominal frequency because of the grid support. As a result, the conventional distance protection cannot accurately measure the impedance, which can result in unnecessary circuit breaker tripping. Therefore, a time-domain distance protection method combined with the least-squares algorithm is proposed to address the problem. The efficacy of the proposed method is validated with real fault cases and simulation.

Keywords: DFIG-based wind farm; transmission line; real fault cases; fault characteristics; distance protection

1. Introduction

In addressing the problems of environmental pollution and energy shortages, the world has been focusing on the development and use of wind energy, and large-scale wind farms are being increasingly integrated into the power grid. Wind turbines based on doubly fed induction generator (DFIG) have been widely used in existing wind farms [1,2] due to the benefits including flexible power control, high efficiency, and small converter capacity [3–5].

However, DFIG is sensitive to grid disturbances, especially to voltage sags. The grid voltage sags cause a large transient overcurrent in the DFIG rotor circuit [6,7]. As a result, the DFIG is disconnected from the grid to avoid damage to the rotor side converter (RSC). With the increasing popularity of wind power, the grid codes [8–10] require the wind farms to remain connected to power grid during grid–voltage sags to ensure system stability. Therefore, low voltage ride through (LVRT) techniques are proposed to protect the converter, meaning that the DFIG does not need to be disconnected from the grid during disturbances [11,12]. A commonly adopted solution to protect the DFIG converter during voltage sags is to use a crowbar circuit to isolate the RSC from the rotor circuit [13,14].

Many studies have been performed to investigate the DFIG fault characteristics considering LVRT capability. Due to the influence of power electronic devices, the DFIG fault characteristics are different from the traditional synchronous generator. The behavior of the DFIG during three-phase voltage sags was analyzed [15]. The authors divided the magnetic flux of the machine into forced flux and natural flux to help understand the causes of the overcurrent that appear during a fault. The physical behavior of the DFIG was presented and later developed with analytical solutions for the generalized DFIG equations operating under symmetrical fault conditions [16]. Comparing the conventional induction machine with DFIG, the short-circuit current expression of a crowbar-protected DFIG and the maximum value of the short-circuit current were outlined [17,18]. The DFIG operation under asymmetrical voltage sags and why such sags are more harmful than symmetrical sags were explained [19]. In addition, the DFIG fault current characteristics under non-severe fault conditions were studied [20–22]. Under non-severe faults, the crowbar protection is not activated and the DFIG fault current characteristics are considerably different than that under severe fault conditions. Notably, the fault characteristics of DFIG with LVRT capability are influenced by the control strategy. A detailed analysis was completed by considering the influence of the controller on short-circuit parameters, such as time constants, initial symmetrical short-circuit current, and the peak short-circuit current [23]. Hence, many researchers have attempted to improve the DFIG control strategy to meet the grid code requirements [24–26].

DFIG pose new problems and challenges for the traditional relaying protection of the power grid. Although the aforementioned DFIG fault characteristics have been scrutinized in many publications, the negative impacts of DFIG on the protection of the transmission system, and particularly distance protection, have not received sufficient attention. Large-scale DFIG-based wind farms are generally integrated into power grids with transmission lines, and distance protection is usually used as either the main or the backup protection for the transmission line [27]. Refs. [28,29] indicated that traditional distance protection and directional elements are unreliable when a balanced fault occurs because the fault current frequency of DFIG significantly deviates from the nominal frequency. The impacts of grounding configurations on ground protective devices used in DFIG-based wind energy conversion systems were studied in [30,31]. However, fault characteristics of DFIG-based wind farms and the performance of distance protection when unbalanced faults occur on high voltage transmission lines should be further studied.

According to analyses in the literature [17–19], the DFIG short-circuit current involves a transient alternating current (AC) component decay at near rotor frequency. As a result, the conventional distance protection based on nominal frequency may fail to measure the correct impedance. To fill the gap, this paper analyzes real fault cases where unbalanced faults occurred at the transmission lines connected to DFIG-based wind farms in China. The fault characteristics and distance protection applicability of the transmission lines connected to DFIG-based wind farms are obtained. Furthermore, a time-domain distance protection method combined with the least-squares algorithm is proposed. The proposed method can overcome the influence of frequency variation as a result that fault distances can be identified accurately and quickly when faults occur on the transmission line connected to a DFIG-based wind farm. With the help of the proposed method, Relay Protection Equipment Manufacturers can improve the equipment level of technology, and the grid companies can improve the safe operation level of power system.

The rest of this work is structured as follows. Section 2 illustrates the composition of DFIG short circuit current and fault characteristics of DFIG-based wind farms connected to transmission lines. Section 3 analyzes the distance protection applicability, and a time-domain distance protection method is proposed for the transmission lines connected to DFIG-based wind farms. The reliability and correctness of the method are validated through real fault cases and simulation tests. Section 4 presents the conclusions.

2. Fault Analysis of DFIG-Based Wind Farm and Its Verification

Areas with abundant wind energy are generally far from load centers. Therefore, large-scale wind farms are established and wind power is transmitted to the power system over long distances. The typical DFIG-based wind farm is shown in Figure 1. The outlet voltage of the DFIG is generally 690 V. The wind power is transmitted through a wind turbine unit transformer, collector system, substation, and transmission lines. The voltage level of the collector system is usually 35 kV and the voltage levels of the transmission line are usually 110 kV or 220 kV. Notably, the 35 kV system is not grounded, and the neutral point of the high-voltage system is directly grounded.

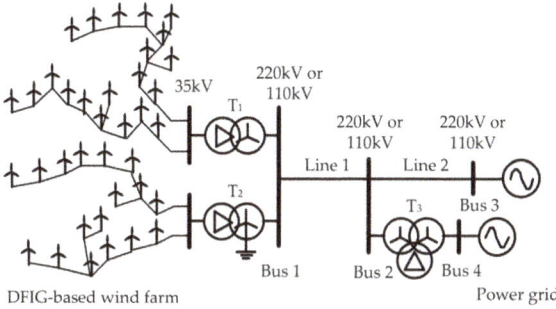

Figure 1. Topology of the doubly fed induction generator (DFIG)-based wind farm.

Figure 2 shows a typical DFIG with a rotor crowbar protection circuit. It mainly consists of a wind turbine, gearbox, asynchronous generator, RSC, grid side converter (GSC), rotor crowbar protection circuit, and control system. The DFIG stator is directly connected to the grid, whereas the rotor is connected to the grid by a back-to-back converter. The converter system enables variable speed operation of the wind turbine by decoupling the power system electrical frequency and the rotor mechanical frequency. A more detailed description of the DFIG system together with its control was previously reported [32].

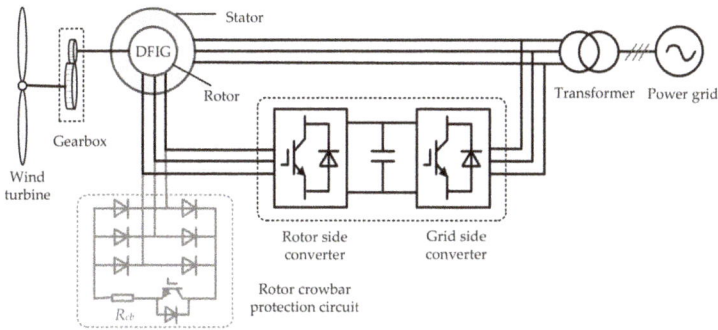

Figure 2. A typical DFIG with a rotor crowbar protection circuit.

2.1. DFIG Short Circuit Current

The transformer connecting the DFIG to the grid is generally a Δ/Y connection; therefore, only the positive and negative sequence components of a DFIG during faults need to be analyzed. A space vector description was used to present the generalized induction machine equations [7,16,17].

Using a generator convention for the stator windings and a motor convention for the rotor windings, the stator and rotor voltage equations and flux linkage equations in a stationary reference frame are expressed as:

$$\begin{cases} \vec{v}_s = -R_s \vec{i}_s + D\vec{\psi}_s \\ \vec{v}_r = R_r \vec{i}_r + D\vec{\psi}_r - j\omega_r \vec{\psi}_r \end{cases}, \tag{1}$$

$$\begin{cases} \vec{\psi}_s = -L_s \vec{i}_s + L_m \vec{i}_r \\ \vec{\psi}_r = -L_m \vec{i}_s + L_r \vec{i}_r \end{cases}, \tag{2}$$

where $\vec{v}$, $\vec{i}$, and $\vec{\psi}$ represent the voltage space vector, current space vector, and flux linkage space vector, respectively; ω is the electrical angular velocity; R is the per-phase resistance; subscripts s and r denote the stator and rotor quantities, respectively; L_s and L_r are the per-phase stator and rotor self-inductances, respectively; and L_m is the per-phase mutual inductance; and D is the time-differential operator.

By solving Equation (2), the current can be written as equation of the flux linkages as:

$$\vec{i}_s = -\frac{1}{L_s'}\vec{\psi}_s + \frac{L_m}{L_r}\frac{1}{L_s'}\vec{\psi}_r, \tag{3}$$

where $L_s' = L_s - L_m^2/L_r$.

Assuming that, at time $t = 0$, an asymmetrical short circuit fault occurs at the DFIG. The post-fault stator voltage is:

$$\vec{v}_s = \vec{v}_{s1} + \vec{v}_{s2} = (1-A)V_{pre}e^{j(\omega_s t+\theta)} + V_2 e^{j(-\omega_s t+\theta')}, \tag{4}$$

where $\vec{v}_{s1}$ and $\vec{v}_{s2}$ are the positive and negative sequence voltages, respectively; A is the voltage sag ratio; V_{pre} is the amplitude of the pre-fault stator voltage; V_2 is the amplitude of the post-fault stator negative sequence voltage; and θ and θ' are the phase angles of $\vec{v}_{s1}$ and $\vec{v}_{s2}$ at time $t = 0$, respectively. The post-fault stator positive and negative sequence flux linkage are obtained in Equation (5). The detailed derivation is presented in Appendix A (i).

$$\begin{cases} \vec{\psi}_{s1} = \vec{\psi}_{sn1} + \vec{\psi}_{sf1} = \frac{AV_{pre}}{j\omega_s}e^{-t/T_s'+j\theta} + \frac{(1-A)V_{pre}}{j\omega_s}e^{j(\omega_s t+\theta)} \\ \vec{\psi}_{s2} = \vec{\psi}_{sn2} + \vec{\psi}_{sf2} = \frac{V_2}{j\omega_s}e^{-t/T_s'+j\theta'} - \frac{V_2}{j\omega_s}e^{j(-\omega_s t+\theta')} \end{cases}, \tag{5}$$

where $T_s' = L_s'/R_s$ is the stator decaying time constant; $\vec{\psi}_{sn1}$ and $\vec{\psi}_{sf1}$ are the natural component and forced component of $\vec{\psi}_{s1}$, respectively; and $\vec{\psi}_{sn2}$ and $\vec{\psi}_{sf2}$ are the natural component and forced component of $\vec{\psi}_{s2}$, respectively.

As for the steady state after a fault in a synchronous rotating reference frame, the positive sequence voltage equations and flux linkage equations are shown in Equations (6) and (7). The negative sequence voltage equations and flux linkage equations are shown in Equations (8) and (9), respectively. The equivalent circuits of DFIG in a synchronous rotating reference frame are shown in Figure 3 [17,33].

$$\begin{cases} \vec{v}_{s1} = -R_s \vec{i}_{sf1} + D\vec{\psi}_{sf1} + j\omega_s \vec{\psi}_{sf1} \\ \vec{v}_{r1} = R_r \vec{i}_{rf1} + D\vec{\psi}_{rf1} + j(\omega_s - \omega_r)\vec{\psi}_{rf1} \end{cases}, \tag{6}$$

$$\begin{cases} \vec{\psi}_{sf1} = -L_s \vec{i}_{sf1} + L_m \vec{i}_{rf1} \\ \vec{\psi}_{rf1} = -L_m \vec{i}_{sf1} + L_r \vec{i}_{rf1} \end{cases}, \tag{7}$$

$$\begin{cases} \vec{v}_{s2} = -R_s \vec{i}_{sf2} + D\vec{\psi}_{sf2} - jw_s\vec{\psi}_{sf2} \\ \vec{v}_{r2} = R_r \vec{i}_{rf2} + D\vec{\psi}_{rf2} - j(2-s)w_s\vec{\psi}_{rf2} \end{cases}, \tag{8}$$

$$\begin{cases} \vec{\psi}_{sf2} = -L_s \vec{i}_{sf2} + L_m \vec{i}_{rf2} \\ \vec{\psi}_{rf2} = -L_m \vec{i}_{sf2} + L_r \vec{i}_{rf2} \end{cases}, \tag{9}$$

where R_r includes the rotor resistance and the crowbar resistance. $s = (w_s - w_r)/w_s$ is the DFIG slip. $L_{s\sigma}$ and $L_{r\sigma}$ are the stator and rotor leakage inductances, respectively, knowing that $L_s = L_{s\sigma} + L_m$ and $L_r = L_{r\sigma} + L_m$.

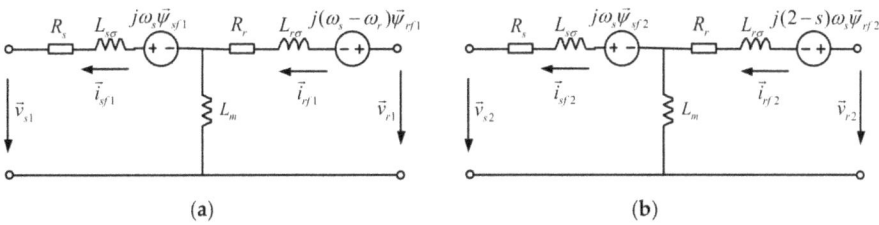

Figure 3. The equivalent circuits of DFIG at the synchronous rotating reference frame: (**a**) the positive sequence equivalent circuit; and (**b**) the negative sequence equivalent circuit.

Then, the post-fault rotor positive and negative sequence flux linkages are obtained in Equation (10), respectively. The detailed derivation is presented in Appendix A (ii).

$$\begin{cases} \vec{\psi}_{r1} = \vec{\psi}_{rn1} + \vec{\psi}_{rf1} = [\vec{\psi}_{r0} - f_1(s)\frac{(1-A)V_{pre}}{jw_s}e^{j\theta}]e^{-t/T_r'+jw_r t} + f_1(s)\frac{(1-A)V_{pre}}{jw_s}e^{j(w_s t+\theta)} \\ \vec{\psi}_{r2} = \vec{\psi}_{rn2} + \vec{\psi}_{rf2} = f_2(s)\frac{V_2}{jw_s}e^{j\theta'}e^{-t/T_r'+jw_r t} - f_2(s)\frac{V_2}{jw_s}e^{j(-w_s t+\theta')} \end{cases}, \tag{10}$$

where $T_r' = L_r'/R_r$ is the rotor decaying time constant; $\vec{\psi}_{rn1}$ and $\vec{\psi}_{rf1}$ are the natural component and forced component of $\vec{\psi}_{r1}$, respectively; and $\vec{\psi}_{rn2}$ and $\vec{\psi}_{rf2}$ are the natural component and forced component of $\vec{\psi}_{r2}$, respectively; $f_1(s) = \frac{R_r L_m/L_s}{R_r + j(w_s - w_r)L_r'}$, $f_2(s) = \frac{R_r L_m/L_s}{R_r - j(2-s)w_s L_r'}$, and $L_r' = L_r - L_m^2/L_s$; $\vec{\psi}_{r0}$ is the pre-fault rotor flux linkage.

Substituting Equations (5) and (10) into Equation (3), the stator positive and negative sequence short circuit currents can be obtained:

$$\begin{cases} \vec{i}_{s1} = -\frac{AV_{pre}}{jw_s L_s'}e^{-t/T_s'+j\theta} + \frac{L_m}{L_s'L_r'}[\vec{\psi}_{r0} - f_1(s)\frac{(1-A)V_{re}}{jw_s}e^{j\theta}]e^{-t/T_r'+jw_r t} + [\frac{L_m}{L_r'}f_1(s)-1]\frac{(1-A)V_{pre}}{jw_s L_s'}e^{j(w_s t+\theta)} \\ \vec{i}_{s2} = -\frac{V_2}{jw_s L_s'}e^{-t/T_s'+j\theta'} + f_2(s)\frac{L_m}{L_s'L_r'}\frac{V_2}{jw_s}e^{-t/T_r'+j(w_r t+\theta')} - [\frac{L_m}{L_r'}f_2(s)-1]\frac{V_2}{jw_s L_s'}e^{j(-w_s t+\theta')} \end{cases}. \tag{11}$$

According to Equation (11), the stator short circuit current consists of a decaying direct current (DC) component, a decaying AC component at rotor speed frequency, and a forced AC component at synchronous frequency. Especially when the voltage sag is deep, meaning A is large, the amplitude of the forced AC component will be small, and the decaying DC component and decaying AC component will be the main components of the short circuit current. The short circuit current of a DFIG is the basis for researching the fault characteristics of the DFIG-based wind farm connected to transmission lines.

2.2. Fault Analysis of the DFIG-Based Wind Farm without LVRT Capability

If a DFIG-based wind farm does not have LVRT capability, the wind farm will immediately be removed from the grid when a fault occurs, and no short circuit current will be provided by DFIGs. When the high voltage side of the main transformer grounded provides a zero-sequence circuit, short circuit current on the transmission line of the wind farm side still exists under a ground fault. Figure 4 shows a real fault case where a single-phase ground fault occurred on the transmission line connected to a DFIG-based wind farm without LVRT capability. The wind power system details are presented in Appendix B.

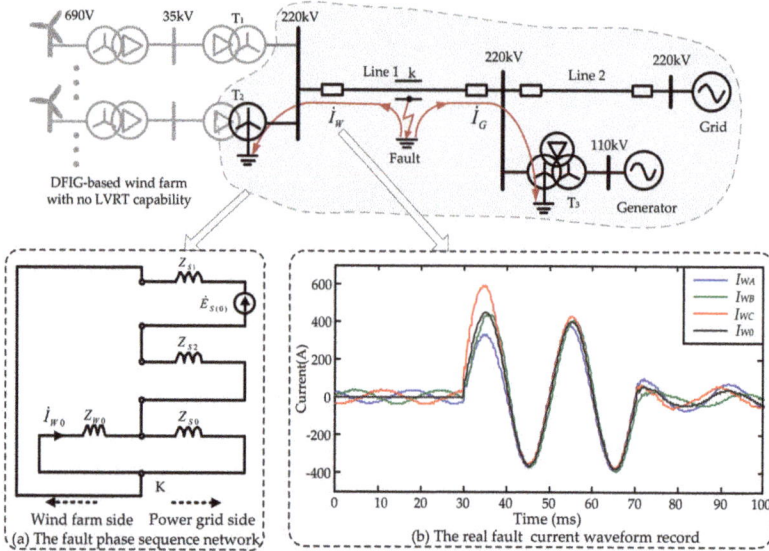

Figure 4. The real fault case when a phase C ground fault occurs on the transmission line connected to DFIG-based wind farm without low voltage ride through (LVRT) capability.

In Figure 4, $\dot{I}_W$ is the wind farm side short circuit current on the transmission line and $\dot{I}_{WA}$, $\dot{I}_{WB}$, $\dot{I}_{WC}$, and $\dot{I}_{W0}$ are phase A, phase B, phase C, and zero sequence current of $\dot{I}_W$, respectively; Z_{W0} is zero sequence impedance of the wind farm side, and it consists of the zero sequence impedance of the transformer grounded and transmission line; Z_{S0}, Z_{S1}, Z_{S2} are the zero sequence, positive sequence, and negative sequence impedance of the grid side, respectively; and $\dot{E}_{S(0)}$ is the grid voltage before the fault. Due to the disconnection of the wind farm, the wind farm side positive and negative sequence circuits are equal to open roads, as shown in Figure 4a. There is no positive or negative sequence current on the wind farm side. However, the ground point of transformer T_2 provides a zero-sequence circuit. The wind farm side three-phase currents on the transmission line are all zero sequence components, as shown in Figure 4b. The fault occurs at 30 ms and the fault duration is about 40 ms; then, the differential protection initiates and the circuit breaker trips. Except for the first few milliseconds of the transient process, $\dot{I}_{WA}$, $\dot{I}_{WB}$, $\dot{I}_{WC}$, and $\dot{I}_{W0}$ are basically the same after the fault. That is to say, the short circuit current on the transmission line connected to the wind farm only contains zero-sequence components.

2.3. Fault Analysis of the DFIG-Based Wind Farm with LVRT Capability

For a DFIG-based wind farm with LVRT capability, the wind farm remains connected to the power grid when a fault occurs on transmission line. The capacity of a wind farm is generally much smaller than that of the power grid. The equivalent impedance of the wind farm converted from 690 V to a high voltage level is much larger than the equivalent impedance of the grid. The activation of the crowbar circuit further increases the equivalent impedance of the wind farm. Therefore, compared with the traditional power system, the wind farm side is a weak power supply system.

According to the above analysis in Section 2.1, the fault current consists of a decaying DC component, a decaying AC component at rotor speed frequency, and a forced AC component at synchronous frequency. Especially when the voltage sag is deep, the AC component at rotor speed frequency is the main AC component at the beginning of the fault. The DFIG rotor speed is generally 0.7 to 1.3 pu. Thus, the main frequency component of the wind farm side current is a 35–65 Hz AC component in the early stage of the fault, although the voltage is mainly 50 Hz due to the grid support. That is to say, the wind farm side shows current frequency deviation during disturbances. Two real fault cases are provided to illustrate the fault characteristics of the wind farm: a single-phase ground fault and a phase-to-phase short circuit fault on the transmission lines connected to DFIG-based wind farms with LVRT capability. As for the balanced fault, the detailed work has been done in papers [28,29].

2.3.1. The Single-Phase Ground Fault Case

The 35 kV system of wind farm is not grounded. The zero-sequence impedance of the wind farm only contains the zero-sequence impedance of transformer grounded and transmission line when a fault occurs on the transmission line. Therefore, the zero-sequence current of the wind farm is relatively large and is not influenced by the DFIG. Due to the weak power supply of the wind farm, the positive and negative sequence currents are small. The zero-sequence current is the main component of the wind farm side current when a ground fault occurs. For the real DFIG-based wind farm shown in Figure 4, the DFIG is equipped with LVRT technology afterwards. A single-phase ground fault occurs on Line 1 and the real fault waveform records are shown in Figure 5.

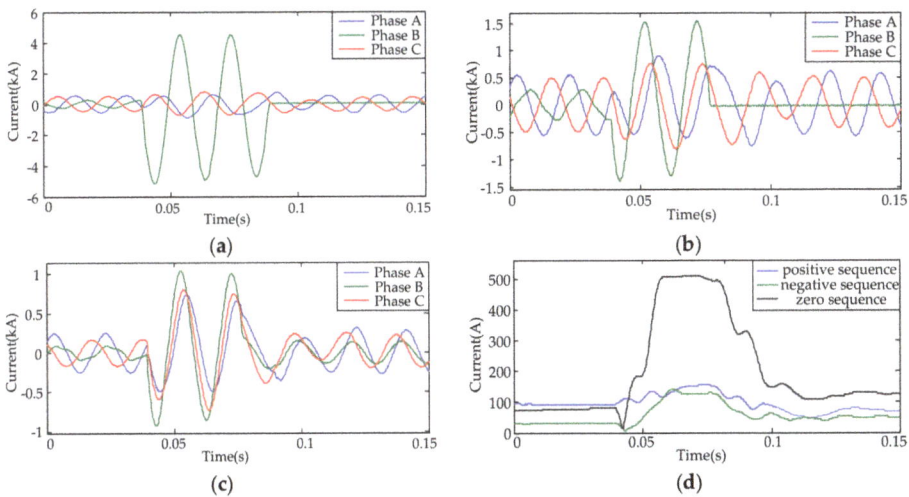

Figure 5. The real fault waveform records when a phase B ground fault occurs on the transmission line connected to a DFIG-based wind farm with LVRT capability: (**a**) currents of the grid side; (**b**) currents of the wind farm side; (**c**) currents of the transformer T$_2$; and (**d**) Root-Mean-Square (RMS) current values of the transformer T$_2$.

For the DFIG-based wind farm power system shown in Figure 4, the fault k occurs at 0.04 s and the fault duration is 40–60 ms; then, the differential protection initiates and the circuit breaker trips. The measured fault currents at the two terminals of the Line 1 are shown in Figure 5a,b separately. Obviously, the phase B current of the grid side is much larger than that of the wind farm side during the fault. Since wind farm is a weak power supply system, the short circuit current of the wind farm is small, hence the differences of the three-phase currents is not obvious. In order to observe the fault currents provided by the wind farm in detail, the measured currents on the high-voltage side of the transformer T_2 are shown in Figure 5c. The RMS values of the positive, negative and zero-sequence currents are shown in Figure 5d. The three-phase short circuit currents are essentially coincident and the zero sequence current is much larger than the positive and negative sequence currents. In summary, for ground faults, the fault current of the wind farm side is dominated by the zero sequence current, the frequency of which is 50 Hz. The frequency deviation is not obvious.

2.3.2. The Phase-to-Phase Short Circuit Fault Case

For the similar wind farm with LVRT capability, the topology is shown in Figure 4, but the voltage level of the transmission lines are 110 kV and the transformer T_3 is out of operation. The wind power system details are presented in Appendix B. Figure 6 shows the real waveform records when a phase-A-to-phase-B fault occurs on Line 2.

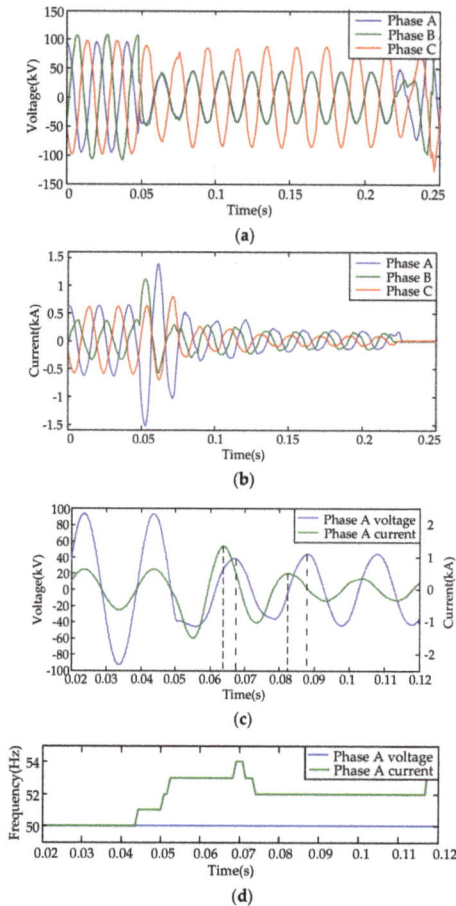

Figure 6. The real fault waveform records when a phase-A-to-phase-B fault occurs on the transmission line connected to a DFIG-based wind farm with LVRT capability: (**a**) voltages of the wind farm side, (**b**) currents of the wind farm side, (**c**) phase A voltage and current of the wind farm side, and (**d**) phase A voltage frequency and current frequency of the wind farm side.

The fault occurs at 0.05 s and the fault duration is about 180 ms; then, the distance protection initiates and the circuit breaker trips. As shown in Figure 6a, the phase A voltage is approximately equal to the phase B voltage after the fault. It can be seen From Figure 6b that the fault currents increased after fault, fault current consists of a decaying DC component, a decaying AC component at rotor speed frequency, and a forced AC component at synchronous frequency according to Equation (11). The natural components of fault currents gradually decrease. At last, the forced component remains and the value is small than load current because of voltage sags. For the phase-to-phase short circuit fault, DFIG-based wind farms only provide positive and negative sequence currents, the amplitudes of which are small, and no zero-sequence current exists. Therefore, the frequency deviation is obvious. Figure 6c shows the wind farm side voltage and the current waveforms of phase A, which is extracted from Figure 6a,b. The frequency of the phase A voltage remains 50 Hz because of the system voltage support. At the beginning of the fault, the frequency of the phase A current is obviously greater than 50 Hz due to the existence of decaying AC component at rotor speed frequency, as shown in Figure 6d.

3. Distance Protection and Its Improvement Method for the Transmission Line Connected to DFIG-Based Wind Farms

3.1. Application of Conventional Distance Protection

Distance protection exploits the fact that the voltage and current change at the same time when a fault occurs. The ratio of the measured voltage to the measured current represents the distance from the fault point to the relay location. The distance protection initiates if the fault distance is less than the setting value. For a conventional power grid with the synchronous generators, the measured voltages and currents in the relay location are mainly at nominal frequency. The Fourier algorithm is usually used to measure voltages and currents correctly. However, the short circuit currents of wind farm side may contain rotor speed frequency AC components, which cannot be filtered accurately. As a result, the calculation result error using the Fourier algorithm may be significantly large [34]. Then, the reliability of the distance protection based on nominal frequency voltage and current cannot be guaranteed. According to the real fault cases for phase-to-ground fault and phase-to-phase fault, the applicability of distance protection of the transmission line is analyzed.

3.1.1. Phase-to-Ground Fault

For the case of the single-phase ground fault outlined in Section 2.3.1, the fault currents of the wind farm side are dominated by the zero-sequence current, the frequency of which is 50 Hz. According to the real fault waveform data, the impedance trajectory calculated by the Fourier algorithm is shown in Figure 7.

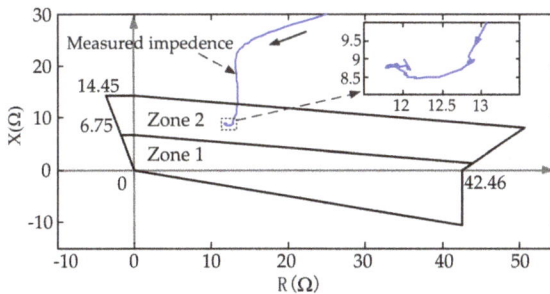

Figure 7. The phase B impedance trajectory of the wind farm side when a phase B ground fault occurs on the transmission line connected to a DFIG-based wind farm.

In Figure 7, the blue line is the impedance trajectory measured by conventional distance protection. Zone 1 and Zone 2 are the polygon action areas of wind farm side instantaneous and time delay instantaneous distance protection on Line 1, respectively. 6.75 and 14.45 Ω are the instantaneous and time delay instantaneous distance protection setting values of the reactance, respectively. 42.46 Ω is the setting value of the ground resistance. Referring to Figure 4, the length of Line 1 is 22 km and the actual fault point is 21 km away from Bus 1 (the corresponding reactance is about 9 Ω). Thus, the impedance trajectory is beyond Zone 1 and lies in Zone 2. From the magnified view of the impedance trajectory, the measured reactance is stable finally and close to 9 Ω during the fault. That is to say, the conventional ground distance protection of the wind farm side is relatively reliable.

3.1.2. Phase-to-Phase Fault

For the real phase-to-phase fault case mentioned in Section 2.3.2, the topology of wind farm is shown in Figure 4, but the voltage level of the transmission lines are 110 kV and the transformer T3 is out of operation. The phase-A-to-phase-B fault occurred on Line 2. However, the wind farm side phase-to-phase instantaneous distance protection on Line 1 maloperated. Obviously, the fault case is a maloperation accident. According to the real fault waveform records shown in Figure 6, the amplitude, phase and trajectories of the measured phase-A-to-phase-B impedance are calculated by the Fourier algorithm, as shown in Figure 8a–c, respectively.

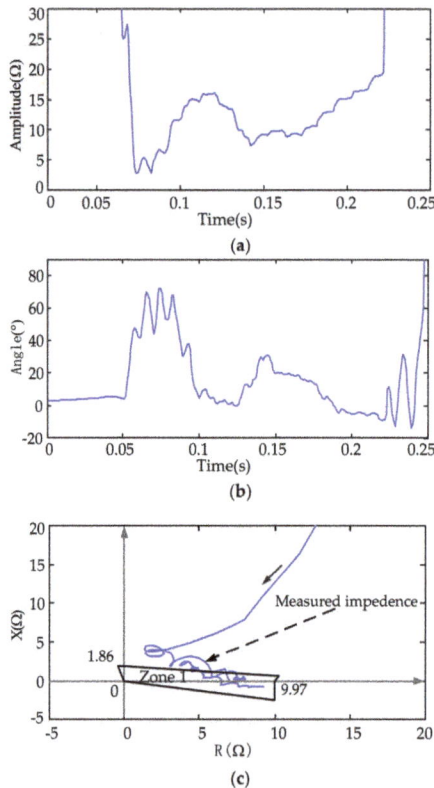

Figure 8. The phase-A-to-phase-B impedance of the wind farm side when a phase-A-to-phase-B fault occurs on the transmission line connected to a DFIG-based wind farm: (**a**) the amplitude of impedance; (**b**) the angle of impedance; and (**c**) the trajectory of impedance.

It can be seen from Figure 8a,b that the amplitude and phase of the measured impedance are both constantly changing. In Figure 8c, the blue line is the impedance trajectory measured by conventional distance protection. Zone 1 is the polygon action area of wind farm side instantaneous distance protection on Line 1. 9.97 and 1.86 Ω are the instantaneous distance protection setting values of the resistance and reactance, respectively. It can be seen that the trajectory of the impedance enters Zone 1 and causes the maloperation of wind farm side phase-to-phase instantaneous distance protection on Line 1. The conventional instantaneous distance protection of the wind farm side is no longer reliable when a phase-to-phase fault occurs on a transmission line.

3.2. Time-Domain Distance Protection Based on the R-L Model

The method used to solve the differential equation based on the R-L time domain model is not involved in the signal frequency domain information. The method establishes the differential equations for the transmission line by using the instantaneous current and voltage values in the relay location. It is not affected by the DC component, the low-frequency component, or the fluctuation in the power grid frequency. Therefore, the differential equation algorithm can overcome the influence of the frequency deviation of the wind power system. If the distribution capacitance of the protected line can be ignored, then the line segment from the fault point to the relay location can be approximated by a resistor and an inductor series circuit. When a fault occurs in a certain point on the line, the differential equation in the relay location is:

$$u = Ri + L\frac{di}{dt}, \tag{12}$$

where R and L are the positive sequence resistance and inductance of the fault point to the relay location, respectively, and u and i are the instantaneous voltage and current values of the relay location, respectively. When using a microcomputer, the derivative of the current can be calculated by the difference equation. Then, the unknown part of the differential equation is only R and L. In theory, the equations can be solved with only three sampling points. To ensure the accuracy of the solution, the voltage and current are sampled multiple times. These values are substituted into Equation (12) to obtain a series of difference equations. Then, the equations are solved with the least-squares method [35]. Set the sampling period to be T_s. t_k and t_{k+1} are two sampling times, and the corresponding measured voltage and current sampling value are, respectively, u_k, u_{k+1}, i_k, and i_{k+1}. For simplicity, set $y_k = (u_k + u_{k+1})/2$, $x_k = (i_k + i_{k+1})/2$, and $D_k = (i_{k+1} - i_k)/T_s$. Substituting them into Equation (12) and N observations are performed, the sum of squares of each error is expressed as:

$$J = \sum_{k=1}^{N} (y_k - Rx_k - LD_k)^2, \tag{13}$$

where N is the length of data window. N can vary with the estimation accuracy requirement. The larger the N, the better the estimation accuracy. The least-squares method involves finding the R and L, which minimize the value of J, so the partial derivative of R and L are equal to zero and two equations are obtained. Solve the two equations, and the calculation formula of the estimated value of the resistance $\hat{R}$ and the inductance $\hat{L}$ are obtained:

$$\hat{R} = \frac{\sum\limits_{k=1}^{N} y_k x_k \sum\limits_{k=1}^{N} D_k^2 - \sum\limits_{k=1}^{N} y_k D_k \sum\limits_{k=1}^{N} x_k D_k}{\sum\limits_{k=1}^{N} x_k^2 \sum\limits_{k=1}^{N} D_k^2 - \left(\sum\limits_{k=1}^{N} x_k D_k\right)^2}, \tag{14}$$

$$\hat{L} = \frac{\sum\limits_{k=1}^{N} x_k^2 \sum\limits_{k=1}^{N} y_k D_k - \sum\limits_{k=1}^{N} y_k x_k \sum\limits_{k=1}^{N} x_k D_k}{\sum\limits_{k=1}^{N} x_k^2 \sum\limits_{k=1}^{N} D_k^2 - \left(\sum\limits_{k=1}^{N} x_k D_k\right)^2}. \tag{15}$$

3.3. Case Studies

According to the real phase-to-phase fault case mentioned in Section 2.3.2, the trajectory of impedance measured by Equations (14) and (15) is shown in Figure 9. The red line and green line in Figure 9 are obtained based on the real resistance and inductance parameters of Line 1 and Line 2, respectively. Zone 1 is the polygon action areas of wind farm side instantaneous distance protection on Line 1, and 9.97 Ω and 1.86 mH are setting values of resistance and inductance, respectively. Actually, the fault occurs on Line 2. Compared with Figure 8c, the measured impedance trajectory does not enter Zone 1 in Figure 9 and the wind farm side instantaneous distance protection on Line 1 will not maloperate. Thus, the reliability of distance protection on wind farm side transmission lines is guaranteed when using the time-domain distance protection method proposed in this paper.

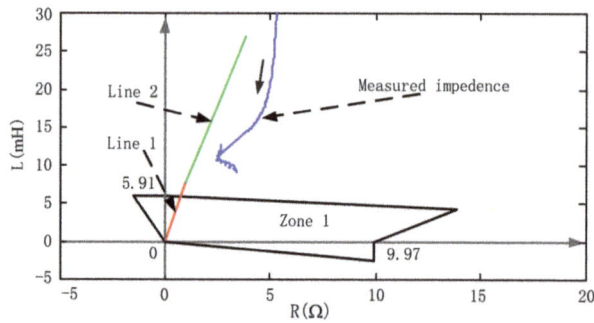

Figure 9. The real phase-to-phase fault case's trajectory of impedance measured with the time-domain method.

According to the real wind farm mentioned in Section 2.3.1, the corresponding wind farm model is established by Matlab/Simulink (R2014a, MathWorks, Natick, MA, USA, 2014). The topology of the wind farm is shown in Figure 4. The wind farm is equipped with 132 DFIGs with LVRT capability, and each main transformer connects 66 DFIGs. The parameters of a real DFIG and transmission lines are shown in Appendix B. The detailed model and control strategy of DFIG can be found in previous studies [14,36].

This paper mainly researched the fault characteristics of DFIG-based wind farms connected to transmission lines and its influence on relay protection. The duration of the fault is relatively short, so the wind speed of the wind farm is considered as unchanged during the research process. For the convenience of research, all wind turbines are set to operate under the same conditions. Additionally, the impedance of the collector system is ignored.

To compare the time-domain distance protection method with the traditional distance protection method based on the Fourier algorithm, the following three conditions are given as typical simulation examples: a phase A ground fault, a phase-A-to-phase-B fault, and a three-phase short circuit fault. The transition resistances are all zero. The fault points are all 11 km away from Bus 1 on Line 1. The wind speed is 11 m/s before the fault. The comparison of two kinds of distance protection methods used on the wind farm side transmission lines connected to DFIG-based wind farms are shown in Figure 10.

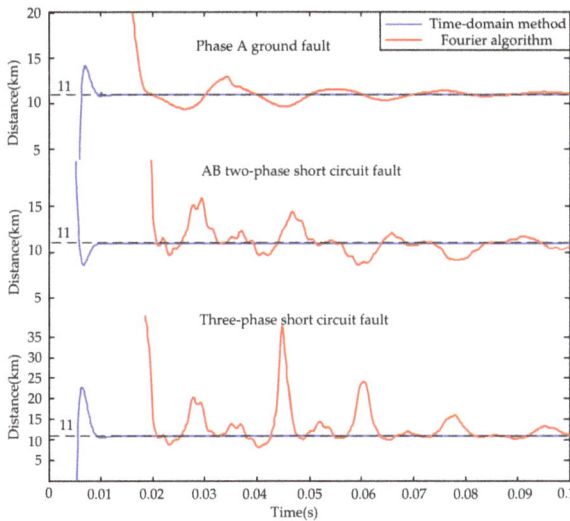

Figure 10. Comparison of two kinds of distance protection methods.

In Figure 10, the red lines are the distance measured by Fourier algorithm and the blue lines are the distance measured by time-domain method. The faults occur at 0 s and the rotor speed before the fault is greater than the synchronous speed. Due to the existence of a rotor speed frequency AC component, the distance calculation results, which are obtained from the traditional distance protection method based on the Fourier algorithm, fluctuate considerably. The error of ground fault is smaller due to the domination of the zero-sequence current. The time-domain method proposed in this paper is not affected by the rotor speed frequency component, the distance calculation results converge to the real fault distance within a dozen milliseconds.

To further verify the performance of the time-domain method, three fault points at 5, 10 and 15 km are selected on Line 1. The following four conditions are given as typical simulation examples: a phase A ground fault, a two-phase ground fault, a phase-to-phase short circuit fault and a three-phase short circuit fault. The transition resistances are all zero. The wind speed is 11 m/s before the fault and the corresponding rotor speed is greater than the synchronous speed. The data of 20–40 ms after faults are selected to calculate the errors of two methods' measurement results. The error calculation formula is shown as:

$$\sigma = \frac{1}{N} \sqrt{\sum_{i=1}^{N} \left(\frac{x_i - x_t}{x_t} \right)^2},$$
(16)

where x_i and x_t are separately the measurement value and true value of fault distance. The errors of fault distance measurement results are separately shown in Tables 1 and 2.

Table 1. Performance of the time-domain method when faults occur.

Fault Type	The Error of Fault Distance Measurement Result		
	5 km	10 km	15 km
Single-phase ground	0.17%	0.20%	0.22%
Two-phase ground	0.23%	0.16%	0.30%
Phase-to-phase short circuit	0.04%	0.63%	0.03%
Three-phase short circuit	0.07%	0.18%	0.03%

Table 2. Performance of the Fourier method when faults occur.

Fault Type	The Error of Fault Distance Measurement Result		
	5 km	10 km	15 km
Single-phase ground	3.67%	3.38%	3.11%
Two-phase ground	3.27%	3.13%	2.94%
Phase-to-phase short circuit	7.24%	7.64%	7.29%
Three-phase short circuit	26.39%	26.07%	25.71%

According to the two tables, the accuracy of the calculation results using the time-domain method is significantly high, and no error is more than 1%. The errors of the calculation results using the Fourier method are significantly bigger than that of the time-domain method. The time-domain distance protection method can accurately and quickly identify fault distances when faults occur on the transmission line connected to a DFIG-based wind farm.

4. Conclusions

This paper analyzes the composition of DFIG short circuit current. Moreover, using data of real fault cases, this paper illustrates the fault characteristics of DFIG-based wind farms connected to transmission lines and the corresponding applicability of the conventional distance protection based on the Fourier algorithm.

The DFIG short circuit current consists of a decaying DC component, a decaying AC component at rotor speed frequency, and a forced AC component at synchronous frequency. Wind farm is a weak power supply system. The short circuit current of wind farm side is much smaller than that of the power grid side. Additionally, the zero-sequence current will be the main component of the wind farm side short circuit current when a ground fault occurs. The wind farm side may show current frequency deviation during disturbances. The wind farm side voltage of transmission line is nominal frequency due to the grid support, whereas its frequency may no longer be nominal due to the existence of rotor speed frequency component. The conventional ground distance protection based on the Fourier algorithm is still relatively reliable, whereas the phase-to-phase instantaneous distance protection is no longer reliable for the transmission line connected to a DFIG-based wind farm. Finally, a time-domain distance protection method based on the R-L model and least-squares algorithm is proposed to solve the problem. The time-domain method accurately and quickly identified fault distances when faults occur on the transmission line connected to a DFIG-based wind farm. The efficacy of the proposed method is validated with real fault cases and simulation. The proposed method is immune to the frequency variation. Thus, it overcomes the defect of traditional distance protection. With the help of the proposed method, Relay Protection Equipment Manufacturers can improve the equipment level of technology, and the grid companies can improve the safe operation level of the power system.

Acknowledgments: This work was supported by the National Science Foundation for Excellent Young Scholars of China (Grant No. 51422703).

Author Contributions: Bin Li conceived the study and wrote the paper; Junyu Liu performed the experiments and analyzed the data and wrote the manuscript; Xin Wang analyzed the topology and fault characteristics of DFIG; Lili Zhao contributed to the data of wind farms and DFIG.

Conflicts of Interest: The authors declare no conflict of interest.

Appendix A.

(i) The derivation of stator flux linkage

From Equation (1), ignore R_s, there is

$$\vec{v}_s = D\vec{\psi}_s. \tag{A1}$$

The solution of Equation (A1) can be written as:

$$\vec{\psi}_s = \vec{\psi}_{sn} + \vec{\psi}_{sf}. \tag{A2}$$

It can be found that the response of the stator flux after the fault consists of two components: natural flux linkage $\vec{\psi}_{sn}$ and forced flux linkage $\vec{\psi}_{sf}$. $\vec{\psi}_{sf}$ can be determined by the post-fault terminal voltage:

$$\begin{cases} \vec{\psi}_{sf1} = \int (1-A)V_{pre}e^{j(\omega_s t+\theta)} = \frac{(1-A)V_{pre}}{j\omega_s}e^{j(\omega_s t+\theta)} \\ \vec{\psi}_{sf2} = \int V_2 e^{j(-\omega_s t+\theta')} = -\frac{V_2}{j\omega_s}e^{j(-\omega_s t+\theta')} \end{cases} \tag{A3}$$

It is well known that the flux can only change continuously. Thus, there is

$$\vec{\psi}_s(0_-) = \vec{\psi}_s(0_+), \tag{A4}$$

and pre-fault stator flux linkage is

$$\vec{\psi}_{s0} = \frac{V_{pre}}{j\omega_s}e^{j(\omega_s t+\theta)}. \tag{A5}$$

Therefore, the stator natural flux linkage is obtained as:

$$\begin{cases} \vec{\psi}_{sn1} = \frac{AV_{pre}}{j\omega_s}e^{-t/T_s'+j\theta} \\ \vec{\psi}_{sn2} = \frac{V_2}{j\omega_s}e^{-t/T_s'+j\theta'} \end{cases} \tag{A6}$$

(ii) The derivation of rotor flux linkage.

Appendix A.1. The Derivation of Pre-Fault Rotor Flux Linkage

Using a generator convention for the stator windings and a motor convention for the rotor windings, the flux linkage equations in a synchronous *dq* reference frame are expressed as [20]:

$$\begin{cases} \psi_{sd} = -L_s i_{sd} + L_m i_{rd} \\ \psi_{sq} = -L_s i_{sq} + L_m i_{rq} \end{cases} \tag{A7}$$

$$\begin{cases} \psi_{rd} = -L_m i_{sd} + L_r i_{rd} \\ \psi_{rq} = -L_m i_{sq} + L_r i_{rq} \end{cases} \tag{A8}$$

The stator active and reactive power output of DFIG can be expressed as [37]:

$$\begin{cases} P_s = \frac{3}{2}(u_{sd}i_{sd} + u_{sq}i_{sq}) \\ Q_s = \frac{3}{2}(u_{sq}i_{sd} - u_{sd}i_{sq}) \end{cases} \tag{A9}$$

Assuming that the *d*-axis of the reference frame is aligned with the stator flux linkage vector,

$$\psi_{sd} = \psi_s, \psi_{sq} = 0 \tag{A10}$$

$$u_{sd} = D\psi_{sd} = 0, u_{sq} = V_s = \omega_s\psi_s, \tag{A11}$$

can be obtained.

Substituting Equation (A10) into Equation (A7),

$$\begin{cases} i_{rd} = \frac{L_s}{L_m}i_{sd} + \frac{\psi_s}{L_m} \\ i_{rq} = \frac{L_s}{L_m}i_{sq} \end{cases} \tag{A12}$$

can be obtained.

Substituting Equation (A11) into Equation (A9),

$$
\begin{cases}
i_{sd} = \frac{2Q_s}{3V_s} \\
i_{sq} = \frac{2P_s}{3V_s}
\end{cases}
,
\tag{A13}
$$

can be obtained.

Based on Equations (A8), (A12) and (A13), the rotor flux linkage equation is expressed as:

$$
\begin{cases}
\psi_{rd} = (L_s L_r - L_m{}^2)\frac{2Q_s}{3L_m V_s} + \frac{V_s L_r}{\omega_s L_m} \\
\psi_{rq} = (L_s L_r - L_m{}^2)\frac{2P_s}{3L_m V_s}
\end{cases}
.
\tag{A14}
$$

If $V_s = V_{pre}$ and the DFIG stator terminal instantaneous complex power before the fault is $S_{pre} = P_s + jQ_s$, then pre-fault rotor flux linkage is

$$
\vec{\psi}_{r0} = (L_s L_r - L_m{}^2)\frac{2Q_s}{3L_m V_s} + \frac{V_{pre} L_r}{\omega_s L_m} + j(L_s L_r - L_m{}^2)\frac{2P_s}{3L_m V_s}.
\tag{A15}
$$

Appendix A.2. The Derivation of Post-Fault Rotor Flux Linkage

After fault, the crowbar circuit is activated and the RSC is shorted so that $\vec{v}_{r1}$ and $\vec{v}_{r2}$ are both zero in Figure 3. Furthermore, the forced ac component of rotor flux rotates synchronously so that its amplitude is constant in the synchronous rotating reference frame. Therefore, $D\vec{\psi}_{rf1}$ and $D\vec{\psi}_{rf2}$ are both zero. Then solving Equations (6)–(9), respectively, the post-fault steady rotor positive and negative sequence flux linkages are expressed as:

$$
\begin{cases}
\vec{\psi}_{rf1} = \frac{R_r L_m/L_s}{R_r+j(\omega_s-\omega_r)L_r'}\vec{\psi}_{sf1} = \frac{R_r L_m/L_s}{R_r+j(\omega_s-\omega_r)L_r'}\frac{(1-A)V_{pre}}{j\omega_s}e^{j(\omega_s t+\theta)} \\
\vec{\psi}_{rf2} = \frac{R_r L_m/L_s}{R_r-j(2-s)\omega_s L_r'}\vec{\psi}_{sf2} = -\frac{R_r L_m/L_s}{R_r-j(2-s)\omega_s L_r'}\frac{V_2}{j\omega_s}e^{j(-\omega_s t+\theta')}
\end{cases}
,
\tag{A16}
$$

where $L_r' = L_r - L_m{}^2/L_s$. For simplicity, set $f_1(s) = \frac{R_r L_m/L_s}{R_r+j(\omega_s-\omega_r)L_r'}$, $f_2(s) = \frac{R_r L_m/L_s}{R_r-j(2-s)\omega_s L_r'}$.

The flux can only change continuously, so there is

$$
\vec{\psi}_r(0_-) = \vec{\psi}_r(0_+).
\tag{A17}
$$

According to Equations (A15)–(A17), the rotor natural flux linkage is obtained as:

$$
\begin{cases}
\vec{\psi}_{rn1} = [\vec{\psi}_{r0} - f_1(s)\frac{(1-A)V_{pre}}{j\omega_s}e^{j\theta}]e^{-t/T_r'+j\omega_r t} \\
\vec{\psi}_{rn2} = f_2(s)\frac{V_2}{j\omega_s}e^{j\theta'}e^{-t/T_r'+j\omega_r t}
\end{cases}
,
\tag{A18}
$$

where $T_r' = L_r'/R_r$ is rotor decaying time constant.

Appendix B.

The real DFIG-based wind farm shown in Figure 4 is equipped with 132 DFIGs and the total capacity is 198 MW. At the beginning, the DFIG is not equipped with LVRT technology. But then, the DFIG is equipped with LVRT technology by technological transformation. The transmission line parameters of the wind farm are shown in Table A1. The main protection of the transmission line is current differential protection and the backup protection is conventional distance protection.

Table A1. The 220 kV transmission line parameters.

Parameter Name	Line 1	Line 2
Length (km)	22.018	70.202
Z_1 (Ω/km)	$0.080 + j0.430$	$0.048 + j0.320$
Z_0 (Ω/km)	$0.360 + j1.000$	$0.317 + j1.000$

The real DFIG-based wind farm mentioned in Section 2.3.2 is equipped with 56 DFIGs and the total capacity is 84 MW. The DFIG is equipped with LVRT technology. The transmission line parameters of the wind farm are shown in Table A2. The main protection of the transmission line is conventional distance protection.

Table A2. The 110 kV transmission line parameters.

Parameter Name	Line 1	Line 2
Length (km)	5.325	16.168
Z_1 (Ω/km)	$0.113 + j0.419$	$0.154 + j0.407$
Z_0 (Ω/km)	$0.871 + j1.085$	$0.593 + j1.216$

The parameters of a real DFIG used in the wind farm which is shown in Figure 4 are shown in Table A3.

Table A3. The parameters of DFIG.

Parameter Name	Value	Parameter Name	Value (p.u.)
Nominal capacity	$P_n = 1.5$ MW	Stator resistance	$R_s = 0.0173$
Rated stator voltage	$V_n = 690$ V	Stator leakage inductance	$L_{s\sigma} = 0.170$
System frequency	$f_n = 50$ Hz	Rotor resistance	$R_r = 0.0120$
Rated voltage of DC-link	$V_{dc} = 1150$ V	Rotor leakage inductance	$L_{r\sigma} = 0.236$
Pairs of poles	$p = 2$	Mutual inductance	$L_m = 10.491$
Nominal wind speed	$v_n = 11$ m/s	Crowbar resistance	$R_{cb} = 0.0432$

References

1. Hossain, M.M.; Ali, M.H. Future research directions for the wind turbine generator system. *Renew. Sustain. Energy Rev.* **2015**, *49*, 481–489. [CrossRef]
2. Li, H.; Chen, Z. Overview of different wind generator systems and their comparisons. *IET Renew. Power Gener.* **2008**, *2*, 123–138. [CrossRef]
3. Muller, S.; Deicke, M.; De Doncker, R.W. Doubly fed induction generator systems for wind turbines. *IEEE Trans. Ind. Appl. Mag.* **2002**, *8*, 26–33. [CrossRef]
4. Hughes, F.M.; Anaya-Lara, O.; Jenkins, N.; Strbac, G. Control of DFIG based wind generation for power network support. *IEEE Trans. Power Syst.* **2005**, *20*, 1958–1966. [CrossRef]
5. Iwanski, G.; Koczara, W. DFIG-based power generation system with UPS function for variable-speed applications. *IEEE Trans. Ind. Electron.* **2008**, *55*, 3047–3054. [CrossRef]
6. Marques, G.D.; Sousa, D.M. Understanding the doubly fed induction generator during voltage dips. *IEEE Trans. Energy Convers.* **2012**, *27*, 421–461. [CrossRef]
7. Ling, Y.; Cai, X.; Wang, N. Rotor current transient analysis of DFIG-based wind turbines during symmetrical voltage faults. *Energy Convers Manag.* **2013**, *76*, 910–917. [CrossRef]
8. The Grid Code 3. Available online: http//www.nationalgrid.com/uk/Electricity/Codes/gridcode (accessed on 1 January 2006).
9. Transmission Code 2007—Network and System Rules of the German Transmission System Operators. Available online: https://www.bdew.de/ (accessed on 1 August 2007).

10. Inspection and Quarantine of China. *Technical Rule for Connecting Wind Farm to Power System*; Technical Report for Inspection and Quarantine of China; Inspection and Quarantine of China: Beijing, China, 2011.

11. Rahim, A.H.M.A.; Nowicki, E.P. Supercapacitor energy storage system for fault ride through of a DFIG wind generation system. *Energy Convers. Manag.* **2012**, *59*, 96–102. [CrossRef]

12. Cardenas, R.; Pena, R.; Alepuz, S.; Asher, G. Overview of control systems for the operation of DFIGs in wind energy applications. *IEEE Trans. Ind. Electron.* **2013**, *60*, 2776–2798. [CrossRef]

13. Akhmatov, V. *Induction Generators for Wind Power*; Multi-Science Publishing Company Ltd.: Brentwood, UK, 2005.

14. Morren, J.; de Haan, S.W.H. Ride through of wind turbines with doubly-fed induction generator during a voltage dip. *IEEE Trans. Energy Convers.* **2005**, *20*, 435–441. [CrossRef]

15. Lopez, J.; Sanchis, P.; Roboam, X.; Marroyo, L. Dynamic behavior of the doubly-fed induction generator during three-phase voltage dips. *IEEE Trans. Energy Convers.* **2007**, *22*, 709–717. [CrossRef]

16. Pannell, G.; Atkinson, D.J.; Zahawi, B. Analytical study of grid-fault response of wind turbine doubly fed induction generator. *IEEE Trans. Energy Convers.* **2010**, *25*, 1081–1091. [CrossRef]

17. Sulla, F.; Svensson, J.; Samuelsson, O. Symmetrical and unsymmetrical short-circuit current of squirrel-cage and doubly-fed induction generators. *Electr. Power Syst. Res.* **2011**, *81*, 1610–1618. [CrossRef]

18. Morren, J.; de Haan, S.W.H. Short-circuit current of wind turbines with doubly fed induction generator. *IEEE Trans. Energy Convers.* **2007**, *22*, 174–180. [CrossRef]

19. Lopez, J.; Gubia, E.; Sanchis, P.; Roboam, X.; Marroyo, L. Wind turbines based on doubly fed induction generator under asymmetrical voltage dips. *IEEE Trans. Energy Convers.* **2008**, *23*, 321–330. [CrossRef]

20. Kong, X.; Zhang, Z.; Yin, X.; Wen, M. Study of fault current characteristics of the DFIG considering dynamic response of the RSC. *IEEE Trans. Energy Convers.* **2014**, *29*, 278–287.

21. Ouyang, J.; Xiong, X. Dynamic behavior of the excitation circuit of a doubly-fed induction generator under a symmetrical voltage drop. *Renew. Energy* **2014**, *71*, 629–638. [CrossRef]

22. Xiao, F.; Zhang, Z.; Yin, X. Fault Current Characteristics of the DFIG under Asymmetrical Fault Conditions. *Energies* **2015**, *8*, 10971–10992. [CrossRef]

23. Naggar, A.E.; Erlich, I. Fault current contribution analysis of doubly fed induction generator-based wind turbines. *IEEE Trans. Energy Convers.* **2015**, *30*, 874–883. [CrossRef]

24. El-Naggar, A.; Erlich, I. Short-circuit current reduction techniques of the doubly-fed induction generator based wind turbines for fault ride through enhancement. *IET Renew. Power Gener.* **2017**, *11*, 1033–1040. [CrossRef]

25. Pannell, G.; Atkinson, D.J.; Zahawi, B. Minimum-threshold crowbar for a fault-ride-through grid-code-compliant DFIG wind turbine. *IEEE Trans. Energy Convers.* **2010**, *25*, 750–759. [CrossRef]

26. Xiao, S.; Yang, G.; Zhou, H.; Geng, H. An LVRT Control Strategy Based on Flux Linkage Tracking for DFIG-Based WECS. *IEEE Trans. Ind. Electron.* **2013**, *60*, 2820–2832. [CrossRef]

27. Piwko, R.; Miller, N.; Sanchez-Gasca, J.; Yuan, X.; Dai, R.; Lyons, J. Integrating large wind farms into weak power grids with long transmission lines. In Proceedings of the Transmission and Distribution Conference and Exposition: Asia and Pacific, Dalian, China, 18–18 August 2005; pp. 1–7.

28. Hooshyar, A.; Azzous, M.A.; El-Saadany, E.F. Distance protection of lines connected to induction generator-based wind farms during balanced faults. *IEEE Trans. Sustain. Energy* **2014**, *5*, 1193–1203. [CrossRef]

29. Hooshyar, A.; Azzous, M.A.; El-Saadany, E.F. Three-Phase Fault Direction Identification for Distribution Systems with DFIG-Based Wind DG. *IEEE Trans. Sustain. Energy* **2014**, *5*, 747–756. [CrossRef]

30. Saleh, S.A.; Aljankawey, A.S.; Meng, R. Impacts of grounding configurations on responses of ground protective relays for DFIG-Based WECSs-Part I: Solid ground faults. *IEEE Trans. Ind. Appl.* **2015**, *51*, 2804–2818. [CrossRef]

31. Saleh, S.A.; Aljankawey, A.S.; Meng, R. Impacts of grounding configurations on responses of ground protective relays for DFIG-Based WECSs-Part II: High-Impedance Ground Faults. *IEEE Trans. Ind. Appl.* **2016**, *52*, 1204–1214.

32. Arachchige, L.N.W.; Rajapakse, A.D.; Muthumuni, D. Implementation, Comparison and Application of an Average Simulation Model of a Wind Turbine Driven Doubly Fed Induction Generator. *Energies* **2017**, *10*, 1726. [CrossRef]

33. Zheng, Z.; Yang, G.; Geng, H. Short circuit current analysis of DFIG-type WG with crowbar protection under grid faults. In Proceedings of the International Symposium on Industrial Electronics, Hangzhou, China, 28–31 May 2012; pp. 1072–1079.

34. Smith, J.C.; Milligan, M.R.; DeMeo, E.A.; Parsons, B. Utility wind integration and operating impact state of the art. *IEEE Trans. Power Syst.* **2007**, *22*, 900–908. [CrossRef]
35. Li, B.; Duan, Z.; Wang, X.; Wu, J. Loss-of-excitation analysis and protection for pumped-storage machines during starting. *IET Renew. Power Gener.* **2016**, *10*, 71–78. [CrossRef]
36. Yin, M.; Li, G.; Zhou, M.; Liu, G.; Zhao, C. Study on the Control of DFIG and Its Responses to Grid Disturbances. In Proceedings of the Power Engineering Society General Meeting, Montreal, QC, Canada, 18–22 June 2006; pp. 1–6.
37. Abad, G.; Rodriguez, M.A.; Iwanski, G.; Poza, J. Direct Power Control of Doubly Fed Induction Generator Based Wind Turbines under Unbalanced Grid Voltage. *IEEE Trans. Power Electron.* **2010**, *25*, 442–452. [CrossRef]

applied
sciences

MDPI

Article

Feasibility Study of Wind Farm Grid-Connected Project in Algeria under Grid Fault Conditions Using D-Facts Devices

Lina Wang [1], Kamel Djamel Eddine Kerrouche [1,5,*], Abdelkader Mezouar [2],
Alex Van Den Bossche [3], Azzedine Draou [4] and Larbi Boumediene [2]

[1] School of Automation on Science and Electrical Engineering, Beihang University, Beijing 100191, China;
    wangln@buaa.edu.cn
[2] Electro-Technical Engineering Lab, Faculty of Technology, Tahar Moulay University, Saida 20 000, Algeria;
    abdelkader.mezouar@univ-saida.dz (A.M.); lboumediene2005@yahoo.fr (L.B.)
[3] Electrical Energy LAB EELAB Sint–Pietersnieuwstraat 41, B 9000 Ghent, Belgium;
    Alex.VandenBossche@ugent.be
[4] Department of Electrical Engineering, Madinah Mounawara University, Madinah 20012, Saudi Arabia;
    a_draou@yahoo.co.uk
[5] Satellite Development Center CDS, BP 4065 Ibn Rochd USTO, 31000 Oran, Algeria
[*] Correspondence: kerrouche20@yahoo.fr; Tel.: +86-131-2191-9077

Received: 21 October 2018; Accepted: 8 November 2018; Published: 15 November 2018

Abstract: The use of renewable energy such as wind power is one of the most affordable solutions to meet the basic demand for electricity because it is the cleanest and most efficient resource. In Algeria, the highland region has considerable wind potential. However, the electrical power system located is this region is generally not powerful enough to solve the problems of voltage instability during grid fault conditions. These problems can make the connection with the eventual installation of a wind farm very difficult and inefficient. Therefore, a wind farm project in this region may require dynamic compensation devices, such as a distributed-flexible AC transmission system (D-FACTS) to improve its fault ride through (FRT) capability. This paper investigates the implementation of shunt D-FACTS, under grid fault conditions, considering the grid requirements over FRT performance and the voltage stability issue for a wind farm connected to the distribution network in the Algerian highland region. Two types of D-FACTSs considered in this paper are the distribution static VAr compensator (D-SVC) and the distribution static synchronous compensator (D-STATCOM). Some simulation results show a comparative study between the D-SVC and D-STATCOM devices connected at the point of common coupling (PCC) to support a wind farm based on a doubly fed induction generator (DFIG) under grid fault conditions. Finally, an appropriate solution to this problem is presented by sizing and giving the suitable choice of D-FACTS, while offering a feasibility study of this wind farm project by economic analysis.

Keywords: wind farm; Fault Ride Through (FRT); Distributed-Flexible AC Transmission system (D-FACTS); Distribution Static VAr Compensator(D-SVC); Distribution Static Synchronous Compensator (D-STATCOM)

---

1. Introduction

In Algeria, the first attempt to connect the wind energy conversion system (WECS) to the electricity distribution network dates back to 1957, with the installation of a 100-kW wind turbine at the Grands Vents site (Algiers) by the French designer Andreau.

Nowadays, the depletion of fossil fuels reserves in Algeria, fluctuations in oil price and the location of energy resources are causing instability in energy policy.

In addition, the use of fossil fuels for conventional power plants triggers alarms of an environmental disaster. Currently, to reduce the harmful impact of conventional resources and improve Algerian energy efficiency, the energy policy program announced by the Ministry of Mines and Energy aims, by 2030, to produce 40% electrical energy from renewable resources [1]. For WECS, the power to be produced over the period 2012–2022 is estimated at approximately 516 MW, of which 10 MW are installed at Kabertene (70 km from Adrar) in the Algerian desert [1–4]. This pilot wind farm consists of 12 wind turbines with a unit capacity of 0.85 MW, the energy produced will be injected to the 30/220 kV step up transformer situated in the same locality [1], as shown in Figure 1. Currently, the Algerian electrical grid code does not consider WECS. In the region of Adrar, the electrical grid is not interconnected with the north; it is a local grid (or micro-grid). Therefore, this program of the energy policy must be accompanied by continual development of wind energy technology and optimization techniques, looking for better options concerning reduced costs, improvement regarding wind turbine performances, reliability of electrical groups and electrical grid integration.

Figure 1. Wind farm at Kabertene in Adrar. Reproduced for reference [5].

In the period of 2009–2010, Sebaa Ben Miloud F et al. [6] and Himri et al. [7] undertook the first study to identify a suitable site in Adrar region for the wind farm installation. In addition, Himri et al. [7] used data of wind speed over a period of nearly 10 years to assess the potential of wind power stations in two southern Algerian regions, namely, Timimoun and Tindouf. In [8], a study of the wind potential in seven southern Algerian sites was undertaken, from west to east, Tindouf, Bechar, Adrar, and Ghardaia, In Amenas and In Salah (Tamanrasset). In [9] wind speed data was collected over a period of almost 5 years, from three selected stations in northern Algeria. Within this context, some studies in the Algerian high plateau region were performed in [10–12]. However, the authors did not consider the integration issue of the wind farm into the electrical grid and it is well known that the electrical grid influences greatly the performance of wind farm installation and production.

At the present time, doubly fed induction generators (DFIG) are the most used in WECS [13,14] and especially in the Algerian wind farm at Kabertene in Adrar. Simple induction generators have some weaknesses such as reactive power absorption and uncontrolled voltage during variable rotor speed. These complications are avoided by the installation of DFIG and power converters or power drives [15,16]. The particular feature of the DFIG is that the injected power by the rotor converter is only a small part from the total provided power with its stator directly connected to the electrical grid [17–19]. Hence, the size, the cost and losses of the power converter are optimized compared to a full-size power converter of the other generators.

One of the most important considerations in a wind farm grid-connected project is fault ride-through (FRT) capability, where the energy grid is often weak and the DFIG is frequently working under grid faults when the wind farm is located relatively far from this electrical grid [20]. Therefore, many research works focus on studying the dynamic behaviors of wind farms during and after the clearance of the grid fault conditions without disconnection from the electrical grid. In [21], several methods employed to improve the FRT capability of the fixed-speed wind turbines are based on induction generators. In [20], an enhanced application to overcome grid fault conditions is studied for a wind farm based on DFIG. FRT control of wind turbines with DFIGs under symmetrical voltage dips is presented in [21]. In [22] a flexible AC transmission system (FACTS) system for DFIG to reduce the effects of grid faults is proposed.

The present paper can extend the aforementioned research works. This paper, shows the feasibility of installing a wind farm in an Algerian highland region, confirmed by some data of wind potential in the selected geographical location in the first part, which is an input for the wind power system. On the other hand, another important aspect is the weakness of the electrical grid, which is often an obstacle in many countries that have established wind energy projects. Consequently, during the feasibility study, some techniques to identify the cost-effectiveness of areas for the wind farms installations, the possible electrical path of distribution lines, and their corresponding estimated cost are used, and the incorporation of electrical devices such as distributed FACTS (D-FACTS) technology is considered. Furthermore, the investors may be confident to fund this possible project when these technical difficulties are taken into consideration. Then, in order to ensure the economic success of the future wind farm project in the highland region, an accurate study by some simulation results, showing the interaction between wind turbine generators and the electrical grid in this region with the impact of the D-FACTS systems, is undertaken, which has not been previously done. In this study, the Algerian electrical grid code could be considered in simulations similar to that of the Spanish grid code as shown in Figure 2.

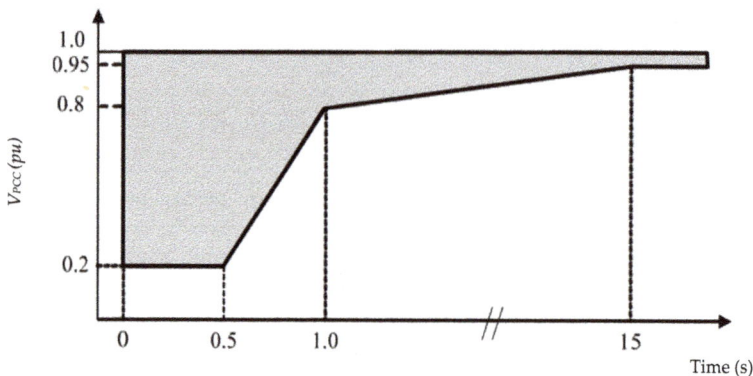

Figure 2. Fault ride-through (FRT) profile according to the Spanish grid code. Reproduced from reference [23].

2. Algerian Wind Potential

This section discusses a method for determining the production of wind energy at different sites in Algeria in order to choose a suitable site for a cost-effective energy installation. Thus, we have both the average wind speed and the power produced by wind turbines; we can combine them to calculate the energy produced by these wind turbines. Furthermore, in this paper, five selected geographical locations (altitude, latitude and longitude) shown in Table 1 and Figure 3 were obtained from the National Meteorological Office (NMO) [24].

Table 1. Coordinates of stations at different Algerian sites.

Station	Coordinates		
	Altitude (m)	Latitude (deg)	Longitude (deg)
South region (Sahara)			
Adrar	263	27°49′ N	00°17′ W
Ghardaia	468	32°24′ N	03°48′ E
Coastal region			
Algiers	24	36°43′ N	03°15′ E
Oran	90	35°38′ N	00°37′ W
Highland region			
Tiaret	1080	35°37′ N	01°32′ E

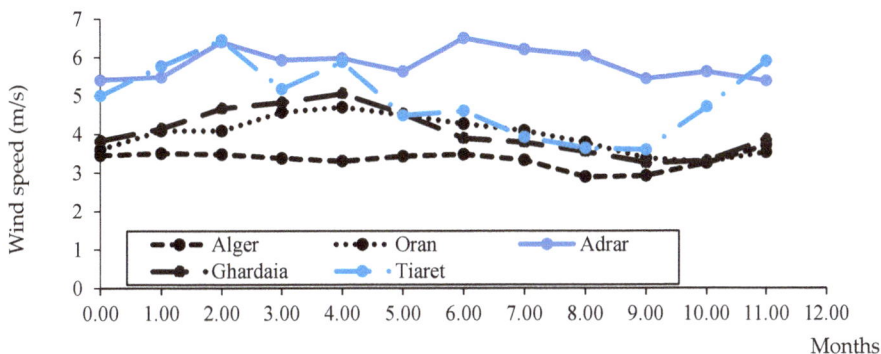

Figure 3. Wind speed variations at different locations in Algeria.

These wind speed data are collected only at 10 m of altitude, measured using a type of anemometer cup and vane. However, the action of the wind speed at the turbine (tower height over 70 m) is very complex, and includes both deterministic effects (wind and shadow average round), and stochastic fast varying wind speed is turbulent. In fact, wind speed describing these variations is usually measured in the lower atmosphere using either instrumented towers or tethered balloons, which have not been available in previous stations [25,26].

The wind speed is the most important aspect of wind potential; in fact, the annual variation of the long-term average wind speed provides a good understanding of the long-term trend of wind speed and gives confidence to investors on the availability of wind energy in the years ahead [27]. Figure 4 provides the average wind speed during five years of data collection at 5 stations in Algeria, which are considered in this study [7,10–34].

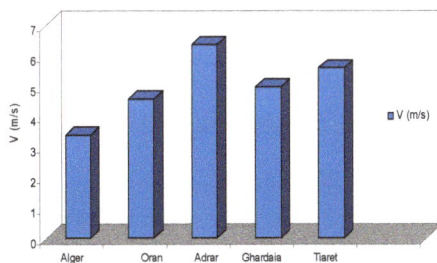

Figure 4. Annual wind speed in Algeria.

3. Connection Issue of a Wind Farm to the Electrical Grid

In this paper, the proposed wind farm in a highland region with average power is considered as a decentralized generator unit, which is most often connected to the distribution network and that differs from centralized generator units. In Algeria, the electrical distribution grids are the most important infrastructure of the whole power system, which is considered as the final interface that leads to most industrial and domestic customers. These distribution grids are operated in ranges of voltages below 50 kV, which is the voltage level of the Medium Voltage (MV) and Low Voltage (LV) ranges. Moreover, in the Algerian distribution grid, the nominal voltage of the MV is 10 kV and 30 kV. These voltage levels allow a good compromise to limit the voltage drops, minimizing the number of source positions (connecting to High Voltage (HV)/MV power station) and reduce the inherent constraints to high voltages (investment costs, protection of property and persons). Moreover, Algerian distribution grids are, in most cases, radially networked. The map of the western Algerian electrical grid is shown in Figure 5. This figure shows the structure of the High Voltage B 220 kV transmission lines, while the substations and power plants are also shown in this figure [1]. The structure of High Voltage A 60 kV distribution lines is shown in Figure 6.

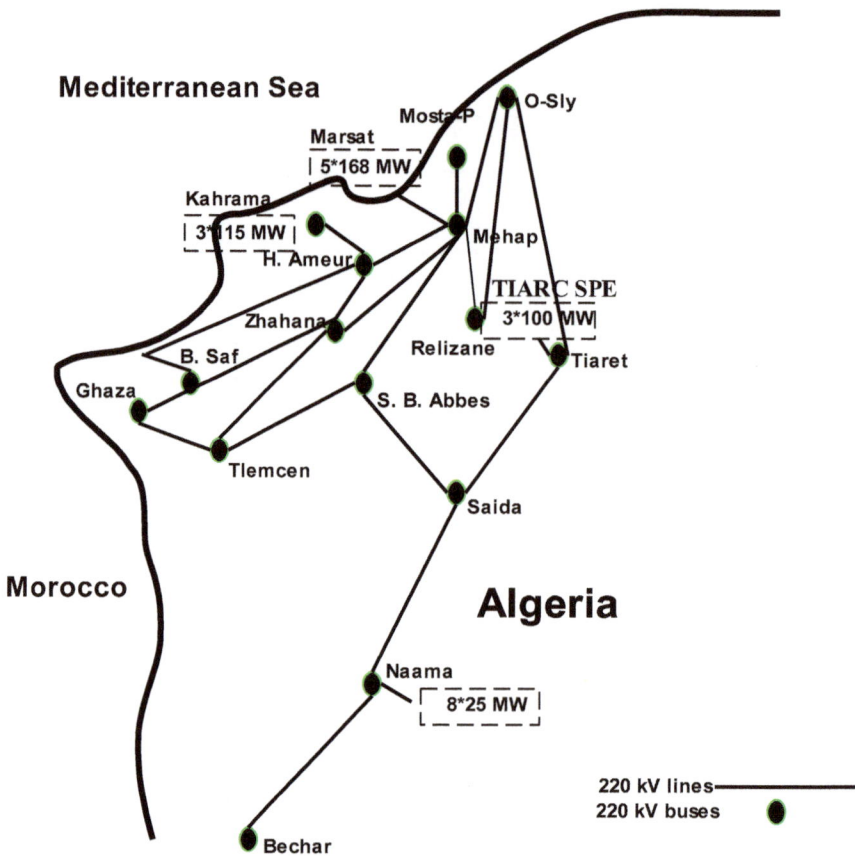

Figure 5. Map of the western Algerian electrical grid. Reproduced from reference [1].

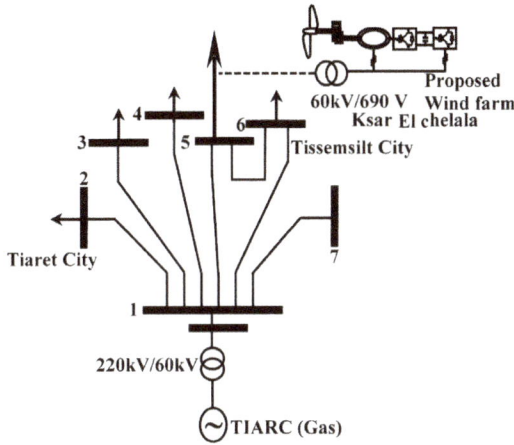

Figure 6. Western electrical grid near to the proposed wind farm.

Based on the structure of the electrical grid on the west side near the proposed wind farm (see Figure 6), this radial electrical grid of 60 kV can be reconfigured; it is then simulated using the Power System Analysis Toolbox (PSAT) software with actual electrical grid parameters and consumer profiles at the peak load of each bus, with a centralized generation source Tiaret City (TIARC) power plant. Simulation of the latter gives the results shown in Figures 7 and 8. More details on the overall simulation of the west Algerian grid can be found in [34].

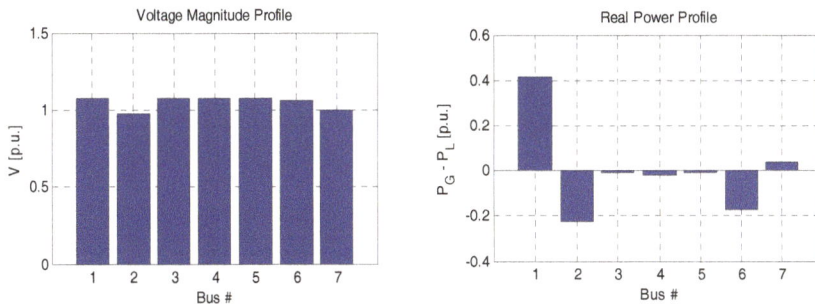

Figure 7. In each bus of the 60-kV electrical grid in Tiaret region: (**a**) voltage amplitudes; (**b**) active powers.

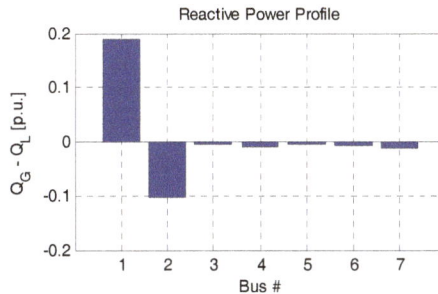

Figure 8. Reactive power in each bus of the 60-kV electrical grid in the Tiaret region.

The existence of an SNVI industrial site in the bus of N 2 between the city of Tiaret and the city of Tissemsilt can justify the presence of the voltage drop across this line as shown in Figure 7a. This resulted in demand of an excessive reactive power and active line losses due to the high-fluctuated demand of energy to the industrial site, as shown in Figures 7 and 8.

4. Distribution Flexible Alternative Current Transmission System (D-FACTS)

Shunt D-FACTS devices can be classified into two main categories, namely the variable impedance type such as the distribution static var compensator (D-SVC) and the switching converter type such as the distribution static synchronous compensator (D-STATCOM).

The configuration of the D-SVC connected to the distribution grid is shown in Figure 9.

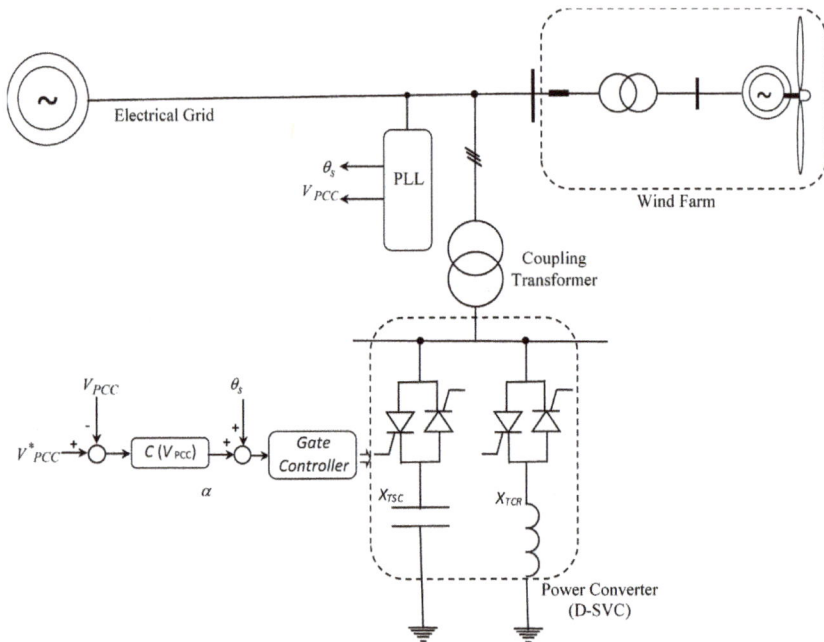

Figure 9. Control configuration of the distribution static var compensator (D-SVC) connected to the point of common coupling (PCC) with a wind farm.

This figure shows a D-SVC consisting of a thyristor switched capacitor (TSC) part composed by two switching thyristors connected with capacitive reactance X_{TSC}; the other part is the thyristor-controlled reactor (TCR) composed by two thyristors connected with an impedance of an inductive reactance branch X_{TCR}. By controlling the angle thyristors (the angle with respect to the zero crossing of the phase voltage), the device is able to control the amplitude of the voltage at the point of common coupling (PCC) due to the changes in the angle resulting mainly in changes of the current. Therefore, the amount of the reactive power consumed by the inductor L, for an angle of $\alpha = 90°$, the inductive circuit is activated, whereas for $\alpha = 180°$, this inductive circuit is off.

The configuration of the D-STATCOM connected to the distribution grid is shown in Figure 10.

This figure shows the different blocks constituting this configuration of the control strategy, which consists of: A phase lock loop (PLL) for synchronization of the component of the positive sequence voltage with the primary voltage of the power distribution grid. An external control loop consists of controlling the DC bus voltage and the grid voltage. The outputs of the voltage controllers are the current references for the current controllers. The internal current control loop consists of the current controllers. The outputs of the current controllers consist in imposing the amplitude and phase of voltages of the D-STATCOM by generating Pulse Width Modulation (PWM) signals.

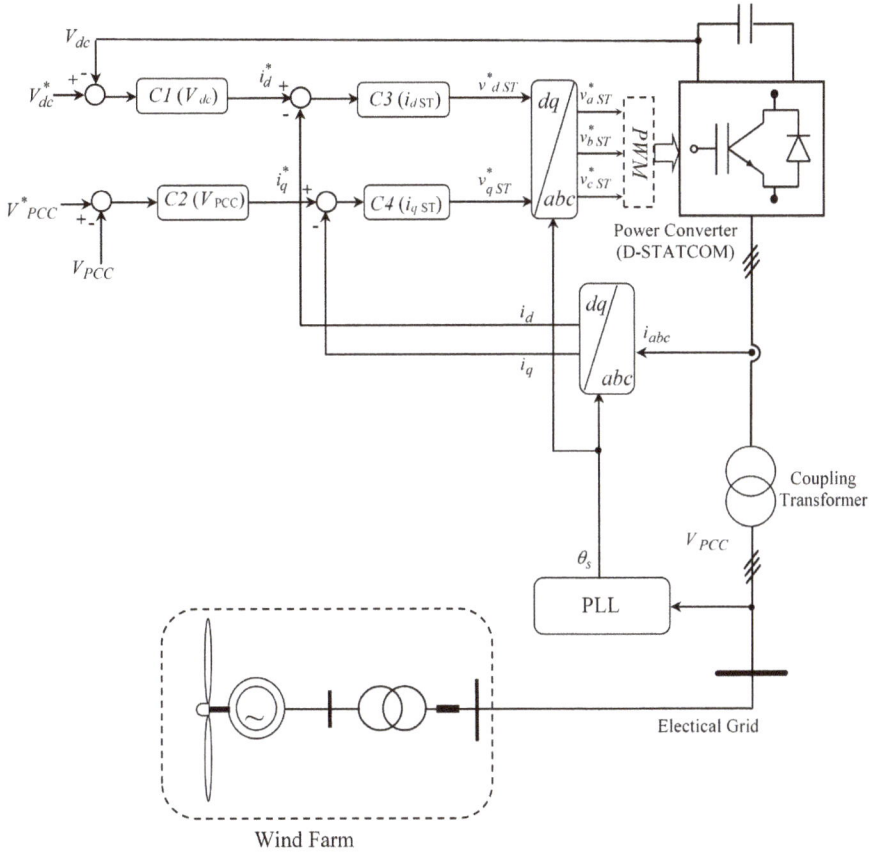

Figure 10. Control configuration of the distribution static synchronous compensator (D-STATCOM) connected to the PCC with a wind farm.

5. Constitution of the Wind Farm and the Location of the Shunt D-FACTS

The wind farm connected to the distribution system proposed in this section is shown in Figure 11, which consists of eight DFIGs with 1.5 MW of power for each wind turbine. These generators are connected between them to a voltage level of 30 kV by a step-up transformer 690 V/30 kV with 4 MVA of power for each generator. Then a 45 km line that is connected to the source substation 60 kV through another step-up transformer 30 kV/60 kV with 47 MVA of power. For this study, these lines are modelled with the π model.

Based on the work done in [21,35] the simulation results obtained with a D-FACTS provides an effective support to the bus voltage to which it is connected. Therefore, in this study, D-FACTS is placed at the PCC for two reasons:

- The location for the reactive power support should be as close as possible to the point at which the carrier is necessary because of the variation in the voltage and, therefore, power loss (Joule loss) in the distribution line associated with reactive power flow,

• In the studied system, the effect of the change in voltage is most common in this bus.

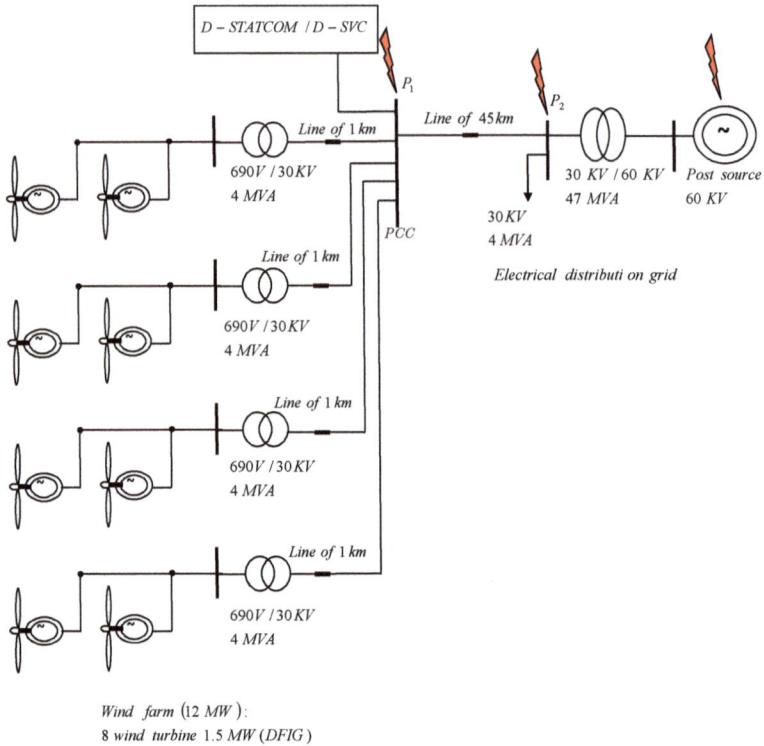

Figure 11. Structure of the studied system based on a wind farm and a distributed-flexible AC transmission system (D-FACTS) connected to the Algerian distribution grid.

6. Simulation Results

In this section, simulations are performed on Matlab/Simulink, to show the impact of D-FACTS on the ability to control the voltage at the PCC between the electric distribution grid and the wind farm, which is described in the previous section (Figure 11). According to the previous section of the wind potential in the Algerian highland region, the considered wind speed in the simulation starts at 8 m/s and then reaches 9 m/s. The parameters of generators and D-FACTS are presented in the Appendix A.

For the more detailed study, the proposed system devices, the D-SVC and D-STATCOM structures used in the context of this paper are the same as the SVC and STATCOM structures, which are presented in [36,37], giving their associated models with their appropriate control schemes. In addition, the detailed model with some simulations for WECS based on DFIG in power system dynamics are described in [38–41]. Generally, a short-circuit fault has a significant effect on the wind farm; a voltage drop is caused even if the fault is located near or far from the PCC or the wind farm. This voltage drop at the PCC leads to an over-current in the rotor circuit of DFIG, and fluctuations in the DC bus voltage. Therefore, the rotor side converter (RSC) of the DFIG should be blocked to avoid being damaged by overcurrent in the rotor circuit.

The block diagram of the simulated wind farm connected to the electrical distribution grid is shown in Figure 12.

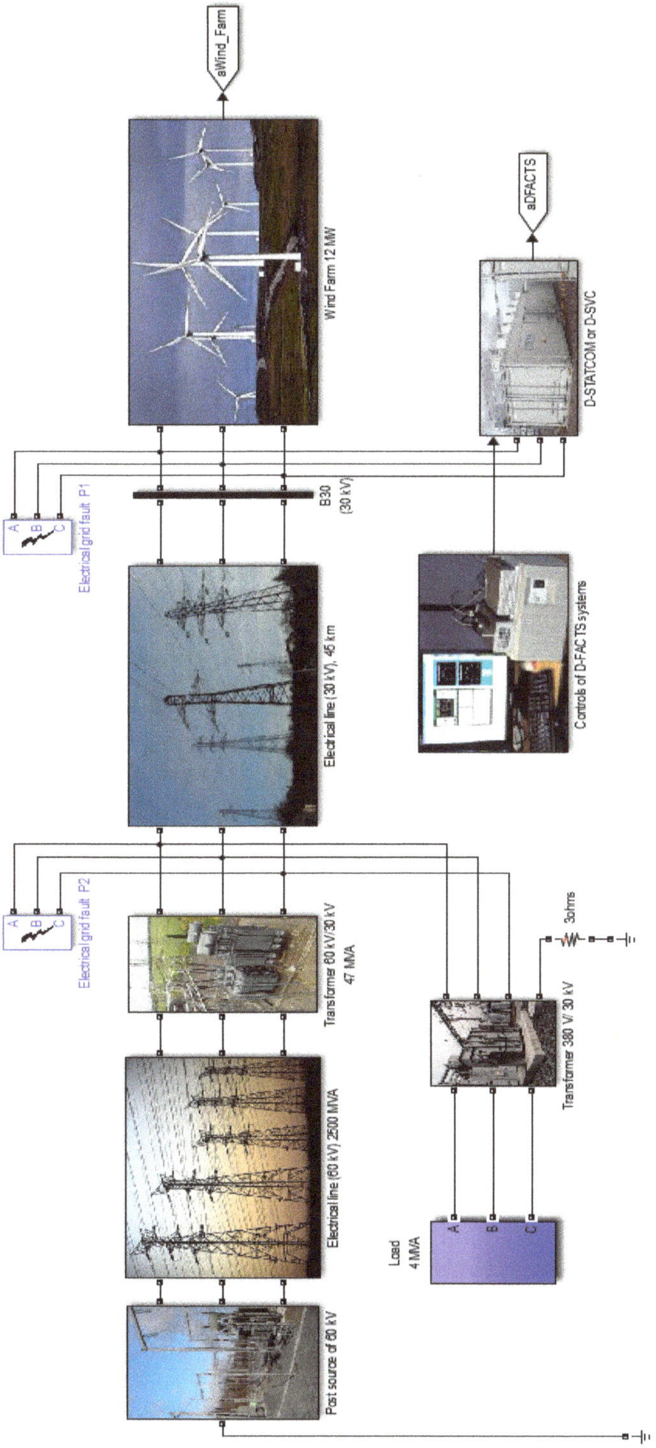

Figure 12. Block diagram of the proposed system on Matlab/Simulink.

In order to study the behavior and the impact of electrical faults in the distribution grid on the wind farm, worst-case scenarios of grid faults were assumed and included in the simulation. Therefore, the entire system was tested under two types of grid faults:

- Line to line electrical grid fault.
- Voltage drop at the 60 kV bus.

The wind speed is considered as constant during the grid fault period, except for this disturbance; and generators and the electrical distribution grid are considered to be working in ideal conditions (no disturbances and no parameter variations in the studied system).

6.1. Simulation Results of the Line-to-Line Electrical Grid Fault

In this section, we consider that the phases "b" and "c" at the PCC at P1 come into accidental contact. Then, the same grid fault is considered at a distance of 45 km from the PCC to the point P2 (Figure 11). Simulation results for this grid fault are shown in Figures 13–15.

The active powers at the PCC with a short temporary grid fault of two-phase to ground are shown in Figure 14. The reactive powers at the PCC with a short temporary grid fault of two-phase to ground are shown in Figure 15.

According to Figure 13, which reveals that without the use of D-FACTS systems the voltage at the PCC exceeds the acceptable voltage level 1 pu due to the voltage swell. However, using D-FACTS devices such as D-SVC and the D-STATCOM these undesirable effects are corrected. In addition, voltages at the PCC during a temporary grid fault presented in this figure show that, without the use of D-FACTS and when the grid fault is at the point P1, the voltage at the PCC drops to the value of 0.48 pu, which is less than the acceptable value. Thus, when the grid fault is at the point P2 without D-FACTS, the voltage drops to 0.52 pu. However, with the presence of D-FACTS devices, when this type of grid fault is at point P1, the voltage at the PCC drops to 0.63 with a slight fluctuation. In addition, when the fault is at the point P2 with D-FACTS, the voltage is maintained at 0.71 pu. Moreover, it is noticed that the voltage at the PCC, when using D-STATCOM many oscillations are mitigated compared to using the D-SVC.

According to Figure 14, which shows that during the occurrence of the same grid fault type in both points P1, P2 and without the presence of D-FACTS systems, no active power is supplied. Then, when the grid fault of 1000 ms duration exceeds the limit (see Figure 2), the wind farm is disconnected from the grid. Moreover, the installation of D-FACTS systems at the PCC guarantees the wind farm commissioning during and after this type of grid fault at these points (P1, P2) without disconnecting from the electrical distribution grid, providing the active power of 10.6 MW. Therefore, in the presence of D-FACTS systems, production of active power by the wind farm is uninterrupted and in the absence of these systems the wind farm is disconnected from the grid by triggering the protection system.

From Figure 15, it is noticed that in the absence of D-FACTS systems and when the grid fault is located at the points P1 and P2, no exchange of reactive power is provided to the electrical distribution grid. However, in the presence of D-FACTS systems, it provides almost the same amount of reactive power, 7.09 MVAr, when the grid fault is located at point P1 and 8.17 MVAr when the grid fault is located at point P2. Indeed, these injected reactive powers are required for compensation to maintain the stability of the wind farm with the voltage at the PCC around the acceptable value. Therefore, the wind farm is kept in service during and after this type of grid fault without disconnecting from the electrical distribution grid. Thus, the use of D-STATCOM has a capacity to compensate faster than using the D-SVC and the peaks of the injected reactive powers are eliminated.

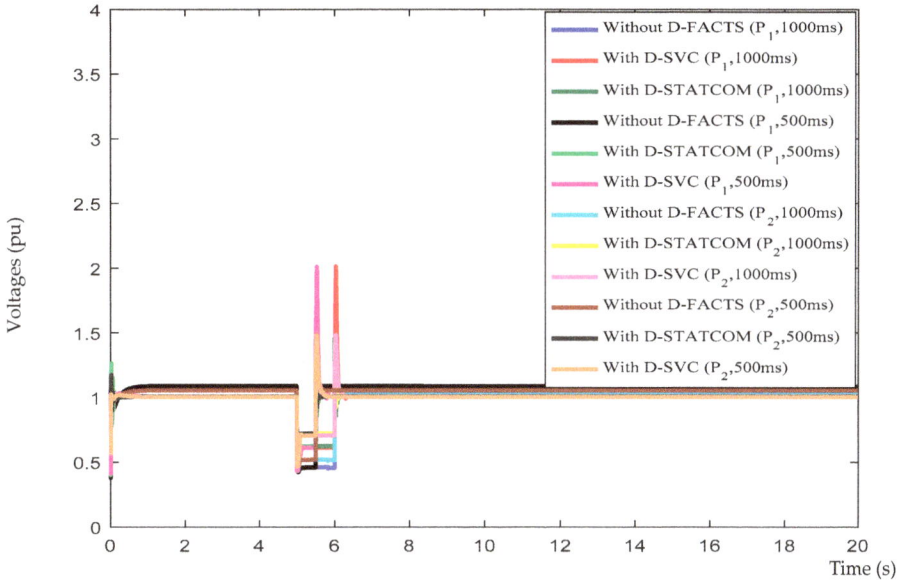

(a)

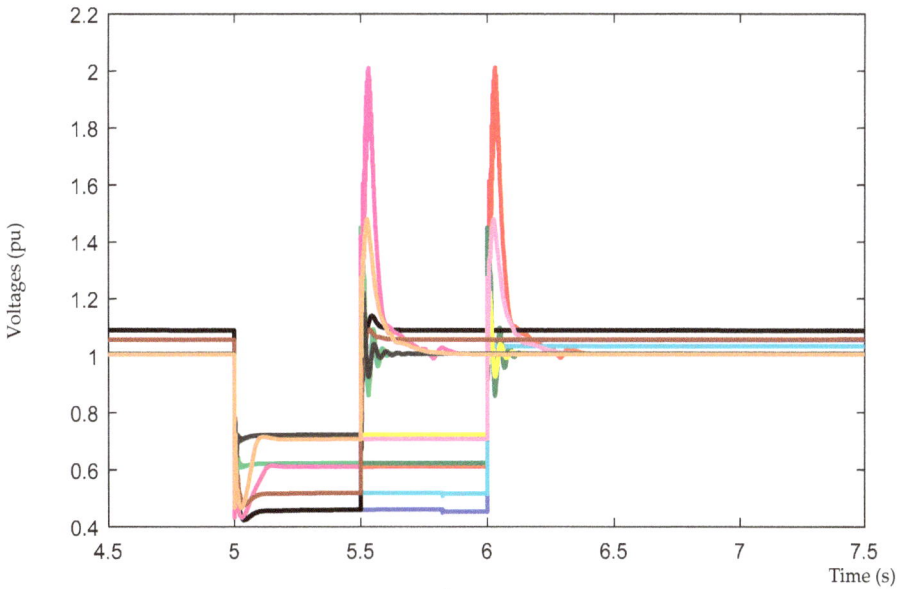

(b)

Figure 13. (**a**) Voltage at the PCC during line-to-line electrical grid fault; (**b**) zoom of voltage at the PCC during line-to-line electrical grid fault.

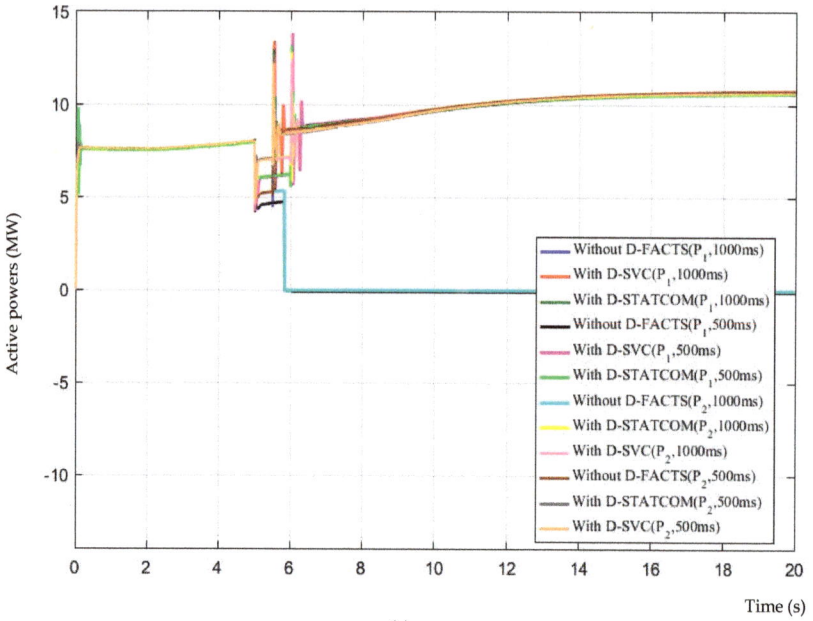

(a)

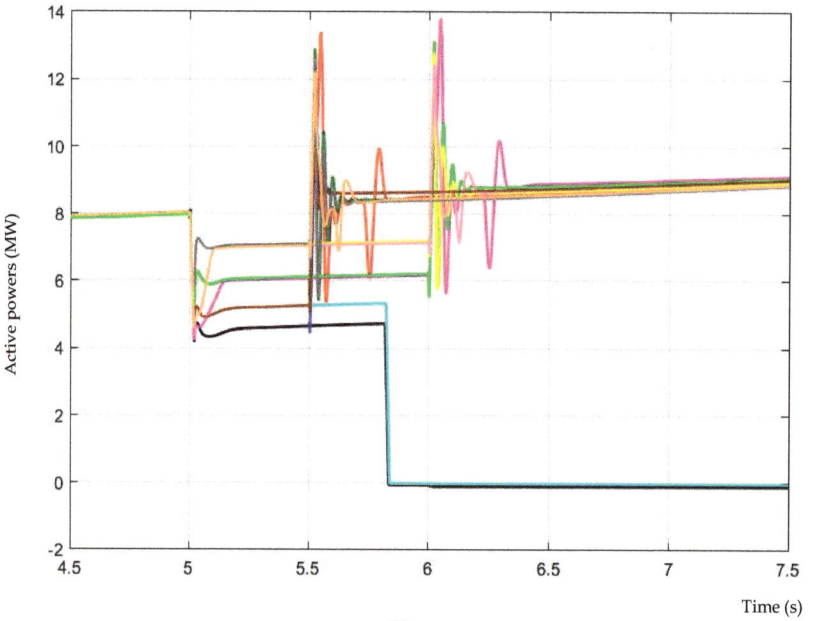

(b)

Figure 14. (**a**) Active power at the PCC during line-to-line electrical grid fault; (**b**) zoom of active power at the PCC during line-to-line electrical grid fault.

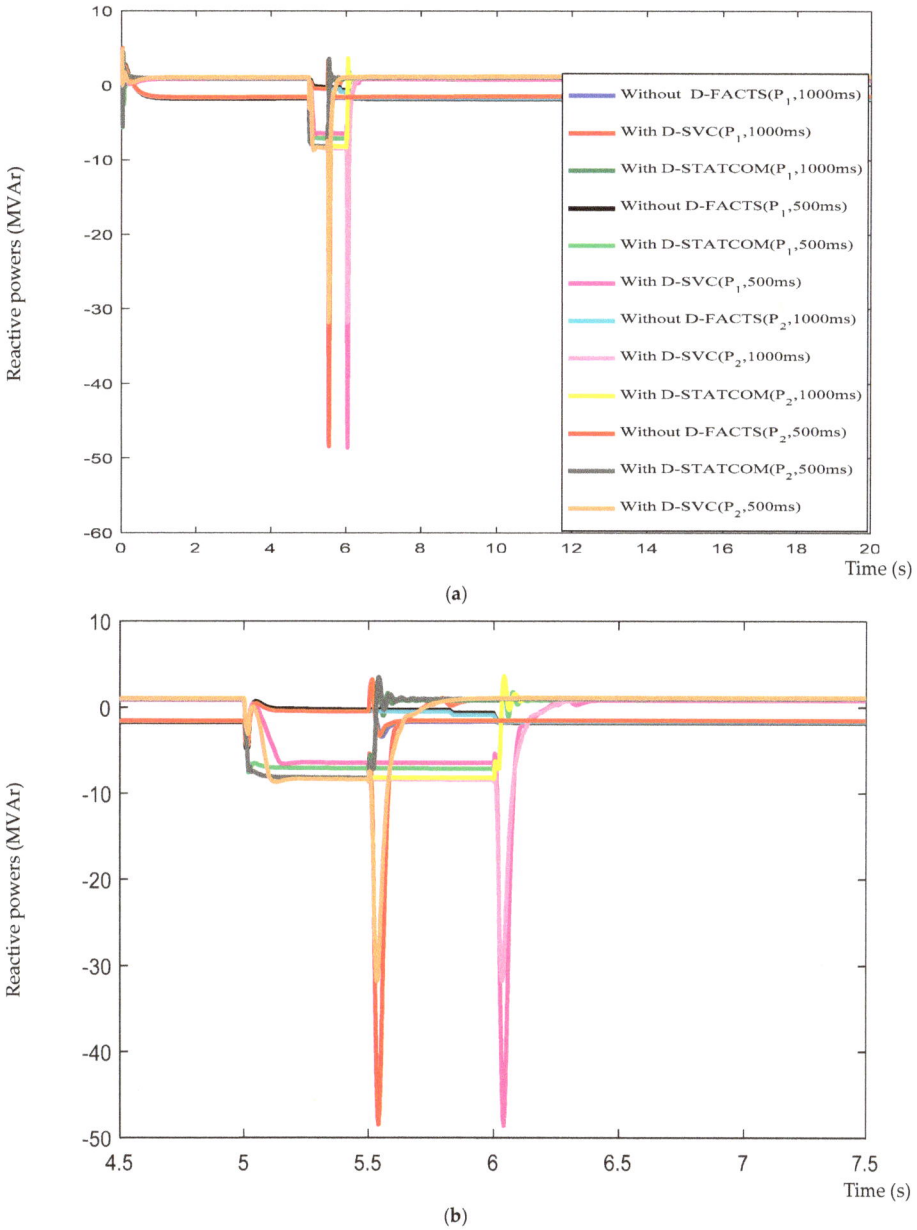

Figure 15. (**a**) Reactive power at the PCC during line-to-line electrical grid fault; (**b**) zoom of reactive power at the PCC during line-to-line electrical grid fault.

6.2. Simulation Results of the Voltage Drop at the 60 kV Bus

The main purpose of this test is to study how a remote grid fault from the PCC may affect the operation of a wind farm; this fault affects the source of 60 kV, which is far from the PCC where the wind farm based on the DFIGs is connected to the grid. Hence, a temporary voltage drop of 50% is applied to the source for the duration of 500 ms at *t* = 10 s, and then the same grid fault for 1000 ms of duration.

Figure 16 shows that during this type of grid fault and without the presence of D-FACTS, the voltage at the PCC drops to 0.44. Therefore, the protection system will be triggered and the wind farm will be disconnected if the fault duration exceeds the electrical interconnection grid code for the wind turbine systems (see Figure 2). However, in the same figure, it is shown that during this grid fault and with the presence of D-FACTS systems, the voltage at the PCC is maintained around 0.88 pu with a transient peak without triggering the protection system. Thus, by using the D-STATCOM, transient peaks are reduced and the time response is faster than by using the D-SVC.

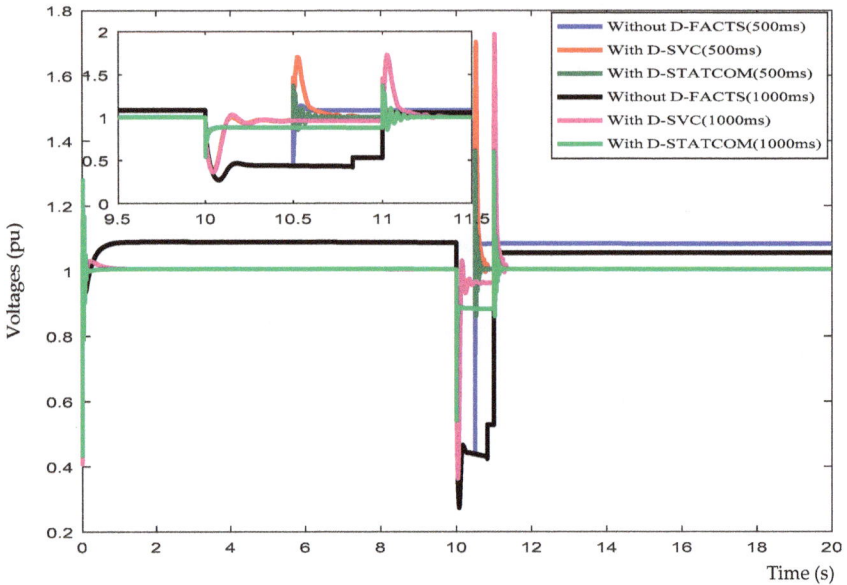

Figure 16. Voltages at the PCC during the voltage drop at the 60 kV bus.

The active powers at the PCC are presented in Figure 17.

Figure 17 shows that, when the D-FACTS devices are not installed at the PCC, the wind farm cannot maintain its connection to the grid during the grid fault that lasts 1000 ms because the protection systems are triggered and the wind farm is disconnected. However, after the installation of D-FACTS devices and with the same grid fault type and the same duration of grid fault, the wind farm can return to the steady state and inject the active power of 10.6 MW to the grid.

From the results shown in Figure 18, it is noticed that without the presence of a compensation system, the wind farm is operating in a weak electrical grid due to its normal behavior and there is no reactive power exchange with the electrical grid. However, in the same situation with the D-FACTS devices, the necessary reactive power is provided to the grid of 12.6 MVAr with D-SVC and 9.7 MVAr with the D-STATCOM. Thus, a very fast and significant fluctuation is observed with the use of the D-SVC compared to the use of the D-STATCOM.

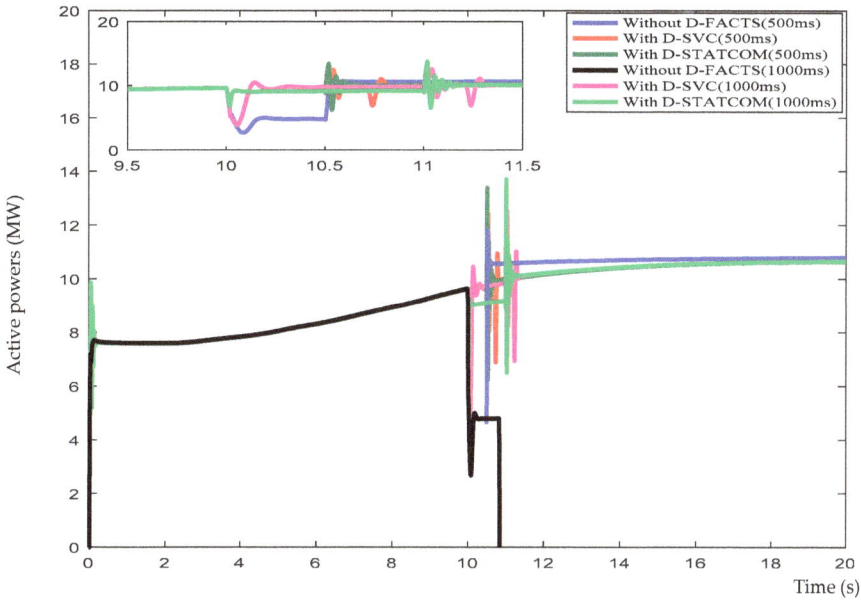

Figure 17. Active power at the PCC during the voltage drop at the 60 kV bus.

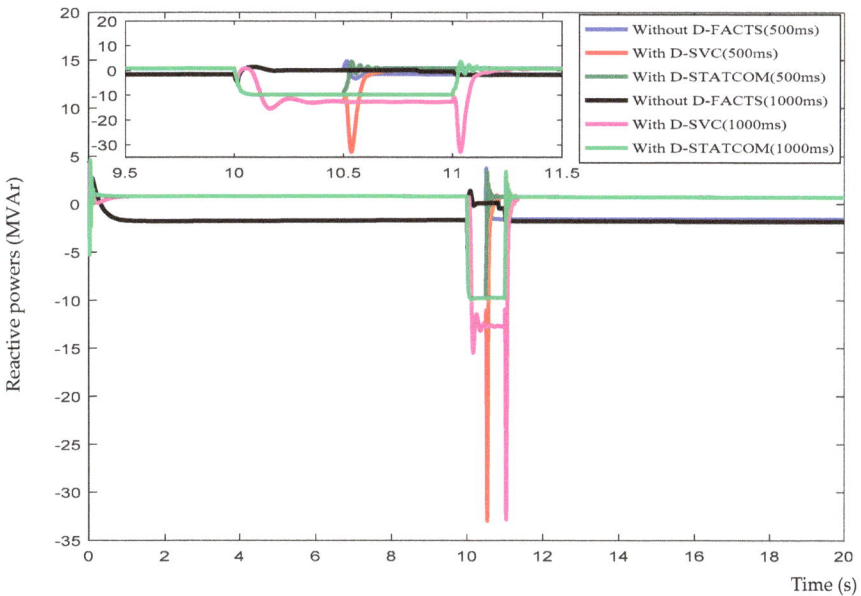

Figure 18. Reactive power at the PCC during the voltage drop at the 60 kV bus.

7. Economic Analysis of D-FACTS Systems

The global market for FACTS and D-FACTS systems is expected to reach $1,386,010,000 in 2018, it had already reached $912,850,000 in 2012 [12]. The D-SVC is the most widely used solution in the world market, followed by fixed capacitor banks. However, devices such as D-STATCOM are one customized solution for specific requirements of the distribution network. Obviously, some D-FACTSs are relatively expensive because they consist of many components such as advanced power electronics components, thyristors, reactors, capacitor banks, switches, protection systems and control systems. In this section, the range of the cost of the key features is often taken from the company Siemens and the Electric Power Research Institute (EPRI) with the database specified in [13], as shown in Figure 19.

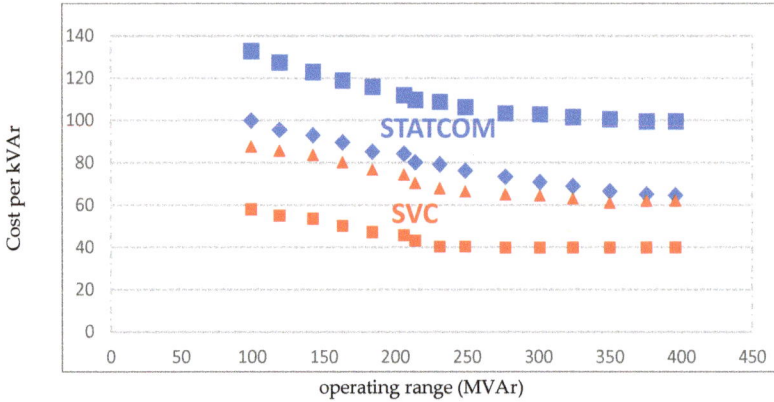

Figure 19. Cost of the operation range for different D-FACTS devices. Reproduced from [14].

Generally, the cost of a D-FACTS system has two components: the installation costs and operating expenses. The total cost of the entire installed systems comprises the equipment price and the delivery and installation of these systems. The operating cost includes the cost of maintenance and service. Specifically, the operating cost of these devices is approximately 5% to 10% of the total installation cost. Therefore, the cost functions for the D-SVC and D-STATCOM are developed as follows [15]:

$$\begin{cases} C_{SVC} = 0.0004\,s^2 - 0.262\,s + 81.5 \\ C_{SVC} = 0.0003\,s^2 - 0.305\,s + 127.38 \\ C_{STATCOM} = 0.0004\,s^2 - 0.3225\,s + 128.75 \\ C_{STATCOM} = -0.0008\,s^2 - 0.155\,s + 120 \end{cases} \tag{1}$$

where s is the operating range of D-FACTS devices kVAr. The marginal cost per kVAr of the installed D-FACTS devices decreases as the operating rate of capacity increases. An overall cost for reactive power 100 MVAr, D-SVC ranges from $60 to $100 per kVAr. Although the D-SVC has sophisticated components such as thyristors, inductors and capacitors, it has a control structure that is relatively simple. Similarly, based on Figure 19, the overall cost of a D-STATCOM varies from $100 to $130 per kVAr and 100 MVAr of operating range. The costs of the installed parallel D-FACTS devices are shown in the Table 2 [16]:

Table 2. Costs of different reactive power compensators.

Parallel Reactive Compensator	Cost (US $/kVar)
Shunt capacitor	8/kVar
D-SVC	60/kVar
D-STATCOM	100/kVar

From the above table, we see that the cost of D-FACTS devices (D-SVC and D-STATCOM) is much more expensive compared to capacitors due to the cost of the control devices and the complexity of the design and application of D-FACTS systems. The D-STATCOM is the source of reactive power compensation that is more expensive because of the used power electronics components like the Insolated Gate Bipolar Transistor (IGBT).

In this study, the Figure 20 summarizes the performance of both D-FACTS types (D-STATCOM and D-SVC), also by comparing the amount of injected reactive power (MVAr) and the installation cost of these devices ($). This comparison provides a basis for system integration D-FACTS in a wind farm consisting of the DFIG type of generator helping to achieve a better balance between performance and cost in the condition of specific defects.

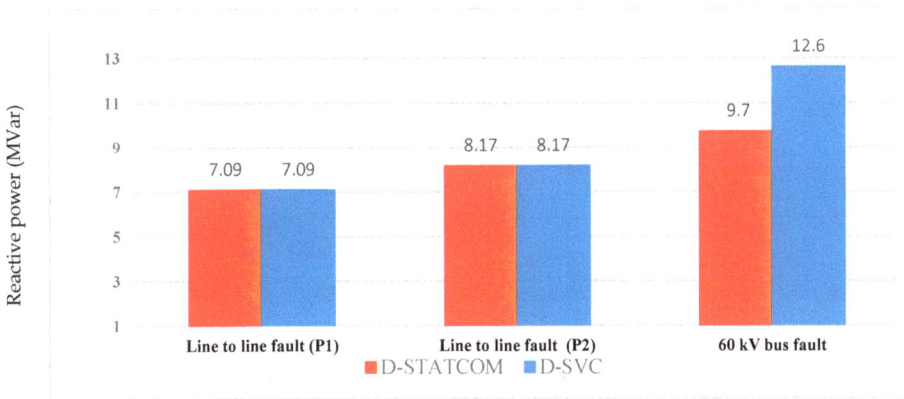

Figure 20. Reactive power injected into the PCC by using D-SVC and D-STATCOM with different types of grid faults.

The D-STATCOM provides a very effective reactive power compensation with respect to the D-SVC for all fault conditions. However, the D-SVC has a capacity option for a large amount of reactive power to be injected during a severe fault condition of the source. In the case of a two-phase ground fault (worst event of the grid fault applied in this study), the cost of installation is an important factor to consider. Therefore, in this paper, the economic analysis aims to compare the total cost of two types of parallel D-FACTS connected to a wind farm based on DFIGs, as presented in Figure 21.

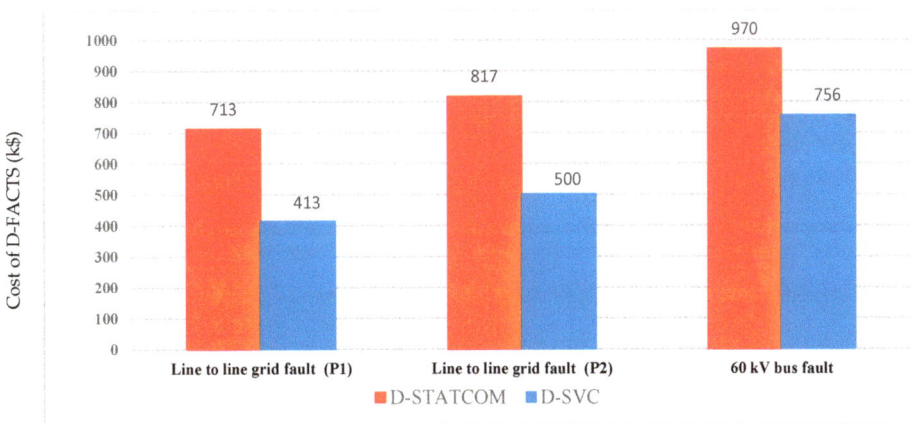

Figure 21. Cost of D-FACTS systems from different types of grid faults.

According to Figure 21, it is clear that the use of D-STATCOM to maintain the wind farm in service during grid fault conditions is an expensive application compared with D-SVC due to the use of the transformer and the cost of power electronics [17,18]. Consequently, one can conclude that the D-STATCOM is more cost-effective compared to D-SVC for voltage support at the PCC and the wind farm connection in the event of the most severe grid fault conditions.

Figure 22 shows the cost breakdown of a proposed 12 MW wind farm installation.

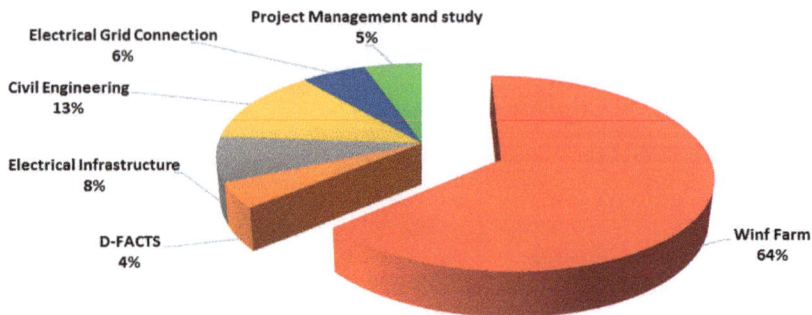

Figure 22. Breakdown of the costs of the wind project.

According to this figure, it can be seen that the proposed solution based on a D-FACTS system represents only 4% of the overall cost of the wind turbine installation. On the other hand, this solution offers good performance at the wind farm, ensuring its connection with the electrical grid and the reliability of the wind energy conversion system.

8. Conclusions

The aim of this paper is to investigate the feasibility study of the wind farm project in an Algerian highland region. Nevertheless, the study in this paper shows that Tiaret's electrical grid is susceptible to host the proposed wind farm in order to benefit from the wind potential. Therefore, in this paper, the application of the D-FACTS systems as the appropriate solution for the electrical grid connection issue and to accomplish the uninterrupted operation of a wind farm based on DFIG during the line-to-line fault and the voltage drop at 60 kV has been investigated. The D-FACTS is connected at the PCC where the wind farm is connected to the grid, to provide necessary reactive power for voltage support of the wind farm. Based on the simulation results, the stability improvement of the wind farm through the incorporation of the D-SVC or the D-STATCOM has been illustrated. In addition, it can be concluded that using D-STATCOM reduces the complexity of controlling the wind turbine-generators, improves the time response of reactive power compensation, and corrects for the lack of a wind turbine.

Future work will evaluate the impact of different scenarios for wind farms integrated into the Algerian electrical grid with other renewable energy sources and their electricity price.

Author Contributions: L.W. and K.D.E.K. contributed to the main idea of this article. K.D.E.K wrote the paper. K.D.E.K. searched literature. K.D.E.K collected data and performed the simulations. L.W., A.V.D.B., A.M., A.D. and K.D.E.K. contributed to the analysis of data. L.W., K.D.E.K., A.V.D.B., A.M., A.D. and L.B. revised the paper critically. All authors approved the final version to be published.

Funding: This research was funded by the National Natural Science Foundation of China grant number 51577005. No. 51877005 and the Aeronautical Science Foundation of China grant number 2015ZC51030.

Conflicts of Interest: The authors declare no conflict of interest.

Appendix A

In this part, simulations are investigated with a 1.5 MW DFIG connected to a 690 V/50 Hz grid [28,40,42,43]. The parameters of the turbine and the generator are presented below:

Parameters	Values
Turbine	
Number of blades	3
Turbine radius	35.25 m
Gear box ratio	90
DFIG	
Power	1.5 MW
Nominal voltage	690 V
Frequency	50 Hz
Number of poles pair	2
Stator resistance	0.012 Ω
Rotorique resistance	0.021 Ω
Stator inductance	0.0137 H
Rotor inductance	0.01367 H
Mutual inductance	0.0135 H
(Turbine + DFIG)	
Generator inertia	1000 Kg.m^2
Friction factor	0.0024 Kg.m/s

The parameters of the D-STATCOM are presented below [44]:

Parameters	Values
Transformer voltage	2.5/30 kV
Nominal frequency	50 Hz
Rated power	3–15 MVA
Resistance	0.22/30 pu
Inductance	0.22 pu
DC-link voltage	4000 V

The parameters of the D-SVC are presented below:

Parameters	Values
Transformer voltage	2.5/30 kV
Nominal frequency	50 Hz
Rated power	3–15 MVA
Rated capacitor power	3–15 MVar
Rated inductance power	1.5–10 MVar

References

1. Djalel, D.I.B.; Bendakir, A.; Metatla, S.; Soufi, Y. The algerian challenge between the dependence on fossil fuels and its huge potential renewable energy. *Int. J. Renew. Energy Res.* **2012**, *2*, 463–470.
2. Souag, S.; Benhamida, F. A dynamic power system economic dispatch enhancement by wind integration considering ramping constraint-application to algerian power system. *Int. J. Renew. Energy Res.* **2015**, *5*, 794–805.
3. Equilibres, La lettre de la Commission de Régulation de l'Electricité et du Gaz, N 12, Mars 2011. Available online: https://www.creg.dz/images/equilibres/Equilibres12.pdf (accessed on 1 Mars 2011). (In French)
4. Programme des Energies Renouvelables et de l'Efficacité Energétique, Mars 2011. Available online: http://www.energy.gov.dz/francais/uploads/2016/Programme-National/Programme-National-Efficacite-Energetique.pdf (accessed on 1 Mars 2011). (In French)
5. La Ferme Eolienne d'Adrar Mise en Service. Available online: http://portail.cder.dz/spip.php?article4098 (accessed on 3 July 2014).

6. Miloud, F.S.B.; Aissaoui, R. Etude du potentiel éolien d'Adrar Sélection de sites pour la ferme éolienne de 10 MW. *Séminaire Méditerranéen En Energie Eolienne* **2010**, *13*, 295–300.

7. Himri, Y.; Himri, S.; Stambouli, A.B. Assessing the wind energy potential projects in Algeria. *Renew. Sustain. Energy Rev.* **2009**, *13*, 2187–2191. [CrossRef]

8. Maouedj, R.; Bouchouicha, K.; Boumediene, B. Evaluation of the wind energy potential in the Saharan sites of Algeria. In Proceedings of the 10th International Conference Environment and Electrical Engineering, Rome, Italy, 8–11 May 2011; IEEE: Piscataway, NJ, USA.

9. Himri, Y.; Stambouli, A.B.; Draoui, B.; Himri, S. Techno-economical study of hybrid power system for a remote village in Algeria. *Energy* **2008**, *33*, 1128–1136. [CrossRef]

10. Kasbadji Merzouk, N.; Merzouk, M.; Abdeslam, D. Prospects for the wind farm installation in the Algerian high plateaus. *Desalination* **2009**, *239*, 130–138.

11. Abdeslam, D.; Kasbadji Merzouk, N. Wind energy resource estimation in Sétif region. In Proceedings of the Conference on the Promotion of Distributed Renewable Energy Sources in the Mediterranean Region, Nicosia, Cyprus, 11–12 December 2009.

12. Dehmas, D.; Kherba, N.; Hacene, F.B.; Merzouk, N.K.; Merzouk, N.; Mahmoudi, H.; Goosen, M. On the use of wind energy to power reverse osmosis desalination plant: A case study from Ténès (Algeria). *Renew. Sustain. Energy Rev.* **2011**, *15*, 956–963. [CrossRef]

13. Zin, A.A.B.M.; Mahmoud, H.A.; Khairuddin, A.B.; Jahanshaloo, L.; Shariati, O. An overview on doubly fed induction generators controls and contributions to wind-based electricity generation. *Renew. Sustain. Energy Rev.* **2013**, *27*, 692–708.

14. Mohseni, M.; Islam, S.M. Review of international grid codes for wind power integration: Diversity, technology and a case for global standard. *Renew. Sustain. Energy Rev.* **2012**, *16*, 3876–3890. [CrossRef]

15. Tazil, M.; Kumar, V.; Bansal, R.C.; Kong, S.; Dong, Z.Y.; Freitas, W. Three-phase doubly fed induction generators: An overview. *IET Electr. Power. Appl.* **2010**, *4*, 75–89. [CrossRef]

16. Tsourakisa, G.; Nomikosb, B.M.; Vournasa, C.D. Effect of wind parks with doubly fed asynchronous generators on small-signal stability. *Electr. Power Syst. Res.* **2009**, *79*, 190–200. [CrossRef]

17. Abdelhafidh, M.; Mahmoudi, M.O.; Nezli, L.; Bouchhida, O. Modeling and control of a wind power conversion system based on the double-fed asynchronous generator. *Int. J. Renew. Energy Res.* **2012**, *2*, 300–306.

18. Kerrouche, K.; Mezouar, A.; Belgacem, K. Decoupled control of doubly fed induction generator by vector control for wind energy conversion system. *Energy Procedia* **2013**, *42*, 239–248. [CrossRef]

19. Wang, Z.; Sun, Y.; Li, G.; Ooi, B.T. Magnitude and frequency control of grid-connected doubly fed induction generator based on synchronised model for Wind power generation. *IET Renew. Power Gen.* **2010**, *4*, 232–241. [CrossRef]

20. Rolán, A.; Pedra, J.; Córcoles, F. Detailed study of DFIG-based wind turbines to overcome the most severe grid faults. *INT J. Electr. Power* **2014**, *62*, 868–878. [CrossRef]

21. Moghadasi, A.; Sarwat, A.; Guerrero, J.M. A comprehensive review of low-voltage-ride-through methods for fixed-speed wind power generators. *Renew. Sustain. Energy Rev.* **2016**, *55*, 823–839. [CrossRef]

22. Ananth, D.V.N.; Kumar, G.N. Fault ride-through enhancement using an enhanced field-oriented control technique for converters of grid connected DFIG and STATCOM for different types of faults. *ISA Trans.* **2016**, *62*, 2–18. [CrossRef] [PubMed]

23. Red Eléctrica de España. REE. P.O.12.3. Requisitos de Respuesta Frente a Huecos de Tensión de las Instalaciones de Producción de Régimen Especial. Available online: http://www.ree.es/sites/default/files/01_ACTIVIDADES/Documentos/ProcedimientosOperacion/PO_resol_12.3_Respuesta_huecos_eolica.pdf (accessed on 8 May 2013). (In Spanish)

24. Hammouche, R. *Atlas Vent de L'Algerie*; Publication Interne de L'office National de Météorologie: Alger, Algiers, 1990.

25. Rachid, M.; Said, D.; Boumediene, B. Wind Characteristics Analysis for Selected Site in Algeria. *Int. J. Comput. Appl.* **2012**, *56*, 450. [CrossRef]

26. Ettoumi, F.Y.; Sauvageot, H.; Adane, A.E.H. Statistical bivariate modelling of wind using first-order Markov chain and Weibull distribution. *Renew. Energy* **2003**, *28*, 1787–1802. [CrossRef]

27. Himri, Y.; Malik, A.S.; Stambouli, A.B.; Himri, S.; Draoui, B. Review and use of the Algerian renewable energy for sustainable development. *Renew. Sustain. Energy Rev.* **2009**, *13*, 1584–1591. [CrossRef]

28. Kerrouche, K.D.; Mezouar, A.; Boumediene, L.; Van den Bossche, A. Speed sensor-less and robust power control of grid-connected wind turbine driven doubly fed induction generators based on flux orientation. *Mediterr. J. Meas. Control* **2016**, *12*, 606–618.

29. Cheggaga, N.; Ettoumi, F.Y. A neural network solution for extrapolation of wind speeds at heights ranging for improving the estimation of wind producible. *Wind Eng.* **2011**, *35*, 33–53. [CrossRef]

30. Merzouk, N.K.; Merzouk, M.; Benyoucef, B. *Profil Vertical de la Vitesse du vent dans la Basse Couche Limite Atmosphérique*; JITH: Albi, France, 2007; pp. 1–5.

31. Cheggaga, N.; Ettoumi, F.Y. Estimation du potentiel éolien. *Revue Des Energies Renouvelables* **2010**, 99–105.

32. Abbes, M.; Belhadj, J. Wind resource estimation and wind park design in El-Kef region, Tunisia. *Energy* **2012**, *40*, 348–357. [CrossRef]

33. Boudia, S.M.; Benmansour, A.; Ghellai, N.; Benmedjahed, M.; Tabet Hellal, M.A. Monthly and seasonal assessment of wind energy potential in Mechria region, occidental highlands of Algeria. *Int. J. Green Energy* **2012**, *9*, 243–255. [CrossRef]

34. Lakdja, F.; Gherbi, Y.; Hocine, G.; Adbsallem, D.O. Impact of STATCOM on a wind farm into the Western of Algerian network. In Proceedings of the International Renewable and Sustainable Energy Conference, Ouarzazate, Morocco, 17–19 October 2014; IEEE: Piscataway, NJ, USA, 2014.

35. Mokhtari, A.; Gherbi, F.Z.; Mokhtar, C.; Kerrouche, K.D.E. Study, analysis and simulation of a static compensator D-STATCOM for distribution systems of electric power. *Leonardo J. Sci.* **2014**, *25*, 117–130.

36. Sharma, P.; Bhatti, T.S.; Ramakrishna, K.S.S. Performance of statcom in an isolated wind–diesel hybrid power system. *Int. J. Green Energy* **2011**, *8*, 163–172. [CrossRef]

37. Karimi, M.; Farhadi, P. Selecting the Best Compatible SVC Type in Power Systems for transient Stability. *Int. J. Green Energy* **2015**, *12*, 87–92. [CrossRef]

38. Kerrouche, K.D.E.; Mezouar, A.; Boumediene, L.; Belgacem, K. Modeling and optimum power control based DFIG wind energy conversion system. *Int. Rev. Electr. Eng.* **2014**, *9*, 174–185. [CrossRef]

39. Kerrouche, K.D.E.; Mezouar, A.; Boumediene, L.; Van Den Bossche, A. Modeling and lyapunov-designed based on adaptive gain sliding mode control for wind turbines. *J. Power Technol.* **2016**, *96*, 124–136.

40. Kerrouche, K.D.E.; Mezouar, A.; Boumediene, L.; Van Den Bossche, A. A comprehensive review of LVRT capability and sliding mode control of grid-connected wind-turbine-driven doubly fed induction generator. *Automatika* **2016**, *60*, 922–935.

41. Okedu, K.E. Stability enhancement of dfig-based variable speed wind turbine with a crowbar by facts device as per grid requirement. *Int. J. Renew. Energy Res.* **2012**, *2*, 431–439.

42. Masaud, T.M.; Sen, P.K. Study of the implementation of STATCOM on DFIG-based wind farm connected to a power system. In Proceedings of the IEEE PES Innovative Smart Grid Technologies, Washington, DC, USA, 16–20 January 2012; IEEE: Piscataway, NJ, USA, 2012.

43. Kerrouche, K.D.E.; Wang, L.; Van Den Bossche, A.; Draou, A.; Mezouar, A.; Boumediene, L. Dual robust control of grid-connected dfigs-based wind-turbine-systems under unbalanced grid voltage conditions. In *Stability Control and Reliable Performance of Wind Turbines*; Intechopen: Rijeka, Croatia, 2018.

44. Kerrouche, K.D.E.; Wang, L.; Mezouar, A.; Boumediene, L.; Van Den Bossche, A. Fractional-order sliding mode control for D-STATCOM connected wind farm based DFIG under voltage unbalanced. *Arab. J. Sci. Eng.* **2018**, *44*, 1–16. [CrossRef]

MDPI

St. Alban-Anlage 66

4052 Basel

Switzerland

Tel. +41 61 683 77 34

Fax +41 61 302 89 18

www.mdpi.com

Applied Sciences Editorial Office

E-mail: applsci@mdpi.com

www.mdpi.com/journal/applsci

www.ingramcontent.com/pod-product-compliance
Lightning Source LLC
Chambersburg PA
CBHW051847210326
41597CB00033B/5806